FROM LOCAL PATRIOTISM TO
A PLANETARY PERSPECTIVE

Science, Technology and Culture, 1700–1945

Series Editors

David M. Knight
University of Durham

and

Trevor Levere
University of Toronto

Science, Technology and Culture, 1700–1945 focuses on the social, cultural, industrial and economic contexts of science and technology from the 'scientific revolution' up to the Second World War. It explores the agricultural and industrial revolutions of the eighteenth century, the coffee-house culture of the Enlightenment, the spread of museums, botanic gardens and expositions in the nineteenth century, to the Franco-Prussian war of 1870, seen as a victory for German science. It also addresses the dependence of society on science and technology in the twentieth century.

Science, Technology and Culture, 1700–1945 addresses issues of the interaction of science, technology and culture in the period from 1700 to 1945, at the same time as including new research within the field of the history of science.

Also in the series

*Acid Rain and the Rise of the Environmental Chemist
in Nineteenth-Century Britain*
Peter Reed

*Matter and Method in the Long Chemical Revolution
Laws of Another Order*
Victor D. Boantza

*Jesse Ramsden (1735–1800)
London's Leading Scientific Instrument Maker*
Anita McConnell

Making Scientific Instruments in the Industrial Revolution
A.D. Morrison-Low

From Local Patriotism to a Planetary Perspective

Impact Crater Research in Germany, 1930s–1970s

MARTINA KÖLBL-EBERT

Jura-Museum Eichstätt, Germany

Routledge
Taylor & Francis Group

LONDON AND NEW YORK

First published 2015 by Ashgate Publishing

2 Park Square, Milton Park, Abingdon, Oxfordshire OX14 4RN
52 Vanderbilt Avenue, New York, NY 10017

Routledge is an imprint of the Taylor & Francis Group, an informa business

First issued in paperback 2020

British Library Cataloguing in Publication Data
A catalogue record for this book is available from the British Library

The Library of Congress has cataloged the printed edition as follows:
Kölbl-Ebert, Martina, author.
    From Local Patriotism to a Planetary Perspective: Impact Crater Research in Germany, 1930s–1970s / by Martina Kölbl-Ebert.
        pages   cm. – (Science, Technology and Culture, 1700–1945)
    Includes bibliographical references and index.
    1. Impact craters – Research – Germany – History – 20th century. 2. Geology, Structural – Germany. I. Title.
    QE612.5.G3K65 2015
    551.3'97072043–dc23   2014024646

ISBN 978-1-4724-3886-7 (hbk)
ISBN 978-0-367-59965-2 (pbk)

Contents

List of Figures   ix
Preface   xvii
    Note on Sources and Translations   xix
    Acknowledgments   xx

1   Introducing the Smoking Gun   **1**
    Geological Context   5
    Early Research on the Ries Basin   9
    Elevation and Subsidence   12
    Central Explosion   14
    Rocks and Minerals   21
    Early Ideas on the Genesis of the Steinheim Basin   22
    Outside the Mainstream   23

2   Early Impactists and their Sources   **27**
    Julius Osvald Kaljuvee   28
    Excursion to Arizona   32
    Importing the Impact Scenario to Germany   35
    Herbert P.T. Rohleder   41
    Otto Stutzer   46

3   Dismissing Impact I   **53**
    Why?   59
    Isolation: Politically, Linguistically and Scientifically   60
    Artificial Boundaries and their Influence on
    Geological Research   62
    Actualism – Method Confused as Fact   75
    Mineralogy and Volcanology   82

4   A Letter from Berlin   **99**
    The Quiring–Kranz Correspondence   101

5   Kaalijärv Crater and Köfels Landslide   **115**
    Walter Kranz as Impactist   120
    Franz Eduard Sueß and the Köfels Landslide   122

6   **Impact Physics – Beyond Human Imagination**   **131**
     Ernst Julius Öpik   134
     Fritz Heide   138
     It Cannot Be   142

7   **'German Geology'**   **145**
     German Geology in the Romantic Period   145
     Geology in Nazi Germany   147
     Defining 'German Geology'   155
     German Geology 'Down the Drain'   163

8   **Setting the Stage**   **171**
     Visit by a 'Crazy' American   173
     Coesite   178
     Meanwhile in Germany   182

9   **The Tide is Turning**   **191**
     Impact in Tübingen   196

10   **Dismissing Impact II**   **205**
     Walter H. Bucher and the Sceptics of Tübingen   213
     Scepticism and Apologetics in Munich   222
     From Scepticism to Chauvinism   230
     Rivalry Between Geology and Mineralogy   234

11   **Testing an Old Theory**   **247**
     No Volcanic Vent at *Altenbürg* Quarry   257
     Gerold Wagner's Tragic Death at Rochechouart   261
     The Problem of Visualisation   268
     The Ries Working Group   273
     Drilling the Ries Crater   279
     Research Drilling at Nördlingen   285
     The Steinheim Basin   290

12   **Ries Crater – A Terrestrial Proxy for the Moon**   **297**
     Shock Metamorphism   299
     Suevite Samples From the Moon   307
     Astronauts at the Ries Crater   310
     Georg Wagner Finally Convinced   319
     A New Routine   324
     Little Green Prussians   329

13 From Local Patriotism to a Planetary Perspective **335**

Glossary *343*
References *347*
Index *373*

List of Figures

1.1   The Ries Crater (image courtesy of Stadt Nördlingen and
      Landesmedienzentrum BW: Fotoarchiv Albrecht Brugger).   1
1.2   The Steinheim Basin with its central mountain (image courtesy
      of Landesmedienzentrum BW: Fotoarchiv Albrecht Brugger).   2
1.3   Suevite (above) and polymictic breccia (below) at Aumühle
      quarry, Ries Basin: During the impact explosion, fractured
      rock material was transported essentially horizontally along the
      ground and deposited as chaotic debris. This material, mainly
      consisting of fragments from Mesozoic rocks, is locally called
      Bunte Brekzie (colourful or polymictic breccia). The grain size
      ranges from house-sized blocks to fine dust. The top of the
      polymictic breccia shows a strong relief. It is overlain by suevite,
      an impact breccia with an appreciable amount of impact melt
      particles. It is characterised by rock fragments mainly from the
      crystalline basement and by vesicular, glassy ballistic bombs,
      locally called *Flädle* (pancakes). Suevite was deposited as hot
      fallout from the impact plume. It is found in small batches, and
      it was not recognised until the 1960s that these batches are only
      the sparse remains of a once continuous ejecta blanket (photo:
      Kölbl-Ebert).   3
1.4   Faults in Ziswingen quarry (Ries Basin): Huge, fractured
      limestone blocks moved along listric faults down into the initial
      crater. These faults are due to the gravitational reorganisation of
      the terrain after the impact (photo: Kölbl-Ebert).   4
1.5   Lacustrine limestone of the post-impact Ries lake at Büschelberg
      quarry (Ries Basin): The limestone is built up by algal reefs. Until
      the early 1960s, these were interpreted as precipitates of thermal
      springs caused by Ries volcanism. Similar deposits also occur at
      the Steinheim Basin (photo: Kölbl-Ebert).   5
1.6   Oblique view over the Rhine-Graben towards the Swabian
      Tableland and the Alps in the far background (drawing by Hans
      Cloos; from Hans Cloos: Gespräch mit der Erde, © 1947 Piper
      Verlag GmbH, München).   6

1.7   The Swabian Volcano: Randecker Maar in the foreground and
        Limburg Mountain in the distance. Due to erosion, the diatremes
        of the 'Swabian Volcano' are cut in different levels. On top of the
        Swabian Alb's surface they appear as shallow depressions. In the
        foreland, the diatreme fillings are harder than the surrounding
        clay-rich country rock and form cone-shaped hills
        (photo: Kölbl-Ebert).   6
1.8   The Swabian Volcano: Chaotic vent breccia in the diatreme at
        Neuffener Steige (photo: Kölbl-Ebert).   8
1.9   Striation at Gundelsheim quarry (east of the Ries Basin): The
        upper surface of Late Jurassic limestone is polished and shows
        numerous parallel scratches resembling glacial action. Polish and
        striae were caused by the horizontal transport of the polymictic
        breccia during impact explosion (photo: Kölbl-Ebert).   11
1.10   Ries model test crater at Swinemünde by Walter Kranz (from
        Kranz 1912: 5).   16
1.11   Reuter's blocks: Reworked distal Ries ejecta (Upper Jurassic
        limestone) within sandy Molasse sediments of the northern
        Alpine foreland (photo: Kölbl-Ebert).   18
1.12   Sketch of the central volcanic Ries explosion (Schuster 1926,
        courtesy of Stadt Nördlingen).   20
1.13   Suevite from Ries Basin (photo: Kölbl-Ebert).   21

2.1   Meteor Crater in the 1930s (Stutzer 1936a: 1).   32
2.2   Otto Stutzer argued in 1936 for an impact origin of the Ries and
        Steinheim Basins (photo from Jurasky 1937: 3).   39

3.1   Richard Dehm denied a central explosion at the Ries Basin
        and opted alternatively for smaller, local volcanic explosions to
        explain the widespread brecciation (photo from Fahlbusch
        1996: 297).   67
3.2   Reinhold Seemann was the main promoter of the idea that the
        features of the Ries Basin evolved through a long history of
        gradual tectonic processes (photo from Hölder 1976: 204).   77

4.1   Drawing of a central mountain in a bomb crater of World War II
        by Ferdinand Trusheim (1940: 3).   100
4.2   Walter Kranz developed, in the 1910s and 1920s, the central
        explosion theory to explain the genesis of the Ries 'volcano'
        (source: Carlé 1987: 3).   106

5.1 Kaalijärv Crater on the Estonian island of Saaremaa in the Baltic
     Sea (photo courtesy of Tõnu Pani, University of Tartu, Estonia).   115
5.2 The Köfels landslide in the Ötztal, Austria, was interpreted by
     Walter Kranz as caused by a volcanic explosion, whereas Franz
     Eduard Sueß opted for a meteorite impact (photo: Kölbl-Ebert).   123
5.3 Köfelsite, a vesicular, slack-like glass from the Köfels landslide.
     The glass was formerly thought to be the product of an obscure
     type of volcanism, while Franz Eduard Sueß interpreted it as
     impact melt (photo: M. Ebert).   123

6.1 The astrophysicist Ernst Julius Öpik published on impact physics
     (photo courtesy of Armagh Observatory).   135
6.2 The German mineralogist Fritz Heide investigated why there are
     no large meteorites preserved on Earth (photo courtesy of Prof.
     Dr Klaus Heide, University of Jena).   139

8.1 Robert Sinclair Dietz promoted shatter cones as macroscopic
     clues to impact processes (photo courtesy of Scripps Institution
     of Oceanography Archives).   174
8.2 *Strahlenkalk*, a shatter cone from the Steinheim Basin
     (photo: M. Ebert).   175
8.3 Phase diagram of SiO_2 showing the temperature and pressure
     conditions under which the high-pressure modifications coesite
     and stishovite form (redrawn using data from <https://www.
     uwgb.edu/dutchs/Petrology/Silica%20Poly.HTM> accessed 12
     November 2013).   182
8.4 Richard Löffler explaining an outcrop at the Ries Basin. After
     Walter Kranz's death, Löffler was the main promoter of a central
     volcanic explosion hypothesis for the Ries Basin (source: Carlé
     1987: 2).   186

9.1 Eugene Merle Shoemaker (right) and Edward Ching-Te Chao
     (left), geologists at the USGS, discovered the high-pressure
     mineral coesite at the Ries Basin in 1960, thus providing the vital
     clue to the impact nature of this structure (photo courtesy of Dr
     Wulf-Dietrich Kavasch).   191
9.2 Coesite from Lake Bosumtwi (Ghana): Thin-section
     photomicrograph (in plane-polarised light) of coesite aggregates
     (dark clusters) within diaplectic quartz glass (Sample LB-44B;
     Suevite from the Bosumtwi crater, Ghana) (Credit: Ludovic
     Ferrière, NHM Wien).   194

9.3   Wolf von Engelhardt (standing), Professor of Mineralogy at
       the University of Tübingen, lecturing about Ries Crater
       geology in front of members of the Apollo 14 crew (photo
       courtesy of NASA).   195

10.1   Georg Wagner was the most vocal critic of the impact theory in
       the 1960s (source: Carlé 1972: 35).   206
10.2   Walter Herman Bucher, impact critic from the USA, joined in
       the Ries debate in 1963 (photo: American Geophysical Union
       (AGU), courtesy AIP Emilio Segre Visual Archives).   214
10.3   Sketch map by Richard Dehm showing the Ries Crater (*Ries*), the
       Steinheim Basin (*Steinhm*), various structural 'lines' (the tectonic
       graben structure of the Upper Rhine Valley (*Oberrheingraben*)
       to the west, the tectonic western boundary of the Bohemian
       Massif (*Fränkische Linie*) to the east, the northern margin of the
       northern Alps (*Nordalpen*) to the south, the tectonic Swabian
       Lineament (*Schwäbisches Lineament*), the northern coastline
       (*Miozäne Klifflinie*) of the Mid-Tertiary Molasse Sea north of
       the Alps (*Molasse-Trog*) and the *Fränkischer Schild Ries-Linie*)
       and volcanically active areas in the Tertiary period (*Kaiserstuhl*,
       Hegau, *Urach* [that is the Swabian Volcano], *Vogelsberg*, *Rhön*
       and unnamed irregular areas surrounded by dashed lines)
       (Dehm 1962: 2).   225
10.4   Sketch of a test explosion of 500 tons of TNT. The developing
       explosion plume consists of several sections, which lead to
       distinct deposits that can also be recognised in the much larger
       Ries impact crater (graphics: Kölbl-Ebert).   244

11.1   The geologist Gerold Heinrich Wagner (source: Carlé 1987,
       detail from 6).   249
11.2   In the quarry of *Altenbürg* a steep contact between suevite and
       a huge allochthonous limestone block can be seen. A second
       such contact with another limestone block is to the left outside
       the image. Formerly, this feature was interpreted as a tuff-filled
       volcanic vent in contact with autochthonous country rock
       (photo: Kölbl-Ebert).   259
11.3   Chaotic fractures in Ries rock drawn by Rudolf Hüttner (side
       length of the cubes: 8 cm; from Hüttner 1969: 147, 8).   283

12.1   PDFs in quartz from the Ries Crater (Germany): Thin-section
       photomicrograph (in crossed polars [CP]) of a shocked quartz
       grain with two sets of planar deformation features (PDFs)
       (Sample NHMW-L4665; Suevite from the Otting quarry, Ries,
       Germany) (Credit: Ludovic Ferrière, NHM Wien).   304
12.2   Astronauts and instructors at field training in the Ries Crater.
       From left to right: M. Abadian, Wolf von Engelhardt, unknown,
       Edgar Mitchell, Joe Engle, Friedrich Hörz, Eugene Cernan,
       Dieter Stöffler, Alan Shepard (photo courtesy of NASA).   311

We are in danger of error, if we seek for slow causes and shun violent agencies further than the facts naturally direct us, no less than if we were parsimonious of time and prodigal of violence. *Time*, inexhaustible and ever accumulating his efficacy, can undoubtedly do much for the theorist in geology; but *Force*, whose limits we cannot measure and whose nature we cannot fathom, is also a power never to be slighted: and to call in the one to protect us from the other, is equally presumptuous, to whichever of the two our superstition leans.

(William Whewell 1858: 592)

Preface

For much of the twentieth century, geologists in Germany struggled to understand two large and conspicuous geological structures in southern Germany: the Steinheim Basin and the larger Ries Basin. What at first glance looked like the product of volcanic activity turned out to be bewilderingly complex and enigmatic. Huge volumes of debris, from finest rock powder to house-sized blocks, littered the landscape, apparently products of two gigantic explosions. Nowhere else on Earth was volcanism known to have achieved such a catastrophic effect; and where were the volcanic rocks anyway? There was no lava, but only a small amount of tuff at Ries Basin resting on top of the blasted debris and therefore obviously deposited later, after the actual explosion. The more people looked, the more they were puzzled. This oxymoron was known as 'The Ries Problem'.

In 1960 – suddenly, according to geohistory – and resulting from the discovery of high-pressure minerals in the Ries Basin by American mineralogists, the volcanic explanation gave way to an interpretation of both structures as impact craters; a shift of paradigm that has been enthusiastically termed a Copernican turning point and hailed as a profound conceptual change by historians of geology (Engelhardt 1994; Marvin 1986: 140).[1] The invention of a new methodology had finally brought progress to geological investigation of

---

[1]   There are several articles or sections in books about the history of the various theories on the formation of Ries and Steinheim Basins (Adam 1980, Carlé 1987, Dehm 1969, 1978, Dorn 1950, Engelhardt 1982, 1987, Engelhardt and Zimmermann 1982, Grau 1978, Grau and Höfling 1978, Hölder 1962, 1971, 1989, Horn 1991, Hüttner 1974, Kölbl-Ebert 2003, 2013, Mark 1995, Pösges and Schieber 1994, Preuss 1965, Reif 1976, Sachs 2011). Most of these give a more or less detailed account of the various historical theories concerning the origin of the two geological structures, classifying these theories into groups (volcanic, glacial, tectonic and so on) and naming their most distinguished promoters in chronological order. Dorn (1950) gave the most detailed overview of older, sometimes forgotten, pre-impact theories, whereas the study by Engelhardt (1982) is recommended as a general treatise up to the recognition of Ries Basin as an impact crater. Carlé (1987) gives a personal narrative of the events around 1960, which is, however, sometimes a little misleading when compared to new evidence from archive material and oral interviews (this study). Apart from Carlé (1987), Hölder (1962, 1971), Kölbl-Ebert (2003, 2013) and Mark (1995) little has been done beyond a systematic approach to research history of the German craters; thus suggesting a smooth story of scientific progress and final success. To overcome this simplified view has been the aim of this research project.

the Ries Basin and solved the problem, while 'irrational resistance' to the new theory and a resulting lack of communication were but unimportant sociological 'phenomena of the contingent history of science, which one has to understand as historian, but which one has always to combat as scientist and philosopher' (Engelhardt and Zimmermann 1982: 341).[2]

According to local historians of geology, an understanding of the research history of the two impact craters in terms of epistemic progress and rationality was important (Engelhardt and Zimmermann 1982: 340):

> We may ... call the change of theory a 'revolution', if this word is not to mean more than the fact of the rejection within a few years of many hypotheses, theories and regulative principles, which have never been doubted over many decades. We may also speak of 'conversion', if the individual researcher suddenly decides to solve his problems in the light of the new theory. All this, however, happened and happens in the course of progress of theoretical knowledge, which can be rationally understood and reconstructed according to the model of 'refined' falsificationism.[3] (Engelhardt and Zimmermann 1982: 348)[4]

Upon closer inspection, the paradigm shift from volcanic explanations of the Ries and Steinheim Basins to the impact theory was indeed not an event in a strictly Kuhnian sense (Kuhn 1962). However, it was neither simply a case of methodological and thus epistemic progress, nor was it as rational and well-reflected as Engelhardt and Zimmermann wished it to be. The epistemic shift was also not that sudden, since it was preceded by some thirty years of argument and followed by at least another decade of discussion – yet it had strong and pervasive repercussions on local scientific culture, which were focused on the early 1960s.

The shift from volcanic to impact theory at the Ries Crater and Steinheim Basin was initiated by a sudden reconnection of German researchers with the international scientific community, after some twenty-five years of scientific

---

2   '... Phänomene der kontingenten Wissenschaftsgeschichte, die man als Historiker verstehen, als Forscher und Philosoph aber immer bekämpfen muß' (Engelhardt and Zimmermann 1982: 341).

3   Lakatos (1970, 1971)

4   'Man mag den ... Theorienwandel eine 'Revolution' nennen, wenn mit diesem Wort nicht mehr bezeichnet werden soll als die Tatsache der in wenigen Jahren vollzogenen Abkehr von vielen über Dezennien hin niemals bezweifelten Hypothesen, Theorien und regulativen Prinzipien. Man mag auch von 'Bekehrung' sprechen, wenn der einzelne Forscher sich mit einem Mal dazu entschließt, seine Probleme im Lichte der neuen Theorie zu lösen. Dies alles geschah aber und geschieht im Zuge eines Fortschritts der theoretischen Erkenntnis, der nach dem Modell des 'verfeinerten' Falsifikationismus rational verstanden und rekonstruiert werden kann' (Engelhardt and Zimmermann 1982: 348).

isolation from the more progressive ideas of their global contemporaries during the build-up to World War II and its aftermath. The reconnection not only introduced a new perspective about the place of the Earth in relation to the cosmos, but also cured German geology of its former self-sufficiency and patriotic arrogance by integrating German geology into an interdisciplinary and international framework. The early 1960s saw a rapid transfer of knowledge and methodology from American mineralogy and physics to German geology, which led to a shift from simple, field-based geology to more chemically and physically oriented geoscience, from pure stratigraphy to a process-oriented view, from local to global interests and, not least, from chauvinism to internationality.

A variety of printed sources, detailed archival materials including letters and personal notes and interviews with veterans of research on the Ries Basin, now provide for a detailed reconstruction of both the historical sequence of events and the personal thoughts, emotions, motives and intentions of the scientists involved. These topics will be explored in the following chapters.

Wolf von Engelhardt, the great old man of German impact research (Chapters 9–12) and historian of science, would have preferred to remain silent about this particular part of the history of geology, which he deemed no 'honourable page in the annals of German geology' (Engelhardt pers. comm. 30 August 2000). The history of science, however, is not only about scientific progress and triumph or about highlighting scientific genius; it is also about how science was actually done, about the circuitous routes that may be needed to achieve what later generations might hopefully call progress, and the vulnerability of science under specific social and historical conditions.

Although 'open-mindedness' is a treasured cliché among scientists to characterise themselves, it must not be forgotten that conservatism (or rather scepticism) against the fanciful and seemingly fantastic is also part of the scientific character – and for good reason. Stubborn resistance to scientific discovery is often explained vaguely as the result of human nature, through a psychological bias or a general *Zeitgeist*, but such 'garbage bin terms' explain everything as well as nothing (Barber 1961: 596–7). Resistance in the face of new scientific evidence is not an embarrassing mental disease that is best not talked about, but instead follows its own logic – an emotional as well as rational logic – whose exploration allows a deeper understanding of the functioning of science and scientists in times of crisis.

Note on Sources and Translations

The most important archives for this study have been the Geologists' Archive of the *Geologische Vereinigung*, which is housed at the library of the University of Freiburg, and the Georg Wagner papers housed at the library of the University of Tübingen. These and other sources have provided a wealth of information, and I have quoted lavishly from them throughout the following chapters as far as the archive laws of Baden-Württemberg allow, because I think it is important not only to provide readers with my thoughts and reasoning, but also to give them all the primary sources they need to evaluate the argument for themselves.

In Baden-Württemberg, however, the use of more recent archival material is restricted to protect people's privacy for up to ten years after their death. For the more recent developments, such as the development of the concept of impact metamorphism, we therefore have less information on personal thoughts and must rely instead on published statements, because the relevant archival material will remain inaccessible for a few more years (or in some cases, decades). This problem was, however, greatly mitigated by the generosity and historical interest of some veterans of Ries Crater research, who either granted me full interviews, spoke to me on the telephone or wrote letters, thus adding their private personal views to the more recent events in Ries Crater research.

The tape-recorded interviews were conducted over the course of several years. They were subsequently transcribed by two assistants and the quoted passages were corrected by the author. The resulting text naturally contains considerable dialect, slang, colloquial language and incomplete sentences. I have tried to keep the English translations as close to the original interviews as possible, except where a literal translation would have obscured the intended meaning. The German original is provided in a footnote that the reader can compare with the translation. Seemingly obsessive shortening of some quotes – signalled in the text by: ... – is no act of quote mining. It is used to eliminate repetitions or digressions that would not help to clarify the narrative. The tapes and transcripts used in this study will be donated to the Geologists' Archive for future research.

Acknowledgments

I am deeply indebted to my interview partners, Helmut Hölder (†), Rudolf Hüttner (Suggenthal), Hans Pichler (Mössingen), Jean Pohl (Munich), Eugen Seibold (†), Peter Signer (Grüt, Switzerland), Rudolf Trümpy (†) and Helmut Vidal (†), for generously sharing their memories. Without their willingness to speak to me and into my microphone, this book would not have been possible. I am likewise grateful to Wolf von Engelhardt (†), Volker Fahlbusch (†) and Wolf-Dieter Grimm (Munich) for providing additional insights via letters or telephone conversations and to Wendy Cawthorne (Geological Society Library, London, UK), Astrid Göbel (Residents' Registration Office, Steinen), Magdalena Kaufmann (Deutsch-Französisches Forschungsinstitut Saint-Louis, France), Halima Khanon (Royal Geographical Society, London, UK), Reinhard Kirner (Bayerisches Hauptstaatsarchiv: Kriegsarchiv, Munich), Marianne

Klemun (University of Vienna, Austria), Ervins Luksevics and Maris Rudzitis (both University of Latvia, Riga), Tõnu Pani (University of Tartu, Estonia), as well as Johannes Seidl (University Archive Vienna, Austria) for providing additional biographical information for hard-to-trace personalities.

I also want to heartily thank the following archives and their staff members for granting me generous access to their repositories: The Geologists' Archive (*Geologenarchiv*) at the University Library of Freiburg i. Br. (Ilse Seibold, Kathrin Lutz, Doris Schweizer), the University Library of the University of Tübingen (Johannes M. Wischnath, Frank Westphal, Irmgard Wagner), the Archive of the Technical University Mining Academy Freiberg (Angela Kießling), the Library of the Geographical Institute of the University of Bonn, the University Archive of the University of Munich, the Archive of the Deutsches Museum (Munich), the Fürst zu Oettingen-Spielberg'sches Archiv (Hartmut Steger), the Centre for Ries Crater and Impact Research Nördlingen (Gisela Pösges) and the American Institute of Physics.

I am also much indebted to Martin Scheuplein (Regensburg) and Monika Boulesnam (Munich) for transcribing the interview tapes.

Sincere thanks are additionally owed to my husband Martin Ebert (Eichstätt), Bernhard Fritscher (†), Martin Guntau (Rostock), Ursula Marvin (Harvard, USA), Peter Schimkat (Kassel) and Hugh Torrens (University of Keele, UK), for helpful discussions, information and encouragement.

Thanks are also due to the following copyright owners for permission to use their images: American Geophysical Union, Armagh Observatory, Bayerisches Landesamt für Umwelt, Bayerische Staatssammlung für Paläontologie und Geologie, Ludovic Ferrière and Christian Koeberl (Natural History Museum Vienna), City of Nördlingen, E. Schweizerbart'sche Verlagsbuchhandlung, Gesellschaft für Naturkunde in Württemberg, Klaus Heide, Landesmedienzentrum Baden-Württemberg, Martin Ebert, NASA, Piper Verlag GmbH, Scripps Institution of Oceanography Archives, Senckenberg Gesellschaft für Naturforschung, Technische Universität Bergakademie Freiberg, Tõnu Pani and Wulf-Dietrich Kavasch.

I am especially grateful to Jennifer Lane (American Museum of Natural History, New York, USA) and David Oldroyd (†) for editing the English of the manuscript.

Finally, I also wish to thank the staff of Ashgate Publishing Ltd for their professional help and friendly support.

MARTINA KÖLBL-EBERT

Chapter 1
Introducing the Smoking Gun

Figure 1.1   The Ries Crater. (Image courtesy of Stadt Nördlingen and Landesmedienzentrum BW: Fotoarchiv Albrecht Brugger).

The Ries Basin (or *Nördlinger Ries* as it is called in Germany, after the town of Nördlingen situated within the structure) and the much smaller Steinheim Basin (*Steinheimer Becken*) some 30 km further to the west-southwest, are geological structures that stand out on any geological map of southern Germany. Both structures are nearly circular. The diameter of the Ries Basin (Figure 1.1) is about 24 km and that of the Steinheim Basin (Figure 1.2) is some 3.4 km. Both structures show an extensive brecciation of the country rock. The Steinheim Basin has a central mountain consisting of rocks, which obviously have been raised for several hundred metres, and within the Ries as well there is a low ring of hills displaying rocks that usually lie at considerable depth. In the area surrounding the Ries Basin, there are patches of a strange rock called suevite, which in the hand-specimen looks similar to a volcanic tuff, whereas in the Steinheim Basin there is no sign of volcanism apart from a limestone resembling the precipitate of a hot spring. Comparable limestone also occurs within the Ries Basin.

These two basins originated some 15 million years ago in one of the most dramatic events geology has to offer: the impact of a small asteroid (a celestial body about one kilometre across) and its tiny moonlet, destroying a lushly vegetated, subtropical countryside and destroying all life for many thousand square kilometres.

* Figure 1.2   The Steinheim Basin with its central mountain. (Image courtesy of
Landesmedienzentrum BW: Fotoarchiv Albrecht Brugger).

The asteroid had an approximate velocity of 70,000 km/h (about 20
km/s). Less than a second before it hit the Earth's surface, the air between
the asteroid and the Earth became highly compressed and heated, causing
the land surface to melt. A jet of melt droplets, which solidified to green
glass during flight, was expelled. These tektites (so-called 'moldavites') can
be found today in the Czech Republic some 400 km from the Ries Basin.
In under a second, the asteroid was halted upon impact. During this time it
penetrated some two kilometres into the Earth's surface. Because of the strong
compression of the rocks upon impact and the braking of the projectile,
the temperature reached several tens of thousands of degrees Centigrade,
leading to near-instant vaporisation and explosion of the projectile and the
target rock.

From the impact site, broken (but also heated and molten) rock was
ejected, creating an explosion crater some 15 km across and 4 km deep.
A mushroom-shaped, incandescent cloud rose up to twenty kilometres
in height. The expelled debris fell down again and covered the ground
to a distance of at least 40 km from the crater (Figure 1.3), changing the
topography so that rivers had to dig new courses.

A pressure wave that rushed from the crater site uprooted trees and
destroyed all life to an even greater distance, permanently deforming and

Figure 1.3   Suevite (above) and polymictic breccia (below) at Aumühle quarry, Ries Basin: During the impact explosion, fractured rock material was transported essentially horizontally along the ground and deposited as chaotic debris. This material, mainly consisting of fragments from Mesozoic rocks, is locally called *Bunte Brekzie* (colourful or polymictic breccia). The grain size ranges from house-sized blocks to fine dust. The top of the polymictic breccia shows a strong relief. It is overlain by suevite, an impact breccia with an appreciable amount of impact melt particles. It is characterised by rock fragments mainly from the crystalline basement and by vesicular, glassy ballistic bombs, locally called *Flädle* (pancakes). Suevite was deposited as hot fallout from the impact plume. It is found in small batches, and it was not recognised until the 1960s that these batches are only the sparse remains of a once continuous ejecta blanket. (Photo: Kölbl-Ebert).

transforming minerals and rocks in the vicinity of the crater. Fine ash must have rained down over half of Europe. The unstable rim slumped into the crater (Figure 1.4), while the rebounding crater floor rose so that the crater became wider and shallower until it was a few hundred metres deep and approximately 24 km wide. The crater eventually filled with water, which was initially rather acidic and brackish because of its high content of carbonate and other solutes (Figure 1.5). Eventually, life returned to this desolate wasteland.

Figure 1.4   Faults in Ziswingen quarry (Ries Basin): Huge, fractured
limestone blocks moved along listric faults down into the initial
crater. These faults are due to the gravitational reorganisation of
the terrain after the impact. (Photo: Kölbl-Ebert).

Although much of the fallout blanket from the asteroid strike has been
eroded since Miocene times, there are still impressive amounts of breccia,
suevite and other impact features, making the Ries Basin a very special place for
geologists.[1]

The Steinheim Basin formed at the same time as the Ries Basin and is the
location at which the 'moonlet' of the Ries asteroid (possibly some 80 m in
diameter) touched down. The centre of the Steinheim Basin is occupied by a hill
called the *Klosterberg* (Figure 1.2). Outside this central mountain is the crater
basin, in which the villages of Steinheim and Sontheim are situated. Still further
to the periphery is a brecciated ring zone containing huge blocks of dislocated
country rock. There is no fallout blanket around the crater, because the material

---

[1]   There are two museums dedicated to the regional impact geology, Rieskratermuseum
in Nördlingen (Ries Basin; <http://www.rieskrater-museum.de/>) and the Meteorkrater-
Museum in Sontheim (Steinheim Basin; <http://www.steinheim.com/meteor/>).
Information relating to the impact scenario, to outcrops and field trips can be found in
BGL (1999), Chao et al. (1978, 1992), Groschopf and Reiff (1993), Hüttner and Schmidt-
Kaler (1999), Kavasch (1997), Mattmüller (1994), Stöffler and Ostertag (1983). General
information on impact processes is provided by French (1998). For information on the post-
impact lake sediments see Arp (2006).

Figure 1.5   Lacustrine limestone of the post-impact Ries lake at Büschelberg quarry (Ries Basin): The limestone is built up by algal reefs. Until the early 1960s, these were interpreted as precipitates of thermal springs caused by Ries volcanism. Similar deposits also occur at the Steinheim Basin. (Photo: Kölbl-Ebert).

constituting the blanket has been eroded since Miocene times. After the impact, the basin filled with water to form a crater lake approximately 55 m deep. Around the central mountain and the basin rim, algal reef deposits are preserved from this time. Numerous other fossils have also been found.

The impact scenario for both the Ries and Steinheim Basins today is common knowledge and taught at schools and universities. However, only a few decades ago, this was completely different. Both structures until quite recently were interpreted as volcanoes, although they did not resemble any other known volcanic edifice.

Geological Context

In order to understand the historical arguments, some background information about the wider geological setting of the Ries and Steinheim Basins is necessary. The two craters are situated at the eastern end of one of the most beautiful landscapes in Germany, formed by the escarpment of the Swabian Alb and the erosional remnants of the so-called 'Swabian Volcano'.

Figure 1.6   Oblique view over the Rhine-Graben towards the Swabian Tableland and the Alps in the far background. (Drawing by Hans Cloos; from Hans Cloos: Gespräch mit der Erde, © 1947 Piper Verlag GmbH, München).

Figure 1.7   The Swabian Volcano: Randecker Maar in the foreground and Limburg Mountain in the distance. Due to erosion, the diatremes of the 'Swabian Volcano' are cut in different levels. On top of the Swabian Alb's surface they appear as shallow depressions. In the foreland, the diatreme fillings are harder than the surrounding clay-rich country rock and form cone-shaped hills. (Photo: Kölbl-Ebert).

For some 160 million years in this area, from the beginning of the Permian to the end of the Jurassic, sedimentary rocks accumulated under diverse depositional and climatic regimes. These rocks include remnants of former seas, with environments ranging from oxygen-poor basins to tropical sponge- and coral-reefs. Other rocks represent a variety of terrestrial depositional basins: barren deserts as well as humid, fluvial landscapes. The rocks forming under these wide-ranging conditions are similarly varied: soft clay and marl alternates with hard sandstone and limestone in a huge layer-cake of rocks.

The Tertiary period saw a major geological transformation of the European Continent. To the south, the collision of the African and European plates led to the formation of the Alps, a process whose influence was also felt much farther north.

The Bohemian Massif was raised along fault lines and some forty million years ago in south-western Germany and eastern France, extensional tectonics led to formation of the Rhine-Graben as the extension and thinning of the continental crust led to a greater transfer of heat from the elevated mantle. Along deep-seated fault zones, magma was able to rise through the Earth's crust and for the first time after an interval of some 200 million years, all of southern Germany was dotted with volcanoes.

The initially horizontal sediments were tilted because of the uplift around the Rhine-Graben by about 2–3°. Erosion occurred along the graben margins (Figure 1.6). Soft beds such as clay deposits were easily eroded, whereas more resistant limestones and sandstones formed steep escarpments, of which the so-called *Trauf* of the Swabian Alb is the most impressive.

Here at the present Alb-escarpment, between the towns of Kirchheim and Urach over an area of some 1,500 km², the Earth's crust is perforated by more than 350 volcanic diatremes, which together have been referred to by local geologists as 'The Swabian Volcano' (Mäussnest 1974, 1978).

Such diatremes are found on top of the Alb-plateau as well as in the foreland. The different erosional levels make it possible to study the internal structure of maar volcanoes, one of the most common types of terrestrial volcano (Figure 1.7).

A maar-type volcano, together with its associated diatreme (the funnel-shaped structure underneath the maar; Figure 1.8), is formed when groundwater gains access to the magma from some tens to as many as three hundred metres below the surface. Violent explosions of water vapour then occur – so-called phreatic (pure vapour explosions) or phreatomagmatic eruptions (juvenile material is also produced). The eruptions are caused by the rapid expansion of vaporising water, which thereby increases from 1,000 to 2,000 times its original volume. During this process, the rock melt (if present) will be cooled down from its initial temperature of about 1,200°C within seconds and fragmented. Entire series of pulsating water vapour explosions, which may last for days, open funnel-shaped diatremes that are further deepened with each explosion pulse.

Figure 1.8   The Swabian Volcano: Chaotic vent breccia in the diatreme at Neuffener Steige. (Photo: Kölbl-Ebert).

During such an eruption, only a small amount of juvenile material actually reaches the surface. The later infilling of the diatreme consists mainly of country rock fragments of all size-ranges, but may also contain volcanic ash and lapilli.

At the surface, the fragmented fallout of the eruption will form a low tuff ring with a bowl- to funnel-shaped depression in the centre (the maar), the bottom of which lies lower than the initial surface of the country rock. In a humid climate these maars fill with water to form circular lakes, which eventually fill up with sediment and mature into a moor.

Since the most recent activity of the Swabian Volcano approximately sixteen million years ago, several tens of metres of the Alb surface have been lost to erosion. Therefore, the initial tuff rings around the maars are no longer preserved, although parts of them have slumped into some of the maar depressions and can still be studied in the field.

For the history of science, it is interesting to note that the diatremes of the Swabian Volcano initially were interpreted by geologists such as Wilhelm von Branca and Eberhard Fraas (early twentieth century) as 'embryonic' volcanoes thought to have been caused by underground explosions of volcanic gases 'punching' their way to the surface but never maturing into a 'true' volcanic edifice. Being roughly contemporary[2] with the impact craters of the Ries and

---

[2]   According to modern radiometric dating, there is about one million years difference.

Steinheim Basins, they too were regarded as (in this case, smaller) examples of an aberrant, gas-driven volcanism. In the 1940s, geologist Hans Cloos (1885–1951) proposed a process of 'tuffitisation' or 'gas-tuff flux' leading to diatreme formation without violent explosions. It was not until the early 1980s that the Swabian diatremes were reviewed and classified within the theoretical framework of modern volcanology (Lorenz 1982; see also Keller et al. 1990).

Early Research on the Ries Basin

Beginning in 1762, Carl von Caspers (1743–1813) had been in military service in the Bavarian Kurpfalz. In 1778 he joined the Kurpfalz-Bavarian corps of military engineers. From 1784 (with an interruption between 1803 and 1806) he was transferred from Kurpfalz on the river Rhine to the building site of the fortress at Ingolstadt, north of Munich. From 1790 to 1799 he was head of the building department, with the rank of Upper Lieutenant in Engineering (Klarmann 1896: 21; pers. comm. Reinhard Kirner, *Bayerisches Hauptstaatsarchiv*, 27 April 2012). During his work in Ingolstadt, von Caspers was concerned by the lack of hydraulic concrete, which in the Kurpfalz had been made from the Eifel ignimbrites and ash tuffs. Consequently, he started to look for similar rocks in the vicinity of Ingolstadt. Soon, he came across a suitable substitute that was to be found in the wider surroundings of Nördlingen and which in 1792 he described in a short publication as volcanic tuff. Based on this conclusion, he quite naturally assumed that there must have been volcanoes in Bavaria (Hüttner 1974; Engelhardt 1982).

In the mid-nineteenth century (1848), systematic geological mapping commenced with the first geological map of the area at a scale of 1:100,000 by Adalbert Schnitzlein (1814–68, apothecary in Feuchtwangen and later Professor of Botany at the University of Erlangen: Freyberg 1974: 140) and Albert Frickhinger (1818–1907, apothecary and politician from Nördlingen: Freyberg 1974: 45–6). These two researchers came to the conclusion that the Ries Basin must have experienced uplift of its rim while the central part had subsided as a result of volcanic action (Pösges and Schieber 1994).

Carl Wilhelm (von) Gümbel (1823–98)[3] mapped in the Bavarian part of the Ries Basin with his Bavarian Geognostic Survey and map sheets were published in 1870, 1891 and 1894. Because all of the tuffs he encountered looked similar,

---

[3]   Carl Wilhelm (von) Gümbel studied sciences, especially mineralogy and geognosy, in Munich and Heidelberg. In 1848, he received a degree in mining in Munich. He then worked as a trainee in the mining and saltern industry. In 1851, he became geognost at the newly founded Geognostic Survey of the Bavarian Kingdom. In 1856, he became its sole head and, in 1879, he also became director of the Bavarian mining administration. Gümbel also lectured at the Munich University (Dehm 1966; Sperling 2001).

Gümbel concluded that all of them must have originated from some eruption centre within the Ries Basin. He distinguished several phases of volcanic activity: first, elevation and brecciation of the country rock took place; then a central volcano spewed out tuff; followed by a phase of subsidence accompanied by the formation of a lake. The smaller Württemberg portion of the Ries Basin was mapped by Oscar Fraas (1824–97)[4] and Carl Deffner (1817–77)[5] in the second half of the nineteenth century, during which time they found evidence for lateral transport of chaotic breccia (Engelhardt 1982; Dorn 1950).

However, with more geological knowledge accumulating, the Ries Basin began to puzzle geologists and quite a number of auxiliary and alternative geological hypotheses were developed about the genesis of the structure. Somewhat resigned, Carl Deffner formulated: 'This Ries is a strange volcanic area, a Sphinx deeply sunk into sand and mud, which poses riddles to the scientist, riddles that are not to be solved in a quick victorious sprint' (Deffner 1870).[6]

Outside the southern and western rim of the Ries Basin in particular, a number of quarries provide access to rocks of the undisturbed Jurassic limestone country rock and overlying blanket of ejecta. The upper surface of the limestone is polished and shows numerous parallel scratches (Figure 1.9) resembling the result of glacial action. Because the contemporary volcanic theories demanded a Ries Mountain[7] anyway, it is no wonder that in 1870 Carl Deffner proposed that glaciers descending from this hypothetical Ries Mountain not only caused the scratches, but also carried imbricated slivers of debris into the Ries Basin's surroundings (Engelhardt 1982; Dorn 1950; Pösges and Schieber 1994).

The idea of glaciers in the Ries Basin was revived around the turn of the twentieth century by Ernst Koken (1860–1912), Professor of Geology, Palaeontology and Mineralogy at the University of Tübingen. He had studied

---

[4]   Oscar Friedrich (von) Fraas studied theology, palaeontology and geology in Tübingen, Paris and Würzburg. He obtained a theologian's degree in 1845 and a doctorate in geology from the University of Würzburg in 1850. In 1848, he became curate in Leutkirch and in 1850 vicar in Laufen an der Eyach. In 1854, he was appointed curator of the Natural History Cabinet in Stuttgart (Quenstedt 1961a; Hantzsch 1904).

[5]   Carl Deffner was the owner of a factory for metalwares and an amateur geologist (<http://bsbndb.bsb.lrz-muenchen.de/sfz9492.html> accessed 5 April 2011).

[6]   'Ein merkwürdiges vulkanisches Gebiet ist dieses Ries, eine tief in Sand und Schlamm versunkene Sphinx, welche dem Forscher Rätsel aufgibt, die nicht in kurzem Siegeslauf zu lösen sind' (Deffner 1870).

[7]   The Ries Mountain was a hypothetical stage in the formation model of the Ries Basin as developed by Gümbel (see above). He assumed that the formation of the Ries Basin started with the up-doming of a mountain leading to the brecciation of the country rock. The mountain then produced suevite tuff. Afterwards the mountain subsided again to form the present basin.

Figure 1.9   Striation at Gundelsheim quarry (east of the Ries Basin): The upper surface of Late Jurassic limestone is polished and shows numerous parallel scratches resembling glacial action. Polish and striae were caused by the horizontal transport of the polymictic breccia during impact explosion. (Photo: Kölbl-Ebert).

geology in Göttingen and Berlin at a time when glacial geology was beginning to be accepted as a powerful explanation for many geological features of northern Germany. Koken worked on fish otoliths[8] for his PhD, and his *Habilitation*[9] in 1888 was on fossil crocodilians. In 1890 Koken became a professor in Königsberg and also undertook excursions to Sweden and Estonia. In 1894, he was called to the chair in Tübingen, where his entrance lecture was about the Ice Age (Hüttner 1974; Hölder 1989: 160; Engelhardt and Hölder 1977: 133–6).

Koken's preoccupation with the idea that glaciers had shaped the Ries Basin did not simply reflect a fashion of his time, but instead resulted from a misinterpretation of fossil shells within the Ries breccias. For some time, these were thought to be late Tertiary in age, introducing a considerable time

---

[8]   A mineralised structure of the inner ear, which can be found isolated in the sedimentary record.

[9]   In Germany, in addition to the doctorate a second thesis is required to qualify for professorship. This qualification is called *Habilitation*.

gap between the Ries lake sediments (Miocene) and the deposition, or at least reworking, of the Ries debris (thought to be Pleistocene) (Engelhardt 1982).

Elevation and Subsidence

Koken's model became less convincing when Wilhelm von Branca and Eberhard Fraas, in a series of papers published between 1901 and 1914 (for example: Branco and Fraas 1905; Fraas 1900, 1903, 1914), were able to demonstrate the wide distribution of the breccias far outside the Ries Basin.

Karl Wilhelm Franz von Branca (1844–1928), known until 1907 as Branco, was initially a military officer and farmer who later studied geology in Halle and Heidelberg, where he obtained his doctorate in 1876. After geological work at the universities of Rome, Strasbourg and Munich, he *habilitated* in Berlin in 1881. For some time he was a geologist at the Prussian Geological Survey. In 1887 he became Professor of Geology and Palaeontology in Königsberg. In 1890 he switched to Tübingen. During the winter of 1894–95, he took a prolonged holiday because of some 'nervous complaint' and then applied for his dismissal. In 1895, however, he regained his health and became a professor at the agricultural high school in Hohenheim (near Stuttgart) and from 1899 to 1917 Branca was a professor in Berlin. Initially interested in palaeontology, he later became interested in the 'volcanic embryos' of the 'Swabian Volcano' and the Ries Basin, on which he published from 1901 onwards. Branca also initiated the Tendaguru expedition to German East Africa (1909–12), today evidenced by the impressive dinosaur skeletons in the Berlin Museum of Natural History (Quenstedt 1955; Engelhardt and Hölder 1977: 131–2).

Eberhard Fraas (1862–1915) studied geology and mineralogy in Leipzig and Munich, where he received his PhD on starfish of the Upper Jurassic in 1886. After his *Habilitation* in geology and palaeontology in 1894 in Munich, where he focused on Alpine geology, he succeeded his father, Oscar Fraas, as curator of the natural history cabinet in Stuttgart, where he had worked as an assistant from 1891. He published on the geology of Württemberg, especially the Triassic formation, but also on the Ries and Steinheim Basins and the palaeontology of Africa and North America. His main expertise was in vertebrate palaeontology (Quenstedt 1961b; Engelhardt and Hölder 1977).

Instead of glacial action, both Branca and Fraas opted for a gravitative sliding[10] of the Ries debris – assisted by volcanic explosions – from the hypothetical Ries Mountain, followed by subsidence that turned the mountain into the present basin, accompanied by eruptions of tuffs (suevite) from numerous eruptive vents in a ring

---

[10]   Rather similar to Alpine nappe tectonics, which had newly been discovered in the decades before (Engelhardt 1982).

outside the basin rim (Engelhardt 1982; Dorn 1950). These local tuff eruptions were thought to have added to the brecciation of the rock in their immediate vicinity (Pösges and Schieber 1994), while they themselves were undisturbed by tectonic movements and thus represented the end of the genetic events at the Ries Basin: 'First of all, we observe here too [at the Ries rim] truly volcanic products in the shape of liparite-tuffs with gorgeous glassy lava-bombs. ... the eruption vents at the rim-zone [are] undisturbed and penetrate the allochthonous Mesozoic cover; that is they are younger than the great dislocations' (Fraas 1903: 10).[11]

These ideas were fuelled not only by contemporary or earlier volcanological concepts, but also by the presence of a ring of brecciated rocks – mostly gneisses and granites – generated from the crystalline basement of the Ries Basin. In the area surrounding the Ries Basin, this crystalline basement lies some 500 m deeper and so the morphological Ries depression in fact had an elevated 'core':

> Let us begin with the granite of the Ries; so nearly all opinions conform that it has experienced an elevation in order to move to the level of the Jurassic at which we find it today. Also the opinions conform that today's Ries Basin is the product of subsidence, which followed the elevation. ... In this, too, there is only one view that these enormous movements were due to volcanic forces, the products of which we recognize in the liparite tuffs. ...
>
> Under the presumption that the Ries has experienced an enormous elevation, the conclusion is appropriate that the slumping from this mountain becomes larger the closer we come to the rim, because there the slope can be assumed to have been steepest. ...
>
> Branco and I and von Knebel[12] as well, go even further and assume that these slumping blocks slid far across the Ries rim onto the neighbouring areas of the Alb, thereby smoothing the underlying surface. ... The question is forced on us: Which power was able to press up a round piston of nearly 25 km diameter from the deep, without blasting and fracturing the overlying cover completely but pushing large blocks to the side? (Fraas 1903: 11–13)[13]

---

[11] 'Zunächst beobachten wir auch hier echte vulkanische Produkte in Gestalt von Liparittuffen mit prächtigen glasigen Lavabomben. ... die Explosionskanäle der Randzone [sind] ungestört und durchsetzen das dislozierte Deckgebirge. Sie sind also jünger als die grossen Dislokationen' (Fraas 1903: 10).

[12] The geologist and volcanologist Walter J.H.C.M.R von Knebel (1880–1907) obtained his PhD in 1902 from the University of Berlin. After a brief employment as an assistant at the University of Erlangen, he became a lecturer at the University of Berlin in 1903. In 1907 he died through an accident during fieldwork in Iceland (Freyberg 1974: 82; Mohr 2010).

[13] 'Beginnen wir mit dem Granit des Rieses, so stimmen fast alle Ansichten darin überein, dass derselbe eine Hebung erfahren hat, um in das Niveau des Jura zu rücken,

This power, both Branca and Fraas claimed, was not tectonics, but volcanic forces of a special kind: 'A simple volcanic explosion, however, does not explain the conditions, because its effect is very different: We must assume tensions that first caused an enormous up-doming, which only then found relief in explosions to finally, in connection with subsidence of the cover, exhaust itself' (Fraas 1903: 13).[14]

The up-doming was supposed to be caused by a 'laccolith', a large intrusive mass. A series of gas explosions or 'local outbursts' from various localities, partly accompanied by tuff eruption, would add to the brecciation, as the granites too were mixed with tuffs: 'Here we do not have, as Branco and I formerly assumed, a uniform crystalline basement that has experienced only vertical movement in the sense of up-doming and later subsidence, but a material that has partly suffered enormous horizontal pushing and in parts shows all signs of explosive volcanic activity' (Fraas 1903: 14).[15]

Central Explosion

While Branca and Fraas in their early papers had opted for a purely nappe-style gravitational sliding or thrusting of the Mesozoic cover from the 'Ries Mountain', they later modified this idea because gravity alone would not provide enough energy to account for the widespread brecciation. They added

---

in welchem wir ihn heute finden. Ebenso stimmen die Ansichten darin überein, dass der heutige Rieskessel das Produkt einer Senkung ist, welche der Hebung nachfolgte. ... Auch darin herrscht nur eine Ansicht, dass diese gewaltigen Bewegungen auf vulkanische Kräfte zurückzuführen sind, deren Produkte wir in den Liparituffen erkennen. ... Unter der Annahme, dass das Ries zunächst eine gewaltige Hebung erfahren hat, liegt der Schluss nahe, dass die Abrutschungen von diesem Berge immer größer werden, je mehr wir uns dem Rande nähern, da ja dort der Böschungswinkel am steilsten anzunehmen ist. ... Branco und ich, und ebenso v. Knebel gehen aber noch weiter, und nehmen an, dass diese abgleitenden Schollen weit über den Rand des Rieses auf die benachbarten Gebiete der Alb hinausrutschten, dabei den Untergrund glätteten. ...[Es] drängt sich uns die Frage auf: welche Kraft war imstande, einen nahezu 25 km messenden runden Pfropfen aus der Tiefe herauszupressen, ohne die darauf lastende Decke vollständig zu zerblasen und zu zerschmettern, sondern in grossen Schollen auf die Seite zu schieben?' (Fraas 1903: 11–13).

[14] 'Eine einfache vulkanische Explosion erklärt aber die Verhältnisse nicht, denn diese wirkt ja in ganz anderer Weise: wir müssen Spannungen annehmen, welche erst eine gewaltige Aufpressung verursachten, und sich dann erst in Explosionen Luft machten, um schliesslich, verbunden mit einem Rücksinken der Decke, sich zu erschöpfen' (Fraas 1903: 13).

[15] 'Hier liegt also nicht, wie Branco und ich früher annahmen, eine einheitliches basales Grundgebirge vor, das nur eine Vertikalbewegung im Sinne einer Aufpressung und späteren Senkung erfahren hat, sondern ein Material, das teils gewaltigen Horizontalschub durchzumachen hatte, teils alle Spuren explosiver vulkanischer Tätigkeit an sich trägt'. (Fraas 1903: 14).

the idea of phreatic water vapour explosions caused by groundwater heated by the underlying laccolith. In the 1900s Eduard Sueß (1831–1914), Professor of Geology in Vienna, pointed to the phreatic Bandai-San eruption (1888; see Chapter 3) in which part of the volcano collapsed accompanied by vapour explosions, without erupting any juvenile material (Kranz 1912: 54; Sueß 1909: 655; Engelhardt 1982).

This, in turn, inspired Walter Kranz to develop the theory of a central explosion, which he published and defended in several papers between 1911 and 1952. His theory became the generally accepted model until 1960, although some details continued to be debated (Engelhardt 1982). Kranz, for example, always denied the existence of the additional explosions initially proposed by Branca and Fraas and heavily defended by the 'Munich School' under Richard Dehm and Joachim Schröder (see Chapter 3).

Major Walter Kranz (1873–1953) (see Chapters 4 and 5) had been an engineering officer and geological consultant to the German military during World War I and afterwards was appointed as state geologist in Württemberg. Kranz undertook a test explosion at the beach of Swinemünde (today Świnoujście in Poland), a test site associated with the local military garrison where Kranz was then stationed.

The results Kranz obtained in his artificial explosion crater (Figure 1.10), thought to represent a model of the Ries Basin at a scale of 1:15,000, proved to be remarkably similar to what field geologists saw at the Ries Basin (Kranz 1912); and Kranz was quite correct in his idea of a central explosion, because a hypervelocity impact is nothing other than a huge explosion from the centre of the structure. Kranz, however, did not shoot at his target from above, but buried a mine that he triggered. Consequently, he searched for the source of the much larger natural explosion within the Earth and proposed a huge explosion of superheated water vapour. A deep-seated magma chamber was postulated as the heat source. For Branca and Fraas a series of small explosions from widely distributed localities had been an auxiliary hypothesis to add to their basically gravitational model. In contrast, Kranz's idea was rather different in hypothesising that the entire Ries area was blasted by a single, gigantic explosion, without requiring that the area be mountainous prior to this event. While Branca and Fraas drew their idea from Alpine nappe-tectonics, Kranz was impressed by the explosive force of volcanoes like Bandai-San, Krakatoa or Tambora, which he knew from the geological literature:

> According to today's state of research, I have tried to explain the origin of the Ries in reference to [Eduard] Suess like this: Shortly before the Upper-Miocene explosion over today's Ries Basin, a horst vaulted ..., which originated because here the general subsidence of the Tertiary land surface between Black Forest and Bohemian Forest was locally inhibited by rising magma chambers. The two

Figure 1.10   Ries model test crater at Swinemünde by Walter Kranz. (From Kranz 1912: Figure 5).

uppermost [magma chambers] ... were situated within the granitic basement, the uppermost [of the two] some 12–1500 m below the surface of the horst, overlain by crystalline rocks, Keuper [Upper Triassic] and Jurassic. In the limestone formation of the Jurassic large water masses had accumulated by subterranean dissolution, which, following a tectonic or volcanic disturbance [an earthquake] of this area, descended suddenly along fissures towards the incandescent, liquid magma chamber. Thereby, a gigantic explosion was caused, the epicentre of which lies in the centre of today's Ries Basin ... It can be compared to some extent with the great eruption of Bandai-San in July 1888, except that there the higher parts of a stratovolcano were destroyed; whereas in the Ries an explosion crater was blasted from the flat-lying Earth crust. By the way, the origin of the Ries provides phenomena that, at least on our planet, generally do not occur in recent volcanoes.

... After the main eruption ... obviously later volcanic eruptions [of suevite] have additionally disturbed the basement and the transported blocks. These younger, but still Upper Miocene utterances of volcanism [emplacing the suevite] stem from a deeper magma chamber, which today is still situated as a coagulated laccolith beneath the Ries. ...

Thus, from the Miocene Ries horst, only the crystalline basement-stump is left today; what formerly rested above is blasted away by the Upper Miocene explosion. The Ries is a depression, a basin, but only morphologically as explosion crater, not according to its inner structure as horst. (Kranz 1911)[16]

Thus, in Kranz's opinion, the general morphology of the Ries Basin and its surrounding breccia blanket was caused by a single highly explosive (volcanic) event, whereas secondary eruptions produced suevite from vents near the rim of the structure.

This idea was further strengthened by the observation that distal Ries ejecta, conspicuous limestone fragments often as large as a fist (so-called Reuter's blocks; Figure 1.11), which were common in a certain stratigraphic level of the molasse basin, were distributed in a strewn field with the Ries Basin at its focal area. This discovery was made by Lothar Reuter (1877–1956) when he was inspired to conduct fieldwork after attending a conference and field trip to the Ries Basin in the spring of 1924. It was a conference of the *Oberrheinischer Geologischer Verein*, an association of amateur and professional geology enthusiasts. There, Reuter heard of Kranz's model and found it convincing. In October of the same year, he submitted his manuscript for publication (Reuter 1925). Reuter had obtained his

---

[16] 'Nach dem heutigen Stand der Forschung habe ich die Riesbildung in Anlehnung an Suess folgendermaßen zu erklären versucht: Kurz vor der obermiozänen Explosion wölbte sich über dem heutigen Rieskessel ein Horst ..., welcher dadurch entstanden war, daß hier das allgemeine Absinken der tertiären Landoberfläche zwischen Schwarzwald und Böhmerwald durch emporgedrungene Magmaherde lokal behindert wurde. Die beiden obersten ... lagen noch im granitischen Grundgebirge, der oberste ungefähr 12–1500 m unter der Horst-Erdoberfläche, überlagert von kristallinem Gestein, Keuper und Jura. Im Kalkgebirge des Jura hatten sich durch unterirdische Auslaugungen größere Wassermassen angesammelt, die bei einer tektonischen oder vulkanischen Erschütterung dieses Gebiets auf Spalten plötzlich zu dem glühend flüssigen Magmaherd hinabdrangen. Dadurch entstand eine ungeheure Explosion, deren Epizentrum in der Mitte des heutigen Rieskessels liegt ... . Sie läßt sich einigermaßen mit der großen Eruption des Bandai-San im Juli 1888 vergleichen, nur daß dort die höheren Teile eines Stratovulkans zerstört wurden, während im Ries ein Sprengtrichter aus flach-gelagerter Erdhaut herausflog. Im übrigen bietet die Riesbildung wenigstens auf unserem Planeten Erscheinungen, welche im allgemeinen bei rezenten Vulkanen nicht vorkommen. // ... Nach der Haupteruption ... haben offenbar spätere vulkanische Ausbrüche den Untergrund und die überschobenen Schollen weiter gestört. Diese jüngeren, aber auch noch obermiozänen Äußerungen des Vulkanismus stammen aus einem tieferen Magmaherd, der heute noch als erstarrter Lakkolith im Untergrund des Ries steckt. ... // Von dem miozänen Ries-Horst existiert sonach heute nur noch der kristallinische Grundgebirgsstumpf; was einstmals darüber lag, ist durch die obermiozäne Explosion fortgesprengt. Eine Einsenkung, ein Kessel ist das Ries aber nur morphologisch als Sprengtrichter, nicht nach seinem inneren Bau als Horst' (Kranz 1911).

Figure 1.11   Reuter's blocks: Reworked distal Ries ejecta (Upper Jurassic
limestone) within sandy Molasse sediments of the northern
Alpine foreland. (Photo: Kölbl-Ebert).

doctorate from the University of Erlangen and until 1945 worked as a geologist
at the Bavarian Water Supply Office in Munich (Freyberg 1974: 126).

After the 1924 conference of the *Oberrheinischer Geologischer Verein*,
research at the Ries Basin intensified and the idea of a central explosion became
predominant (Dehm in AG Ries 1963: 8), except among researchers in Munich.

In discussion with Richard Löffler (1886–1967) the central explosion model
was somewhat modified. Löffler had studied science for *Gymnasium* teachers at
the Technical University of Stuttgart, where he finished his teaching exam with
a thesis describing the crystalline basement rocks of the Ries Basin. He further
extended this work in a mineralogical PhD from the University of Tübingen,
which he finished in 1910. From 1911 to 1913, Löffler was private tutor to the
grandsons of the King of Württemberg. He was Professor of Mathematics and
Science at the teacher's seminary in Saulgau from 1913 to 1928. In 1928, he
switched to the teacher's seminary in Schwäbisch Gmünd, which was closer to
the Ries Basin, because he wanted to continue his scientific research. When the
teachers' seminaries in Württemberg were disbanded by the Nazi administration
in 1936, Löffler taught for some time at a *Gymnasium* for girls. When the
Pedagogical Institute in Schwäbisch Gmünd was refounded in 1946, he became
its rector until his retirement in 1952 (Carlé 1987: 77–9; Engelhardt and
Hölder 1977: 244).

Based on the eruption of Krakatoa, Löffler now thought that what had exploded at the Ries Basin was not water vapour but juvenile gases, which accumulated at the top of a giant magma chamber (Figure 1.12) (Löffler 1926; Dorn 1950; Pösges and Schieber 1994).

The general idea of a gigantic central explosion, however, remained: 'The principal funnel is so flat that it could only be created by an exceptionally huge explosion. As old pioneer, I shouldn't know how else that could be explained if not by the [central] explosion theory' (letter from Walter Kranz to Ernst Carl Kraus, 2 April 1952, GA19395).[17]

The joint Kranz–Löffler model was popularised and promoted by Löffler's close friend, Professor Georg Wagner (1885–1972) of Tübingen (Wagner 1931: 540–46).[18] Thus, the central explosion became the established paradigm at the Tübingen institute for geology and palaeontology, where it was supported by the institute's patriarch (Wagner) until the mid-1960s (Wagner 1960: 596–604; see also Chapters 9 and 10).

In Munich, however, the Kranz–Löffler model did not enjoy the same patronage. Instead, scientists there felt 'that the supposition of 'suddenness' [which the central explosion theory claimed for the Ries event] was something forced and improbable' (Schnell 1926). Until the early 1960s, the 'Munich school' (under Richard Dehm, Professor of Palaeontology in Munich: see Chapter 3) denied a single central explosion and instead defended the existence of several much smaller explosive vents to account for the brecciation. As such, the Munich model was considerably closer to Branca's and Fraas's ideas (see also Schröder and Dehm 1950).[19]

---

[17] 'Der Haupttrichter ist so flach, dass er nur durch eine ganz gewaltige Explosion entstanden sein kann. Als alter Pionier wüsste ich nicht, wie man das anders als durch die Sprengtheorie erklären kann' (letter from Walter Kranz to Ernst Carl Kraus, 2 April 1952, GA19395).

[18] Although Löffler was Wagner's friend, Walter Kranz throughout his life was the 'intimated enemy number 1 of Georg Wagner, who always spoke only ironically of 'Major Kranz'' (Carlé 1987: 81).

[19] Whereas the palaeontologist Dehm possibly did not truly understand the principal difference between the two models, Branco and Fraas versus Kranz and Löffler: 'I have not known Walter Kranz personally, and did not try to [meet him] either, because – in my youthful zeal – his conduct against Branca, the true founder of the 'central explosion' seemed incomprehensible'. ['Walter Kranz habe ich persönlich nicht gekannt, habe es auch nicht versucht, da mir – in jugendlichem Eifer – sein Verhalten gegen Branca, den eigentlichen Begründer der 'zentralen Explosion', unverständlich erschienen war'] (letter from Dehm to Carlé, 3 March 1988, GA11/0637).

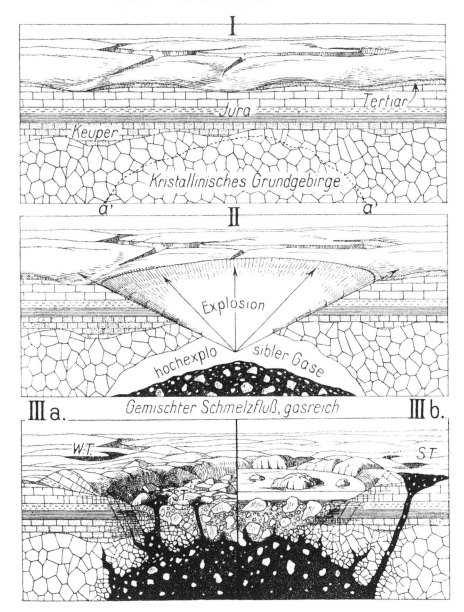

Figure 1.12   Sketch of the central volcanic Ries explosion. (Schuster 1926, courtesy of Stadt Nördlingen).

Figure 1.13   Suevite from Ries Basin. (Photo: Kölbl-Ebert).

Rocks and Minerals

Although ample macroscopic fieldwork had been done in the Ries Basin, considerably less was known about the actual rocks, especially the unusual suevite. Mat(t)hias Bartholomäus von Flurl (1756–1823) in 1805 interpreted it as 'truly volcanic but porous lava'. Others followed him in this interpretation. Friedrich August von Alberti (1795–1878), for example, referred to the suevite (in 1826) as a trass-like mass with inclusions of vesicular slaggy basalt (Figure 1.13; Engelhardt 1982). All such interpretations had been based on examining the hand-specimens. In 1849, Karl E. F. (von) Schafhäutl (1803–90) performed the first chemical analyses of suevite and was impressed by its high SiO_2 content. He imagined that a subterranean water-rich SiO_2 gel might have shrunk as a result of water loss and caused the depression of the Ries basin, while the gel had risen in clefts where it solidified as tuff, which today looked much like disintegrated granite (Engelhardt 1982).

Richard Oberdorfer, a student at the University of Tübingen, wrote a mineralogical PhD thesis on suevite that was published in 1904. For the first time, thin-sections were investigated. Oberdorfer noted strange features such as feldspar and quartz, which must have turned into glass while retaining the solid state. Chemical analyses showed an extremely heterogeneous chemistry. Oberdorfer assumed that the original magma was strongly mafic, but having assimilated large amounts of granite and gneiss (Oberdorfer 1904; see Chapter 3).

Early Ideas on the Genesis of the Steinheim Basin

The very first genetic interpretations of the much smaller Steinheim Basin have been less spectacular than for the Ries Basin. Ami Boué (1794–1881) in the 1820s assumed that the Steinheim structure, because of its similarity to a Meander cut-off, had been caused by water excavation in Tertiary times (Reif 1976; Adam 1980). The basin is tangentially cut by a river valley and today we visit the Steinheim Basin in a privileged time, since much of the basin filling has indeed been eroded by river action, restoring much of the original crater morphology.

Ernst Koken did not rely on glacial action to explain the Steinheim Basin, but assumed some sort of up-doming force for the Klosterberg, the central mountain within the Steinheim Basin (Reif 1976), as here Mid-Jurassic beds had been elevated to the level of the Upper Jurassic.

Theodor Engel (1842–1933), a protestant vicar and amateur geologist who had attended geological lectures by Friedrich August Quenstedt (1809–89) in Tübingen, interpreted the Steinheim Basin in 1881 as a volcanic depression comparable to the summit caldera of a volcano. He interpreted the algal limestone on top of the central mountain as a precipitate of CO_2-oversaturated water (Reif 1976; Adam 1980).

In 1899, Eberhard Fraas gave a lecture on the Steinheim Basin for the *Oberrheinischer Geologischer Verein*, a society of geology enthusiasts in the German southwest. Several papers on both the Ries and Steinheim Basins followed, in which Fraas transferred his Ries model to the Steinheim Basin as well. Here, the slumping of beds from the hypothetical 'Steinheim Mountain' exposed Lower- to Mid-Jurassic beds on top of the mountain. Upon cooling, the subterranean 'laccolith' shrank and the up-doming was transformed into a depression, leaving an elevated piston as the central mountain. After that stage, Fraas believed, volcanic eruptions occurred at the Ries Basin and produced suevite; but not so in Steinheim, which remained 'cryptovolcanic' (Reif 1976).

Kranz, too, recognised the similarities between the Ries and Steinheim Basins and used essentially the same model of a central, two-stage explosion to explain the other structure (Kranz 1914a, b, 1915–16, 1926; Kranz and Gottschick 1925). First the wide funnel or crater, together with thrusting at the rim, was created by a 'super-charged' water vapour (and volcanic gas) explosion from a shallow centre and:

After a pause, the length of which cannot yet be determined with certainty, gases rose again, now perhaps also magma but only up to the level of the Lower Jurassic, where then the second weaker (gas?) explosion with an effect between a 'squeezer' and 'normal mine' tumbled together and built up the underground of the *Klosterberg-Steinhirt* [the central mountain] and also threw fragments of polymictic breccia towards the rim of the basin. (Kranz 1926: 97)[20]

According to a correspondent of Kranz, the Swiss–American Professor Walter H. Bucher (1888–1965) from Cincinnati (see Chapter 10), the conditions at the Steinheim Basin, were quite similar to other 'cryptovolcanic' structures in the USA, such as the 'Serpent Mound disturbance'[21] in Ohio (Kranz 1926: 97).

The Geological Survey in Baden-Württemberg drilled into the lake sediments of the Steinheim Basin in the 1950s. The borehole supplied gravel, freshwater limestone and underneath it bituminous muds. Volcanic glass was not detected, neither in thin-sections nor among the sieved heavy minerals. Instead, quite a lot of titanite, fresh alkali-feldspar, baryte and opal were found, a mineral assemblage alien to the usual Steinheim Basin rocks (mostly limestone and shale), but reconcilable with a volcanic source albeit not truly convincing[22] (letter from Paul Groschopf to Walter Kranz, 11 April 1952, GA19370).

This was enough to be communicated in a less uncertain tone to colleagues: 'You may have heard that [our] colleague Groschopf has encountered volcanic minerals upon drilling in Steinheim?' (Letter from Manfred Bräuhäuser to Walter Kranz, 30 March 1951, GA19364).[23]

Outside the Mainstream

Outside the mainstream of various volcanic theories, several alternative explanations were proposed occasionally (see Dorn 1950; Engelhardt 1982; Pösges and Schieber 1994). The historically most important of these alternatives were hypotheses of a purely tectonic origin of the Ries and Steinheim Basins and the idea of meteorite impacts.

---

[20] 'Nach einer Pause, deren Länge noch nicht sicher bestimmt werden kann, drangen erneut Gase, jetzt vielleicht auch Magma empor, aber höchstens bis in den Lias, wo dann die zweite schwache (Gas?) Explosion mit einer Wirkung zwischen 'Quetscher' und 'normaler Mine' den Untergrund im Klosterberg-Steinhirt durcheinander warf und aufhäufte, sowie Fetzen von Bunter Brekzie an den Beckenrand warf' (Kranz 1926: 97).

[21] Meanwhile also confirmed as an impact crater.

[22] The source may have been Ries ejecta, that is suevite.

[23] 'Sie haben wohl erfahren, dass Herr Kollege Groschopf bei einer Bohrung in Steinheim vulkanische Mineralien erfasst hat?' (Letter from Manfred Bräuhäuser to Walter Kranz, 30 March 1951, GA19364).

Already in the mid-nineteenth century, Friedrich August Quenstedt (1809–89) recognised the parallels between the Steinheim and Ries Basins. He remained cautious as to their genetic interpretations, but favoured tectonic processes over volcanism, especially for the Steinheim Basin (Reif 1976).

In 1932, the teacher Armin Seidl (born 1868), who obtained his doctoral degree in Munich in 1894, proposed that radial tectonic tension would be capable of tearing holes in the Mesozoic cover. The plastically deformable crystalline basement then supposedly migrated into these holes (Seidl 1932; Dorn 1950; Freyberg 1974: 147).

Compression rather than tectonic tension was envisioned as the driving force in 1909 by C. Regelmann (1842–1923), geologist at the Württemberg Office of Statistics (which included a Geological Survey; Freyberg 1974: 123).

Reinhold Seemann (1888–1975), curator at the Stuttgart Museum of Natural History, proposed a similar explanation in 1939 (Dorn 1950). The broken and fragmented debris at the Ries and Steinheim Basins was regarded as tectonic breccias, crushed by a hypothetical subterranean 'spike' of rock moving north, pushed by pressure built up by the alpine orogenesis and surfacing in the Ries area (Pösges and Schieber 1994; see Chapter 3).

> Regarding the divergence of opinions concerning the Ries phenomenon, it is understandable that voices also rose seeking the cause of the Ries formation not on Earth but in cosmic forces. Already in 1904, E. Werner has uttered such an assumption. Stutzer then, in 1936, has tried to further the meteor hypothesis in comparison with the 'Meteor Crater' of Arizona. Against this meteor hypothesis... speak unequivocal geological and geophysical reasons. (Dorn 1950: 358)[24]

Stutzer and Werner were not the only renegades. There were also Kaljuvee (1933) and Rohleder (1933a, b) (see Chapter 2). Nevertheless, this alternative never attained enough status for serious discussion until the 1960s (Chapter 9).

In 1950, Joachim Schröder and Richard Dehm summarised the contemporary ideas on the genesis of the Ries Basin as follows:

> If we disregard some ideas generally recognized as wrong (meteor crater, pull-apart hole, glacier action), then the focus is on the question of whether we have a central [volcanic] explosion as the only cause ([Eduard] Sueß, [Walter] Kranz), as

---

[24] 'Bei dem Auseinandergehen der Meinungen über das Riesphänomen ist es verständlich, daß auch Stimmen laut wurden, welche die Ursache der Riesbildung nicht auf Erden selbst suchten, sondern in kosmischen Kräften. Schon 1904 hat E. Werner eine derartige Vermutung ausgesprochen. Stutzer hat dann 1936 für das Ries die Meteorhypothese unter Vergleich mit dem 'Meteor Crater' von Arizona näher auszubauen versucht. Gegen diese Meteorhypothese ... sprechen eindeutige geologische wie geophysikalische Gründe' (Dorn 1950: 358).

a partial cause in addition to upheavals and local extrusions ([Karl Wilhelm Franz von] Branca and [Eberhard] Fraas) or as completely unreasonable assumption within a mainly tectonic interpretation ([Reinhold] Seemann). That we may safely restrict ourselves to one of these three possibilities and do not need to introduce a completely new theory, seems to be certain to us as well as to previous authors. (Schröder and Dehm, 1950)[25]

Thus, just ten years before the impact theory gained recognition in a spectacular shift of paradigm, a genetic cause from outside the Earth was considered unthinkable by most geoscientists (and especially the declared experts of the Ries and Steinheim Basins).

---

[25] 'Sieht man von einigen allgemein als abwegig erkannten Deutungen des Rieses (Meteorkrater, Zerreißloch, Gletscherwirkung) ab, so steht im Mittelpunkt die Frage nach einer zentralen Sprengung als alleiniger Ursache (SUESS, KRANZ), als Teilursache neben Aufpressung und örtlichen Aufbrüchen (BRANCA & FRAAS) oder als gänzlich unberechtigter Annahme bei einer überwiegend tektonischen Deutung (SEEMANN). Daß man sich auf eine dieser drei Möglichkeiten beschränken darf und nicht etwa eine völlig neue Theorie des Rieses einführen muß, scheint uns wie den bisherigen Autoren gewiß' (Schröder and Dehm 1950).

Chapter 2
Early Impactists and their Sources

First, tentative words in favour of a cause from outside the Earth were uttered by the merchant Ernst Werner (1837–1910), who came from the town of Gmünd (Adam 1980). In 1904, he published a 'review paper' in a popular journal referring to previous work by Carl Deffner, Oscar Fraas, Wilhelm von Branca, Eberhard Fraas and others. In that paper, he extensively quoted from and commented on a field trip guide by Eberhard Fraas for an excursion in April 1903 of the *Oberrheinischer Geologischer Verein*. It seems that Werner attended this field trip and that his paper was meant as a report of the excursion. In his report, Werner stressed that the Ries Basin showed no relevant external features of volcanism and in its 'strangeness' it resembled more the 'curious circular mountains on the Moon, invoking the thought that these as well as the Ries Basin owe their genesis to the same cause' (Werner 1904, column 156).[1] Although lunar craters at that time were mostly interpreted as volcanic, the metaphors Werner used in his paper suggest that he was one of the few advocates for an impact origin of lunar craters.

Werner gave a summary of various contemporary volcanic interpretations for the genesis of the Ries Basin (which, however, did not seem to convince him) and it was in this context that he wrote the single sentence that later apologetics would style as the beginning of impact geology in Germany (see Chapter 10). When explaining the emplacement of a potential laccolith underneath the Ries Basin without volcanic eruption, Werner added: 'One could also pose the question of whether such foreign bodies [laccoliths] could not be of cosmic origin and thus intruded from outside' (Werner 1904, column 159).[2]

Werner's approach was more intuitive than guided by geological learning. However, his metaphors stick to the image of some cause from outside: 'Just as a stone thrown into water creates concentric circles, the Ries collapse has happened in five concentric zones' (Werner 1904, column 164).[3] Again he evoked the similarity of the craters of the Moon and concluded:

---

[1] 'In seiner Fremdartigkeit erinnert es an die merkwürdigen Ringgebirge auf dem Mond, und der Gedanke liegt nicht ferne, daß diese wie das Riesbecken gleichen Ursachen ihre Entstehung verdanken' (Werner 1904, column 156).

[2] 'Man könnte auch die Frage stellen, ob solche Fremdkörper nicht kosmischen Ursprungs und deshalb von außen herein eingedrungen sein könnten' (Werner 1904, column 159).

[3] 'Wie ein ins Wasser geworfener Stein konzentrische Kreise erzeugt, so ist der Rieseinbruch in 5 konzentrischen Zonen vor sich gegangen' (Werner 1904, column 164).

While reviewing what scientific research has established concerning the events in the Ries area, I cannot fend off the impression that the catastrophe broke into the area suddenly and as if from outside and that there must have been an unthinkably long developmental state on our planet, during which time it has been subject to even more severe catastrophes. The nature and origin of the so-called laccoliths is not yet recognized; but their existence is confirmed, even though they are stuck at depths to which no human eye can pierce.

Thus, the Ries is extremely important for geology and receives a fundamental relevance for many previously unresolved geological features on Earth and even for the formation of the world. (Werner 1904, column 168)[4]

While Werner would later be cited as an important forerunner of the impact theory, his influence in the early twentieth century was negligible. The issue of a potential impact at the Ries and/or Steinheim Basins was, however, revived in the 1930s by three independent authors: the Estonian schoolteacher Julius Osvald Kaljuvee in 1933, who again compared the Ries and Steinheim Basins with lunar craters; the British geological prospector Herbert Rohleder, also in 1933; and the German Otto Stutzer, Professor of Fossil Fuels at the Mining Academy in Freiberg, in 1936. Both Rohleder and Stutzer compared these German structures to Meteor Crater in Arizona, USA, although Stutzer considered both the Steinheim Basin and the Ries Basin to be impact craters, whereas to Rohleder only the Steinheim Basin was an impact crater.

Julius Osvald Kaljuvee

Julius Osvald Kaljuvee (formerly Kalkun; 1869–1940) from Estonia worked as a private tutor in natural sciences from 1914 onwards. Later, he became a teacher at a succession of village schools, followed by schools in St Petersburg

---

4   'Beim Überblick auf die durch die wissenschaftlichen Forschungen festgestellten Vorgänge in der Riesalb kann ich mich des Eindrucks nicht erwehren, daß die Katastrophe plötzlich und wie von außen über das Gebiet hereingebrochen ist und daß es einen undenkbar langen Entwicklungsabschnitt auf unserem Planeten gegeben haben muß, während dessen er unzähligen und noch heftigeren derartigen Katastrophen ausgesetzt war. Die Natur und die Herkunft der sog. Lakkolithe ist noch nicht erkannt; aber ihr Vorhandensein ist nachgewiesen, auch wenn sie in Tiefen stecken, in welche kein menschliches Auge dringt. //So ist das Ries für die Geologie ungemein wichtig und erhält eine typische Bedeutung für viele bis jetzt unaufgeklärte Gebirgserscheinungen auf Erden, ja wohl auch für den Weltenbau' (Werner 1904, column 168).

and Tallinn. From 1920 to 1922, he taught geology and mineralogy at the Tallinn Technical School (today the Technical University).[5]

In 1933, when Kaljuvee published, the origin of lunar craters was still far from resolved. In 1886, the American geologist Grove Karl Gilbert (1843–1918) had made simple impact experiments to elucidate the potential impact origin of the craters of the Moon (Mark 1995: 28–9). Alfred Wegener (1880–1930), of Continental Drift fame, did the same in the early twentieth century in Germany. Wegener had studied astronomy in Berlin. From 1909 onwards, he lectured on meteorology, astronomy and astrophysics at the University of Marburg. In 1916, he was an eyewitness to a meteorite fall. In 1919, when research money was sparse following World War I, Wegener began a series of simple impact experiments. In 1921–22, he published several papers about the genesis of lunar craters (Greene 1998; Wutzke 2000; Mark 1995: 106), which in turn seemed to have captured the imagination of Kaljuvee who declared himself to be an ardent admirer of Wegener.

Kaljuvee's ideas on impacts were dealt with in Chapter 3 of his book *Großprobleme der Geologie* [Major Problems in Geology] as a hitherto little-regarded geological factor. He mentioned Wegener's experiments concerning the genesis of lunar craters and also regarded Meteor Crater in Arizona (with its associated Canyon Diablo irons) as well as the Kaalijärv Crater on Saaremaa (then Ösel; Estonia) to be of undoubtable impact origin (Kaljuvee 1933: 100–08).

With regard to Kaalijärv Crater, Kaljuvee reported excavations on Saaremaa by Ivan Reinwaldt (see Chapter 5) and assumed that the size of the projectile should be about equal to the crater size. An explosion, Kaljuvee assumed, led to the destruction of the impacting body and distribution of the material in the surrounding area. As for the cause of the explosion, he quoted the expertise of Reinwaldt, who had attributed it to shock waves reflected backwards into the projectile, when it was about to hit the target. Such shock waves had been observed 'by cinematographic shooting' to destroy cannon balls. The velocity of the projectile was determined by Kaljuvee to be about 50 to 70 km/s within the realistic range (Kaljuvee 1933: 105).

Kaljuvee then, however, failed to consequently apply known physics. Although he mentioned the great velocity of the incoming meteorite, he did not think about the consequences of the energy release upon braking of such a fast projectile, essentially in a mere instant. He only thought it probable (again citing Reinwaldt) that the effect of the shock waves was enhanced by phreatic explosion of water vapour created by contact of the incandescent meteorite with groundwater. Ironically, Kaljuvee expected the destructive effect on the projectile to be worse in smaller rather than larger ones:

---

[5]   Biographical information pers. comm. Tõnu Pani, Estonia, 2 November 1999.

From what was said, it becomes sufficiently clear that when meteorites hit the ground, in many cases one has to expect explosive phenomena. But it is equally clear that this does not always need to happen. ... By the way, a complete devastation of the meteoritic mass by the sudden braking of the rapid velocity of movement and the explosive effect of the vapours from targeted water masses can happen only with small meteorites up to a certain size. Meteorites larger than this may remain intact in their main proportions even when they fall into the ocean. Also, cases can be imagined in which, upon a water-vapour explosion, the meteorite concerned is reflected by the most elastic vapours and falls down elsewhere at a greater distance. It can be deemed by no means impossible that, for example, the meteorites of the Kaalijärw have been reflected in such a way and that they lie somewhere, possibly in the area surrounding Ösel beneath the surface of the Baltic Sea. (Kaljuvee 1933: 107)[6]

As examples of impacts, Kaljuvee described the Tunguska event (Siberia, 1908), the Henbury craters (Australia), the Hoba meteorite (Namibia), Lake Bosumtwi (Ghana) and the Ries and Steinheim Basins (Kaljuvee 1933: 108–17).

According to Kaljuvee, the destruction of the target rock (its extensive brecciation) was again due to the destructive force of shock waves created by the rapid approach of the meteorite: Assuming an incoming velocity of 50 km per second would give the minimum velocity of atmospheric gases forced to escape in front of the projectile. This velocity, according to Kaljuvee, would rapidly increase immediately before the impact, when the atmospheric gases were rapidly squeezed out of the way. An increase in gas velocity up to only six times the initial value would surpass the initial velocity of gases created by the detonation of chemical explosives such as dynamite (275 km/s, according to Kaljuvee):

Therefore, it is no wonder that a meteorite was needed under these conditions to form the Ries – with the ordinary brecciation of the target-rocks (due to the

---

6   'Aus Angeführtem erhellt gewiss zur Genüge, dass man beim Aufschlagen der Meteoriten auf die Erdoberfläche in vielen Fällen mit Sprengungserscheinungen rechnen muss. Dass aber solches nicht immer einzutreten braucht, ist ebenfalls klar. ... Eine totale Zertrümmerung der Meteoritenmasse durch plötzliche Hemmung der rapiden Schnelligkeit der Bewegung und Sprengwirkung der Dämpfe getroffener Wassermassen kann übrigens nur bei kleineren Meteoriten bis zu einer gewissen Grösse eintreten. Diese übersteigende Meteoriten dürften selbst beim Einsturz in den Ozean in ihren Hauptmassen ganz bleiben. Auch wären Fälle denkbar, dass bei einer Wasserdampfexplosion der betreffende Meteorit an den äussert elastischen Dämpfen zurückprallt und in grösserer Distanz anderweitig niederfällt. Es ist durchaus nicht als unmöglich zu erachten, dass z. B. die Meteoriten des Kaalijärw in solcher Weise abprallten und irgendwo vielleicht in der Umgebung Oesels unter dem Ostseespiegel lagern' (Kaljuvee 1933: 107).

rending effect of the escaping gases from the meteorite front together with the displacement by the sheer mass of the meteorite) and the special radial thrusts (stressed by E. Kayser, including the wedgings into neighbouring layers mentioned by Lindeman plus the accompanying friction breccias). If we finally consider that a cosmic body worthy of the Ries basin must also have been able to penetrate the Earth's crust, so that the lava flowing back into the opening created by the cosmic body also was able – in sparse quantities [as suevite?] – to break through to the surface of the depression, we find that only with the assumption of a meteoritic catastrophe, comprising in this case a double hit, can the riddles of the Ries and Steinheim Basins be explained without any unsolved factors remaining. (Kaljuvee 1933: 118)[7]

Kaljuvee saw the (nearly) contemporary volcanism on the Swabian Alb (see Chapter 1), some 40 km west of the Steinheim Basin, as representing the hydrostatic release of the impact pressure at the Ries and Steinheim Basins (Kaljuvee 1933: 120).

Kaljuvee then warmed to his topic, extending its influence to nearly every possible and impossible geological feature: for example, he regarded the Pannonian Basin as a large impact crater with the Carpathians as crater rim and viewed the Czech tektites ('moldavites'; now attributed to the Ries crater) as a consequence of this hypothetical 'Pannonian impact' (Kaljuvee 1933: 121–4).

Kaljuvee also thought that impacts were the driving force behind Wegener's polar wander. For Kaljuvee, impacts explained everything from orogenesis and plutogenesis to large volcanic eruptions. Impacts, he imagined, pushed the continents around and forced up mountain chains such as the Alps, the Atlas and the Pyrenees. Grazing shots altered the axis of the Earth and thus caused ice ages. Kaljuvee regarded the Mediterranean Sea as the trace of such a grazing shot. This bombardment was also considered by Kaljuvee the origin of the biblical Deluge story. Impacts for Kaljuvee also explained various sediments, such as bonebeds or vast accumulations of rock debris. Kaljuvee believed that

---

7   'Es ist daher nicht zu verwundern, dass einem zur Erzeugung des Rieskessels unter solchen Umständen notwendigen Meteoriten ausser der gewöhnlichen Vergriesung des vorliegenden Gesteines durch die auseinanderzerrende Wirkung der ausweichenden Stirngase, vereint mit der Verdrängung durch die blosse Masse des Meteoriten, auch die besonders von E. Kayser hervorgehobenen radialen Schollenüberschiebungen und von Lindeman erwähnten Einkeilungen in die benachbarten Schichten nebst den dazugehörigen Reibungsbrekzien zu verdanken sind. Wenn wir endlich erwägen, dass ein des Rieskessels würdiger kosmischer Körper auch die Erdrinde zu durchschlagen imstande gewesen sein muss, wobei die in die entstandene Öffnung zurückströmende Lava in spärlichen Quanten auch bis zur Oberfläche der Senke hinaufschlagen konnte, so finden wir, dass nur durch Annahme einer Meteoritenkatastrophe und zwar eines Doppelschlages die Rätsel des Nördlinger Rieses und Steinheimer Beckens restlos erklärt werden können' (Kaljuvee 1933: 118).

the enormous bombardment of the Earth was responsible for a vast increase in its mass, which was responsible for the fact that the continents would no longer fit together in a closed shell (Kaljuvee 1933: 123–62).

It seems that Kaljuvee was read by his contemporaries, but it is not surprising that his largely esoteric ideas were mainly disregarded.

Excursion to Arizona

Meteor Crater in Arizona (Figure 2.1), which inspired the European impactists of the 1930s, received its name officially in 1946 (Hoyt 1987: 334). It had previously been known by various other names such as Coon Mountain, Barringer Crater and Arizona Crater. For convenience, we refer to it here by its present name, Meteor Crater, which derives from the nearest post office of Meteor, which in turn was named after the Canyon Diablo iron meteorites discovered in its vicinity.

Figure 2.1   Meteor Crater in the 1930s. (Stutzer 1936a: Figure 1).

Meteor Crater is a simple bowl with a diameter of some 1,200 m, a depth of about 175 m, and a crater rim elevated some 45 m above the surrounding plain. As early as 1891, fragments of meteoritic iron had been found in the surrounding areas and after polishing and etching these fragments showed Widmanstätten figures. From scratches inflicted on polishing machines, it was also quickly determined that the irons contained tiny, impure diamonds (Mark 1995: 25–6).

The crater was the property of the Standard Iron Company, founded c. 1903 by Daniel Moreau Barringer (1860–1929), who had secured the mining patents for the crater and its surroundings, together with a companion, the mathematician and physicist Benjamin Chew Tilghman. Barringer had initially been a lawyer, graduating from the University of Pennsylvania Law School in 1882. Having found life as a lawyer boring, he took a geology course at Harvard University, but dropped out again 'when the instructor asked him a question he considered "childish"' (Rabinowitz n.d.). However, he later studied mineralogy at the University of Virginia and worked as a mining engineer in his own company (Hoyt 1987; Rabinowitz n.d.). In 1906 Barringer discussed the impact origin of Meteor Crater at the International Geological Congress and at the annual meeting of the Geological Society of America. In the same year, unusual metamorphic features were discovered within the crater rocks: pulverised rocks, broken and deformed quartz crystals, pumice-like sandstone containing iron and nickel, and sandstone intermingled and 'baked' together with silica glass similar to fulgurites, those tube-like structures which were known to be generated by the intense heat of lightning striking sand dunes. To melt the quartz, the temperature must have been at least 1,600°C, far more than ordinary volcanic temperatures.[8] And then, there were the meteorites, as well as iron-rich magnetic 'shales' possibly derived from weathering of even more meteoritic irons (Mark 1995: 34–5).

For Barringer, it seemed quite clear that a huge meteorite must have hit the ground and formed the crater, and he assumed that this meteorite must still lie buried somewhere in the crater: a huge mass of iron and nickel, an immensely valuable treasure. Even professors of physics were prepared to accept this thrilling conclusion. Over the course of about twenty years, every effort was made to recover the iron mass, but little evidence was found (Mark 1995: 36–8). While Barringer heavily promoted the impact idea for his crater, geologists of the USGS remained sceptical.

In the USA, unlike in Germany, the impact promoters were physically stationed at the crater and kept the conflict simmering, whereas in Germany, until the paradigm finally shifted in 1960, impactists published their ideas and then dropped off again.

The controversies[9] at Meteor Crater between Barringer and his supporters on the one hand and the USGS geologists on the other, 'revolved around such

---

[8]   Silica-rich melts usually have a temperature of some 750 to 800°C, whereas basaltic lavas reach about 1,150°C.

[9]   The history of these controversies is fully recounted in the detailed and comprehensive book by Hoyt (1987) and the general history of impact research by Mark (1995). A brief

issues as authoritarianism within disciplines, communication barriers between disciplines, professionalism versus dilettantism, questions of priority, the roles of government and commerce in science, and the processes by which the conventional wisdom of science is changed' (Hoyt 1987: 5).

Because Barringer and his mining company were unable to find the supposed iron-nickel mass in Meteor Crater, the conflict remained unresolved for some time. To Barringer's dismay, in the late 1920s physicists and astronomers began to suggest that the impacting body might have been much smaller than previously assumed and that it had exploded upon impact (Mark 1995: 38–9) – a mechanism that had already been suggested for lunar craters as early as 1878 by the British astronomer Richard Antony Proctor (1837–88) (Hoyt 1987: 128).

Mineralogists had also realised and reported to Barringer that a temperature of at least 1,600°C, suggested by the occurrence of silica glass, could not possibly be reached 'during any part of a steam explosion even if the steam were superheated' (Rogers, quoted by Hoyt 1987: 249), implying that the explosion was not caused by exploding groundwater alone.

Although this insight crushed Barringer's hope for a bonanza, it explained many strange geological features, and thus an impact origin for Meteor Crater became more acceptable for geologists. In addition, other possible impact localities had been found and investigated in the meantime:

> As these localities were examined, much of the early skepticism disappeared. By the end of the 1920s the existence of meteorite craters on the surface of the earth was accepted by some people as an established fact, and the controversy concerning the Arizona crater, which by that time was known as Meteor Crater, had almost disappeared. ... In the early 1930s, nine probable meteorite craters, randomly scattered over the globe, were authoritatively listed. Twenty years later the number had doubled. (Mark 1995: xiv)

And so, around 1930, 'there was indeed a general consensus, if not unanimity, among scientists who had considered the matter at all, that Meteor Crater had been created by the impact of a large meteorite' (Hoyt 1987: 318; see also Mark 1995: 89).

In the mid-1930s, even larger and more complicated craters were about to be understood. In 1936 J.D. Boon and C.C. Albritton, Jr interpreted central mountains as due to elastic rebound:

> and that as a result of impact and explosion, a series of concentric waves would go out in all directions, forming ring anticlines and synclines. These waves would

introduction into the history of impact crater research is also provided by Marvin (1986, 1999).

be strongly damped by the overburden and by friction along joint, bedding, and fault planes. The central zone, completely damped by tension fractures produced by rebound, would become fixed as a structural dome. The general and simplest type of structure to be expected beneath a large meteorite crater would, therefore, be a central dome surrounded by a ring syncline and possibly other ring folds, the whole resembling a group of damped waves. (Mark 1995: 71)

Ernst Werner had intuitively realised the same for Ries Basin: It showed concentric rings, like water rippled by a tossed stone (see above).

The experience of World War I, when for the first time large bombing craters were observable from above, inspired new physical research on explosion effects and also invited the comparison of these man-made atrocities with natural crater-shaped geological features of a strikingly similar morphology on Earth (Hoyt 1987: 196) as well as on the Moon. In 1919, the American physicist Herbert Eugene Ives (1882–1953) remarked that the crater-pocked landscape of a bombing test-site showed 'very striking similarity between the craters made by the explosion of bombs and the craters of the moon'; and most relevant, he found central elevations in the man-made craters too. This, he realised, 'formed apparently by a species of rebound' (Ives quoted after Hoyt 1987: 196).

Around 1930, the American physicist and Nobel laureate Percy Williams Bridgman (1882–1961) noticed for the first time that the velocity of an incoming large meteorite would be faster than the velocity of sound in rock, which consequently did not allow the energy to be dispersed with adequate speed. Bridgman thus realised the immense power released by the impact and proposed the forceful explosion of a comparatively small body (Hoyt 1987: 314–17).

Barringer, however, did not want to accept this and give up his dream of a minable iron and nickel mass, after he and his shareholders had spent so much money seeking it in vain. Therefore, while more and more evidence in the USA was accumulating for an explosive and consequently destructive interaction between projectile and target, Barringer still argued against this scenario. He would not believe in such high velocities, because then, to Barringer, more heat and thus melt would be expected (Hoyt 1987: 281).

Importing the Impact Scenario to Germany

Barringer, seeking support for his business enterprise, tried to interest geologists in visiting his crater. In 1912 he managed to get Meteor Crater on 'the itinerary of a transcontinental excursion by some ninety geographers and geologists, many of them distinguished European scientists' (Hoyt 1987: 147); but when Eduard

Brückner (1862–1927,[10] Professor of Geography at the University of Vienna) was interviewed by the *New York Times* after the field trip, he at least was not convinced. Brückner opted for a volcanogenic gas explosion at Meteor Crater and compared it to 'an enormous pit or caldera at Ries, in Wurtemburg [sic]'. According to Brückner, the brecciation of the country rock in the Ries Basin 'indicate pretty clearly that the force causing their displacement must have come from below and was probably due to a geyser-like steam explosion' (quoted after Hoyt 1987: 149).

On the other hand, Rudolf Richter (1881–1957), director of the Senckenberg Museum in Frankfurt am Main, Germany, accepted the impact scenario without hesitancy, but also without familiarising himself more closely with its geology. Richter participated in a transcontinental field trip of the International Geological Congress in 1933, which provided a glimpse of Meteor Crater in the evening-light of a long day that led onwards from the Grand Canyon to volcanic Sunset Crater to Arizona's meteorite crater. To Richter, Meteor Crater seemed like just another tourist curiosity (Richter 1936).

Other Europeans were also quite willing to follow Barringer in the assumption of an impact at Meteor Crater. Max Wilhelm Meyer (1853–1910), astronomer and author of popular scientific literature as well as of poems, novels and theatrical plays, mentioned the Arizona crater briefly in his popular book *Kometen und Meteore* (~1906) and described it as a 'moon-crater-like hole we at least cannot explain in any other way except by the impact of a cosmic mass' (quoted after Hoyt 1987: 192; <http://www.urania-potsdam.de/texte/seite. php?id=354> accessed 6 November 2013).

Richard Hennig (1874–1951), a meteorologist and geographer (Hennig 1969), reported in a popular geographical journal that the existence of meteorite craters was only known with certainty since 1927, when descriptions of the destruction related to the Tunguska event[11] were published. Previously, there had been only speculation concerning Meteor Crater in Arizona. Meanwhile, however, as awareness had increased, a total of eleven meteorite crater sites were known: Tunguska (Siberia), Kaalijärv (Estonia), Meteor Crater, Odessa (Texas), a crater field in South Carolina, Campo del Cielo (South America), the site where the desert glass was found in Libya, Lake Bosumtwi (Gold Coast, Africa), East Persia, Wabar (Arabia) and the Henbury craters in Australia – all

---

[10]   Brückner was born in Jena, Germany. In 1888 he became Professor of Geography at the University of Bern (Switzerland). In 1904, he moved to Halle (Germany) and in 1906 to Vienna (Austria) (Freyberg 1974: 22).

[11]   On 30 June 1908, a meteoroid exploded within the atmosphere some six to ten kilometres above the Podkamennaya Tunguska river in Siberia. The explosive force was equivalent to somewhere around ten to fifteen megatons of TNT. The explosion killed some 80 million trees in an area of more than 2,000 square kilometres (<http://en.wikipedia.org/wiki/Tunguska_event> accessed 4 October 2011).

of them, to Richard Hennig's knowledge, a few thousand years old at the most (Hennig 1934).

A list of nine impact sites was provided by Kurd von Bülow (1899–1971)[12] in another popular journal (Bülow 1937). Bülow cautioned his readers:

> Since science has first known about the Siberian case and the meteor basin of Arizona, it has become nearly a fashion to interpret similar depressions in the Earth's surface as traces of the impact of celestial projectiles as well. However, if one reviews all these 'meteor craters' unbiased, there remain only very few apart from those two [Tunguska and Meteor Crater in Arizona] that cannot be explained by other, 'earthly' means. (Bülow 1937: 23)[13]

Thus, the impact idea was permissible for Bülow only as a last resort. This implied that he did not know of any positive evidence for the impact phenomenon. However, because the absence of evidence is not evidence of absence, such an interpretation of a location as an impact site could hardly be called 'scientific' by common standards. The author showed this same caution regarding possible impact sites in Europe:

> Maybe meteor craters exist in Europe too. At least the investigations on the Estonian island of Ösel in the Baltic Sea by Alfred Wegener showed that here,

---

[12]   Upon finishing his schooling, Bülow took part in World War I and was assigned as helpmate to a military geologist. After the war, he studied geology in Breslau, Berlin and Greifswald until 1921 and finished with a PhD. From 1921 to 1935, he was a geologist at the Prussian Geological Survey (later Reichsamt für Bodenforschung). From 1935 to 1939, he had a leading position at the Geological Survey of Mecklenburg in Rostock, where he also lectured on geology in 1935–36. In 1937, he became a professor of geology at the University of Rostock, but was dismissed in 1946 because of his membership of the Nazi party. In 1939–40 he had taken a hiatus from his professorship to work as a military geologist and from 1943 onwards he commanded the military geologists of the German army. In 1945, Bülow was detained for some time as a prisoner of war by the British authorities. In 1946–47, he worked for a peat company. In 1947, he returned to the Geological Survey in Mecklenburg, also lecturing from 1950 onwards at the University of Rostock. From 1952 until his retirement in 1964, he again worked as Professor of Geology in Rostock (GDR) (<http://cpr.uni-rostock. de/metadata/cpr_professor_000000002320> accessed 4 October 2011).

[13]   'Seit die Wissenschaft den sibirischen Fall kennt und den Meteorkessel von Arizona, ist es beinahe Mode geworden, ähnliche Vertiefungen in der Erdoberfläche ebenfalls als die Spuren von Einschlägen himmlischer Geschosse zu deuten. Prüft man jedoch unvoreingenommen alle diese 'Meteorkrater', so bleiben als sicher außer den beiden genannten eigentlich nur ganz wenige übrig, also solche, die auf andere, 'irdische' Weise nicht erklärt werden können' (Bülow 1937: 23).

too, the possibility must be reckoned with. But here, as well as in numerous other cases, the last word has not yet been spoken by far. (Bülow 1937: 24)[14]

However, it was not only popularisers of science who fell for the charms of the dramatic impact scenario, which made a thrilling and therefore saleable story. Otto Stutzer (1881–1936) (Figure 2.2), Professor of Fossil Fuels at the Mining Academy in Freiberg, Saxony, visited Meteor Crater on 29 February 1936. In his notebook, he described the visit:

> From the plain formed by red rock now rose a small hill that we approached, the Meteor Crater. It is elevated only 50 m above the plain. When we came closer, we saw that its slopes were formed by rock debris, and when we left the car at its upper rim we looked into a 200 m deep crater cauldron. The diameter of this crater is more than 1,300 m. It was an impressive view. But the wonderful thing now is the origin of this great ~~crater~~ hole. It is not a volcanic crater, because volcanic rocks are completely absent in the closer surroundings of this mountain. ~~To the contrary~~ It came into existence because a meteor,[15] a celestial body, slammed into the Earth and created this big hole. ~~One must imagine this~~ It was probably similar to the impact of an enormous bomb shell that created a large shell funnel. That this theory is correct follows from the numerous meteoritic irons that have been found in and around the mountain. They surround the mountain for up to six miles. ... I have bought two pieces as big as a fist and polished somewhat in one place. One is also supposed to contain a diamond. As the meteoritic iron contains nickel, an effort was made to find the meteor at depth thirty years ago. At that time, 27 boreholes were drilled and nine shafts were dug, but in vain. Then, in recent time, the place was investigated geophysically and thereby it was found out that the meteor must lie under the southern rim, so that it must have shot in obliquely. 3 ½ years ago it was drilled anew and the meteorite was hit at a depth of 450 m. Then a shaft was begun to be sunk alongside, but it only reached a depth of 240 m, ~~when~~ possibly because the people ~~possibly~~ ran out of money. It is assumed that the meteorite is not a compact mass, but lies there shattered into many pieces. ... Whatever the case, this deposit is something very curious. (Travel journal of Otto Stutzer concerning his journey to the USA, p. 22a–c: Winslow

---

14   'Vielleicht sind auch in Europa Meteorkrater vorhanden. Jedenfalls zeigte die Untersuchung auf der estnischen Ostseeinsel Ösel durch Alfred Wegener, daß auch hier mit der Möglichkeit gerechnet werden muß. Doch ist so wie hier, auch in zahllosen anderen Fällen wohl noch lange nicht das letzte Wort gesprochen worden' (Bülow 1937: 24).

15   A meteor is a phenomenon of the Earth's atmosphere (the fireball, produced by a meteorite travelling at high speed through the atmosphere, which is heated by friction to a glowing plasma). Nevertheless, the German scientists quoted in this book often use the term 'meteor' to describe the cosmic body, the meteorite itself. This may result from confusion with 'Meteor Crater', the name of the meteorite crater in Arizona.

Figure 2.2   Otto Stutzer argued in 1936 for an impact origin of the Ries and Steinheim Basins. (Photo from Jurasky 1937: 3).

(Arizona). 29 February 1936: Archive of the TU Bergakademie Freiberg, Stutzer-Nachlass)[16]

Stutzer published his travelogue in a short paper, together with some photographs of the crater. Here he reported that the idea to visit the crater came to him because he saw meteorites and other rock samples from Meteor Crater at the University of Princeton. He also stated that the meteoritic origin of the crater had been clarified within the last two decades. In addition to the presence of meteoritic iron and the lack of volcanic material, drillings had shown that the disturbance became less pronounced the deeper the drill bored. Following Barringer, Stutzer assumed that the projectile rested beneath the southern crater rim. The origin of the crater, according to Stutzer, was an explosion that was not further specified (Stutzer 1936a: 442, 447).

It was such personal experience with Meteor Crater that led to the introduction of this general idea at geological conferences in Germany. There were, however, fundamental differences between Meteor Crater and the German sites: the much larger size of the two German craters, which also resulted in a much more complex geology, the lack of accompanying meteorites, the greater

---

[16] 'In der aus rotem Gestein aufgebauten Ebene erhob sich nun ein kleiner Hügel, auf den wir zufuhren, der Meteor Crater. Er erhebt sich nur 50 Meter über die Ebene. Als wir uns ihm näherten, sahen wir, dass seine Abhänge aus Gesteinsschutt aufgebaut waren, und als wir oben an seinem Rand das Auto verliessen, sahen wir in einen 200 Meter tiefen Kraterkessel hinab. Der Durchmesser dieses Kraters ist mehr als 1300 Meter. Es war ein imponierender Anblick. Das Wunderbare ist nun aber die Entstehung dieses grossen K̶r̶a̶t̶e̶r̶s̶ Loches. Es ist kein vulkanischer Krater, denn vulkanische Gesteine fehlen in der näheren Umgebung dieses Berges vollkommen. Er ist v̶i̶e̶l̶m̶e̶h̶r̶ dadurch entstanden, dass ein Meteor [sic], also ein Himmelskörper auf die Erde aufschlug und dieses grosse Loch schuf. M̶a̶n̶ ̶m̶u̶s̶s̶ ̶e̶s̶ ̶s̶i̶c̶h̶ ̶v̶o̶r̶s̶t̶e̶l̶l̶e̶n̶ Es war wohl ähnlich wie das Aufschlagen einer gewaltigen Granate, die einen grossen Granattrichter schuf. Dass diese Theorie richtig ist, ersieht man dadurch, dass zahlreiche Meteoreisen in und um den Berg herum gefunden sind. Bis 6 Meilen liegen sie um den Berg herum. ... Ich habe zwei faustgrosse und an einer Stelle etwas angeschliffene Stücke gekauft. Einer soll auch einen Diamanten enthalten. Da Meteoreisen Nickel enthält, so hat man vor 30 Jahren versucht den Meteoriten in der Tiefe zu finden. Man hat damals 27 Bohrlöcher und neun Schächte niedergebracht, aber ohne Erfolg. Dann hat man in letzter Zeit die Stelle geophysikalisch untersucht und hierbei festgestellt, dass der Meteor unter dem Südrand liegen muss, dass er also etwas schief hineingeschossen ist. Vor 3 ½ Jahren hat man von neuem gebohrt und den Meteoriten in 450 Meter Tiefe getroffen. Man hat dann seitwärts einen Schacht abzuteufen begonnen, der aber nur 240 Meter tief wurde, w̶o̶r̶a̶u̶f̶ wohl weil den Leuten w̶o̶h̶l̶ das Geld ausging. Man nimmt an, dass der Meteorit nicht in einer kompakten Masse, sondern in vielen Stücken zersprasst dort liegt. ... Es ist dieses Vorkommen jedenfalls etwas ganz Eigenartiges'. (Travel journal of Otto Stutzer concerning his journey to the USA, p. 22a–c: Winslow (Arizona). 29 February 1936: Archive of the TU Bergakademie Freiberg, Stutzer-Nachlass).

loss of information by erosion during the last 15 million years and of course, the dense vegetation cover compared to the Arizonan semi-desert environment.

Herbert P.T. Rohleder

The bilingual British geologist Herbert Paul Theodor Rohleder (later Herbert P.T. Hyde, born 1902) held PhD degrees in geology from London and from Munich University. From about 1921 onwards, he worked as a mineral prospector and surveyor for several colonial companies in various African countries such as Abyssinia, Egypt, Rhodesia, South Africa, the Gold Coast and Cameroon. He perceived himself as a 'globe trotter & adventurer with an occasional touch for science' (letter from Herbert Rohleder to Carl Troll, 23 June 1938, BGIB).[17]

Late in 1936, Rohleder was unemployed for some time. During this period, Rohleder travelled around Europe, visiting the UK, Germany, Scandinavia and Russia. In August 1938, he attended the Geologists' Meeting in Munich and commented in a letter on the narrow-mindedness of German geologists and the racist abuse of Eduard Sueß (see Chapter 7). By then, Rohleder had already decided to leave Nazi Germany. He returned to London, where he adopted the surname Hyde, possibly his mother's maiden name.

In 1950, he continued to write letters from London and renewed his old connections with the German colonial geographer Carl Troll (1899–1975). It seems that in the meantime, Rohleder had married.

He worked as a consulting geologist in Ankara (Turkey) in 1951 and in Portuguese Timor in 1957. Between 1955 and 1960 he also spent time in Australia. For the following eight years, he was a visiting professor at the Humboldt University in East Berlin (GDR).

Because of increasing restrictions in the German Democratic Republic (for example, his English journals were confiscated, and after the Berlin Wall was erected he needed a visa to cross between East and West Berlin), he decided to leave this job late in 1968 and seek retirement in Belfast (Northern Ireland), where he seems to have had family connections. In 1962, his wife had left him to settle in Bournemouth, because she no longer wished to stay in the GDR.[18]

In 1933, Rohleder attended the annual meeting of the German Geological Society (DGG). On 5 July, he gave a talk on terrestrial meteorite craters. He

---

[17] 'unsereiner ist und bleibt ein Weltenbummler & Abenteurer mit einem gelegentlichen Einschlag von Wissenschaft' (letter from Herbert Rohleder to Carl Troll, 23 June 1938, BGIB).

[18] Biographical information compiled from Rohleder's letters at the *Bibliothek des Geographischen Instituts der Universität Bonn*, Carl Troll papers; Rohleder 1936: 64; pers. comm. Wendy Cawthorne, Geological Society, London, by e-mail on 14 November 2011; pers. comm. Halima Khanon, Royal Geographical Society, London, by e-mail on 3 January 2011.

first introduced his audience to Meteor Crater in Arizona, as well as to the Pretoria Salt Pan[19] (then Transvaal, today South Africa), only to proceed to the Steinheim Basin. He did so, in accordance with his brusque temperament, in a less than diplomatic way:

> Because of the short time available, I cannot elaborate on the various genetic theories that may be known to every German geologist in their basic features anyway. And this is just as well: Because it has always been customary to head for the solution to the problem by way of a preformed opinion, and finally find that which one wanted to find. (Rohleder 1933a)[20]

Rohleder then pointed out several contradicting statements he had found in the relevant publications: for example, the bedding of sediments in the central mountain of Klosterberg had been called chaotic; nevertheless, the accompanying drawings showed symmetric features resembling the layers in an onion or straightforward horst-structures. Rohleder saw no necessity for a common origin of the Steinheim and Ries Basins. The latter he considered to be clearly volcanic in nature, among other things because of the tuffs (suevite) and the volcanic bombs embedded in them. But the Steinheim Basin, some 30 km from the Ries Basin, was too far away to allow a common volcanic origin and showed completely different features according to Rohleder. First of all, there were none of the 'well-known pressure sutures' or 'ray-shaped pressure figures'[21] in the Ries Basin (Rohleder 1933a; Figure 8.2).

Nothing at the Steinheim Basin, Rohleder claimed, contradicted an origin by meteorite impact. In Meteor Crater too, there was 'an elevation of foreign material',[22] that is some sort of central mountain. The aragonitic travertine as witness to thermal springs, Rohleder explained as being the result of 'heat released from the slowly cooling meteorite at depth'.[23] But Rohleder also had no problem with an impact crater lacking preserved meteoritic material, preferably when the projectile had been a stony meteorite, the substance of which 'can be

---

[19]   The impact origin of the Pretoria Salt Pan was confirmed by drilling in 1989, when shock-metamorphosed minerals and impact melts were found (Reimold et al. 1992).

[20]   'Der knappen Zeit wegen kann ich auf die einzelnen Entstehungstheorien nicht eingehen, die im übrigen wohl jedem deutschen Geologen in den Grundzügen bekannt sein dürften. Und dies ist auch gut so: Denn es ist von jeher ebenso üblich gewesen, mit vorgefaßter Meinung an die Lösung des Problems zu treten und schließlich das zu finden, was man finden wollte' (Rohleder 1933a).

[21]   'die bekannten Drucksuturen' or 'strahlenförmige Druckfiguren', that is *Strahlenkalk*, which today is interpreted as shatter cones, a macroscopic indicator for impact craters.

[22]   'eine Erhebung von ortsfremdem Material' (Rohleder 1933a).

[23]   'durch die Hitzeabgabe des allmählich in der Tiefe erkaltenden Meteoriten' (Rohleder 1933a).

blown much more easily into dust than metals' (Rohleder 1933a).[24] Rohleder concluded:

> that in a meteorite impact too, a secondary explosion cannot be ruled out. When the white-incandescent body touches in a deeper bed a strongly water-impregnated horizon, then so much water vapour must be released that an explosion and thus blasting of the overlying rock-layers must occur. That is, in this case, the mechanism of the meteorite impact would become extraordinarily similar in its effects to those of the true surface-volcanism. (Rohleder 1933a)[25]

Rohleder finished his talk with the remark that his presentation had only been an abstract of a longer English manuscript, but he nevertheless wanted to use the opportunity to inform the German geologists of this new explanatory trial (Rohleder 1933a).

Rohleder's longer English manuscript was published the same year in the *Geological Magazine* (Rohleder 1933b). Here, too, Rohleder opted for a volcanic origin of the Ries Basin, contrary to that of the Steinheim Basin:

> [An] important fact has been deliberately ignored, viz. that only the Steinheim basin is actually cryptovolcanic, whereas the Ries of Nördlingen exhibits undeniable proofs of extensive volcanic activity. There are rhyolitic tuffs, weathered-out lava cords and lava slabs, partly absorbed crystalline rocks, *et alia*. In the case of the Ries of Nördlingen the explanation of the vast caldera and the various geologic abnormalities must obviously be sought in volcanic activity. (Rohleder 1933b: 489)

Whereas many of the factual arguments by conference members against Rohleder's talk centred on the alleged predisposed position of both the Steinheim and Ries Basins, which made it improbable that a meteorite would hit just there (see Chapter 3), Rohleder in his English paper quoted Walter Kranz regarding the Steinheim Basin: 'The lines of disturbing influences simply ignore the existence of our basin' (Rohleder 1933b: 492).

In his English paper too, Rohleder attributed the formation of aragonite to thermal springs caused by a cooling meteoritic mass, but with less emphasis,

---

[24] 'viel leichter zu Staub zerblasen werden kann als Metalle' (Rohleder 1933a).

[25] 'daß auch bei einem Meteoriteneinschlag eine sekundäre Explosionserscheinung nicht von der Hand zu weisen ist. Berührt der weißglühende Körper in einer tieferen Lage einen stark wasserdurchtränkten Horizont, so muß hierdurch soviel Wasserdampf frei werden, daß eine Explosion und damit eine Zerspratzung der hangenden Gesteinsschichten stattfindet. In diesem Fall also würde der Mechanismus des Meteoriteneinschlags und seiner Folgen dem des echten Oberflächen-Vulkanismus außerordentlich ähnlich werden' (Rohleder 1933a).

downplaying the importance of this feature (as he, quite correctly, saw no need for very high temperatures in aragonite formation):

> Tufa with occurrences of aragonite show that the temperature of the water within the basin must at times have exceeded 30°C. So far, all German geologists have regarded this fact as strong evidence for the volcanic origin. Needless to say aragonite does occur in many sedimentary rocks without any connection whatever with volcanic activity. But if this occurrence within the Steinheim basin has so far been regarded as so important, one is surely equally justified in assuming that the source of heat which gave rise to this mineral was not volcanic activity but the meteorite cooling in depth. The amount of aragonite is scanty, and we know that the cooling of igneous bodies can take a considerable time. (Rohleder 1933b: 493)

Finally, Rohleder made reference to Kaljuvee's book:

> It was only after I had finished this manuscript that Professor Keilhack[26] showed me in Berlin a new publication by J. Kaljuvee of Tallinn (Reval). This Estonian teacher, as clever as bold, tries to explain many geological features of the earth's crust through impacts of large meteoric bodies. As such he not only regards the Steinheim basin, but also the Ries of Nördlingen, in spite of its vast size – the diameter of the crater being 15 to 20 miles. He goes even so far as to interpret the volcanic rocks of Swabia as effects of this vast Ries meteorite impact, the volcanic material having been squeezed to the surface in places of minor resistance.
>
> Although I personally regard this latter hypothesis as extremely bold, it seems only fair to state the fact that Kaljuvee's ideas, as far as the origin of the Steinheim basin is concerned, are identical with mine. (Rohleder 1933b: 494)

Rohleder's interest was not just a brief 'flash in the pan', but lasted for some years and was fuelled by his own fieldwork as a prospector in Africa, where he also undertook some research regarding the supposed meteorite crater Lake Bosumtwi, 'an isolated crater lake of 5 miles diameter, which shows astonishing parallels to the Ries' (letter from Herbert Rohleder to Carl Troll, 14 September

---

[26]   Konrad Keilhack (1858–1944), since 1881 geologist at the Prussian Geological Survey, in addition to his practical work, taught at the Berlin mining academy since 1896, from 1900 onwards as professor, applied geology and hydrogeology (Engelmann 1977).

1934, BGIB).[27] Again, he planned to write a brief report in German, whereas the main paper was to be in English.[28]

Lake Bosumtwi had been interpreted as a meteorite crater by James Malcolm Maclaren (†1936) in 1931 (Rohleder 1936: 51). However, based on the close similarity of Lake Bosumtwi to the Ries Basin, which Rohleder considered to be volcanic due to the presence of suevite, he finally argued for an equally volcanic origin of the African structure: 'In my opinion the key to the correct interpretation of the Bosumtwi depression is to be found in the Ries of Nördlingen' (Rohleder 1936: 53).[29]

Rohleder felt himself justified in his volcanic interpretation by the heavily disturbed basement rocks, which he assumed were formed by terrestrial tectonic forces rather than impact:

> The mere fact that the pre-existing metamorphic rocks around Lake Bosumtwi exhibit signs of great tectonic dislocations in a country which is regionally uniform suggests that whatever causes the lake depression also caused these dislocations. This means incidentally further evidence against the meteoritic theory, for the impact of a meteorite would truncate the country rock but not cause complicated structures. (Rohleder 1936: 55)

Although fieldwork was made difficult by dense tropical vegetation and intense weathering of rocks, Rohleder nevertheless found quite a number of astonishing parallels not only to the Ries Basin but also to the Steinheim Basin. There were highly porous 'volcanic' rocks; intense brecciation with components sometimes reduced to fine fragments; surface structures on quartzite that resembled the *Strahlenkalk* of Steinheim, later interpreted as shatter cones by Robert Dietz (see Chapter 8) (Rohleder 1936: 57–61): 'In constantly comparing Lake Bosumtwi with Steinheim basin, the author relies on the morphological resemblance. Contrary to the general German point of view he is not convinced of the volcanic origin of the Steinheim basin, as any definite proofs are so far missing' (Rohleder 1936: 63).

This statement betrays the fact that for Rohleder too, just as for Kurd von Bülow (see above), an impact-based interpretation was permissible only as a last resort. It also shows that in the case of missing associated meteorites, he

---

[27] 'Zur Zeit bin ich mit der Untersuchung des Lake Bosumtwi beschäftigt, eines isolierten 5 Meilen durchmessenden Kratersees, der eine erstaunliche Parallele zum Ries bildet' (letter from Herbert Rohleder to Carl Troll, 14 September 1934, BGIB).

[28] Letter from Herbert Rohleder to Carl Troll, 14 September 1934; the crater is again mentioned in a letter on 18 April 1935, BGIB; Rohleder 1934.

[29] 'In 1961 the suevite-like rock from the Lake Bosumtwi crater was shown to contain both lechatelierite and coesite. This definitely established the Bosumtwi depression as a meteorite crater, and as the fourth location in which coesite had been found' (Mark 1995: 122).

had no other positive evidence to offer. However, for Rohleder, the occurrence of melts automatically pointed to volcanism. This led to some inconsistency in his thinking. Although in 1933 he mentioned ray-shaped *Drucksuturen* [pressure sutures] as being characteristic for the Steinheim Basin impact crater, distinguishing it from the 'volcanic' Ries (Rohleder 1933a), he seemed to be prepared to give up this idea now that he had found similar features at the 'volcanic' Lake Bosumtwi: 'Although the origin of these so-called *Drucksuturen* [at the Steinheim Basin] has never been explained in a satisfactory way, they are generally regarded as being due to pressure-effects of volcanic explosions' (Rohleder 1936: 61).

Apart from the discovery of seemingly volcanic rocks at Lake Bosumtwi, detailed fieldwork had also revealed the intricacies of this geological structure and Rohleder drew the same conclusion for the African crater as the Ries experts did for their field of expertise: 'The more one is able to drag the apparently sudden process of formation of the Bosumtwi depression into different phases, the less the meteoritic theory can be upheld' (Rohleder 1936: 55).

Otto Stutzer

The next scientist who dared to publicly argue for an impact crater in Germany was Otto Stutzer (1881–1936), whom we have already accompanied on his field trip to Meteor Crater. He was born in Bonn as the son of a university professor and studied geology in Königsberg (now Kaliningrad), Munich, Tübingen and Heidelberg. He obtained his PhD in 1904 at the University of Tübingen, with a dissertation on the geology in the vicinity of Gundelsheim, a village close to the eastern rim of the Ries Basin. In 1905, he became an assistant at the Mining Academy in Freiberg (Germany) and from 1913 to 1936 he was Professor of Fossil Fuels in Freiberg.

Stutzer became an exceptionally widely travelled geologist, whose field trips, expeditions and prospecting work led him to various places in Europe, northern Africa, the Near East, the Belgian Congo, Canada, Alaska, Colombia, Venezuela and Russia. His initial interest was in ore deposits, but later shifted increasingly to the geology of oil and coal.

In 1935, Stutzer was elected as the first European to become president of the Society of Economic Geologists and travelled to its annual meeting in the USA, followed by a lengthy private field trip that also led him to Meteor Crater in Arizona in February 1936 (see above). He died on 29 September 1936 from heart failure (Engelhardt and Hölder 1977: 242; Jurasky 1937; Künstner 1981).

On 27 August, half a year after his visit to Meteor Crater in Arizona and a month before his sudden death, Stutzer gave a talk at the annual meeting of the German Geological Society (DGG) in Kassel on Meteor Crater and the

Ries Basin. First he gave a short description of Meteor Crater, claiming that 'An explosion must have created this hole' (Stutzer 1936b).[30]

Stutzer founded his claim on several arguments: as drilling campaigns had shown, the deeper underground part of the crater was undisturbed and thus the explosion must have been situated near the surface. Also there were fragments of meteorites to be found. Stutzer, like Barringer, assumed that the meteorite might still rest within the crater, although it had not yet been found by drilling. Once found, the meteorite could be expected to have a high content of nickel as well as platinum. Silica-glass had also been found in Meteor Crater, containing nickel that for geochemical reasons must stem from the meteorite.

As to his impact mechanism, Stutzer was guided by a simple field observation. Upwardly bent beds of rocks at the crater rim suggested that the explosion took place not at the point of contact at the Earth's surface, but somewhat below:

> It shows that the force of explosion was greater than the force of impact, because the beds around the crater hole are not bent downwards but upwards. The explosion was possibly caused when the incandescent hot meteorite came into contact with groundwater, which led to the sudden creation of a great amount of water vapour. (Stutzer 1936b)[31]

As Stutzer reported to his colleagues, other meteorite craters were also known: in Odessa (Texas), Henbury (Australia), Wabar (southern Arabia), Ösel (Kaalijärv, Estonia), the Tunguska event, Gran Chaco (Argentina), a crater field in North and South Carolina and 'at this point, we may refer to the craters of the Moon as well, which nowadays are interpreted as impact craters by most astronomers' (Stutzer 1936b).[32]

At the Ries Basin too, there had been an explosion:

> The explosion theory, elaborated by Kranz, explains the geological setting of these rocks in a lucid way. It is possibly accepted by most geologists. ... The following is conspicuous: Among the ejecta volcanic rock is missing. This always has astonished everyone who has worked on the Ries. The tuffs ('suevite') that occur

---

[30] 'Eine Sprengung muß dieses Loch geschaffen haben' (Stutzer 1936b).

[31] 'Es zeigt dies, daß die Kraft der Explosion größer war als die Kraft des Aufschlages, denn die Schichten sind um das Kraterloch herum nicht abwärtsgebogen, sondern aufgerichtet. Die Sprengung wurde wohl dadurch hervorgerufen, daß der glühend heiße Meteor in Berührung mit Grundwasser kam, was zur plötzlichen Bildung einer großen Menge Wasserdampf führte' (Stutzer 1936b).

[32] 'Es sei hier auch auf die Mondkrater verwiesen, die heute von den meisten Astronomen als Aufsturzkrater gedeutet werden'. (Stutzer 1936b); but compare Chapter 8.

in the area are younger than the debris cover and therefore also younger than the explosion. (Stutzer 1936b)[33]

Stutzer, like Kranz, assumed a shallow explosion centre. He referred to an explosion funnel of 21 kilometres width – assuming the whole of Ries Basin in its present morphology to represent the explosion funnel:[34]

> And how easy is everything, if one assumes a meteorite impact. Then the fantastic assumption of a water-vapour concentration in solid crystalline rocks as a consequence of melting of the basement rock in unknown depth is unnecessary, because the hot meteor falling from above suddenly evaporated the groundwater, which is present in the Jurassic beds in ample amounts. Then, one is no longer astonished that volcanic material is missing. For anyone who knows the Nördlinger Ries and has seen Meteor Crater in Arizona, a veil has fallen. (Stutzer 1936b)[35]

That meteoritic material was missing did not in Stutzer's opinion speak against this hypothesis, as no Tertiary or older meteoritic iron was known. It was oxidised and disintegrated too fast to be expected in a 'fossilised' state. If the impacting body had been a stony meteorite, which was ten times more common anyway, it would have disintegrated even faster or might even have been pulverised upon impact.

Stutzer suggested that the local magnetic anomalies should be checked to determine if they might point to meteoritic material underground. Stutzer cautioned against overly high expectations concerning 'large nickel and platinum masses', but nevertheless, once he had thus set out his bait, he added: 'Should,

---

[33]   'Die durch Kranz ausgearbeitete Sprengtheorie erklärt die geologische Lagerung dieser Gesteine in klarer Weise. Sie dürfte von den meisten Geologen angenommen sein. ... Auffallend ist nun folgendes: Unter den Auswürflingen fehlt jedes vulkanische Gestein. Dies hat stets das Erstaunen aller, die das Ries bearbeiteten, hervorgerufen. Die im Gebiet auftretenden Tuffe ('Suevite') sind jünger als die Schuttdecke, demnach auch jünger als die Sprengung'. (Stutzer 1936b).

[34]   The explosion of the projectile in fact creates a transient crater, which is reshaped within seconds to minutes by elastic rebound of the strongly compressed target rock underneath the crater and further – within minutes to months – by slumping of the unstable rim. Thus, the initial smaller and deeper crater becomes wider and shallower (see Chapter 1).

[35]   'Und wie einfach ist alles, wenn man Meteor-Einschlag annimmt. Dann ist die phantastische Annahme einer Wasserdampfansammlung im festen kristallinen Gestein als Folge einer in unbekannter Tiefe erfolgten Aufschmelzung des Grundgebirges unnötig, da der von oben niedergestürzte heiße Meteor das in den Juraschichten reichlich vorhandene Grundwasser plötzlich verdampfte. Dann wundert man sich auch nicht mehr erstaunt, daß vulkanisches Material fehlt. Wer das Nördlinger Ries kennt und den 'Meteor Crater' in Arizona gesehen hat, für den ist ein Schleier gefallen' (Stutzer 1936b).

however, the state or a private company undertake drillings because of such an idea, all Ries geologists would welcome it' (Stutzer 1936b).[36] He also continued to suggest a drilling campaign for the smaller Steinheim Basin.

Stutzer cited Wilhelm Ahrens (1894–1968), according to whom suevite consisted exclusively of melted basement rock and did not contain any juvenile material (Ahrens 1929). Stutzer claimed that the necessary heat originated in the meteorite:

'When meteorites impact on Earth with high velocity (at least 2.5 km per second),[37] they develop heat' (Stutzer 1936b).[38] Stutzer continued to estimate the temperature reached by the impact by considering the melting and evaporation temperatures of material that had been found in meteorite craters: silica glass 1400–1800°C, nickel 3377°C, iron 3200°C. 'At any rate, it is no utopia to assume that upon the impact of the large meteorite of Nördlingen so much heat was created that shattered crystalline rock could melt at a depth of 1000 m' (Stutzer 1936b).[39]

A physical reasoning about mass and energy balance (relating the assumed size of the meteorite to the amount of molten material), was, however, missing from Stutzer's considerations and what happened to the projectile's kinetic energy upon impact was not mentioned. The only hint in that direction was his mention of the meteorite's velocity, which Stutzer, however, assumed to be about one order of magnitude too small (see above), consequently assuming two orders of magnitude less energy than in reality.

Stutzer must have known that this was the weakest point in his presentation, or he reacted to his critiques, because in the printed version he added in smaller print that he did not want to fix his mind concerning suevite. Perhaps it was due to secondary volcanism – that is, Stutzer did not want to exclude the possibility that the impact had allowed magma, present deep underneath the Ries Basin, to rise and erupt (Stutzer 1936b).

As to the Steinheim Basin, Stutzer referred to Herbert Rohleder:

---

[36] 'Sollten der Staat oder ein Privatunternehmen auf Grund einer solchen Anschauung hin aber Bohrungen ansetzen, so werden dies alle Riesgeologen begrüßen' (Stutzer 1936b).

[37] This is not close enough for a large impacting body. A large projectile with a diameter of a few hundred metres to several kilometres contains so much kinetic energy that the braking effect of the Earth's atmosphere becomes negligible. Thus, the minimum velocity of a large incoming body corresponds to the Earth's escape velocity of 11.2 km/s. However, it may well reach 30 to 60 km/s without having to be considered extraordinary (cf. Chapter 6).

[38] 'Wenn Meteoriten mit großer Geschwindigkeit (mindestens 2,5 km Sekundengeschwindigkeit) auf die Erde aufschlagen, entwickeln sie Wärme' (Stutzer 1936b).

[39] 'Jedenfalls ist es keine Utopie anzunehmen, daß beim Aufschlag des großen Nördlinger Meteoriten soviel Wärme entstand, daß zerschmettertes kristallines Gestein in 1000m Tiefe schmelzen konnte' (Stutzer 1936b).

Unfortunately, his elaborations have not been especially happy. They have been assailable in details and thus did not find the attention that the underlying idea of his communication should have found. It is not understandable why Rohleder interpreted the Ries as volcanic, but the Steinheim Basin as the result of meteorite impact. Equal origin must of course be assumed for Ries *and* Steinheim, as it is otherwise done, I think, by all geologists, who have concerned themselves with this area. (Stutzer 1936b)[40]

In 1926, Walter Kranz had assumed two explosions at the Steinheim Basin, by means of water vapour caused by plutonism, in order to explain the central mountain: first shallow in the karst-water horizon to form the basin, then weaker and deeper (in Lower Jurassic beds) to raise the central mountain. Stutzer now correctly pointed out that karst water cannot be highly pressurised because fissures and clefts are open to the surface; that is the necessary heat transfer cannot be reached slowly – volcanically – but instead must have happened suddenly. Everything became clear to Stutzer once a heat source from above, the 'incandescent hot' meteorite was assumed. The explosive force would be even stronger, because apart from water vapour, oxyhydrogen gas would be formed:

How easily a falling hot meteorite penetrating into the karst horizon explains everything: here, blasted broken rock explained by sudden vapour release, which shortly afterwards and somewhat deeper led to a weaker second explosion. ... It had been my purpose in this lecture to provide an impetus to reflect on the Ries-problem for once earnestly in the light of the meteorite impact, instead of passing over this idea with empty talk or with a shaking of the head. For many it will be difficult because of persistence to give up the crypto-volcanism! But there is no crypto-volcanism, just pseudo-volcanism. Meteor craters are not fashion, but facts. And the Nördlinger Ries and Steinheim Basin are meteor craters! (Stutzer 1936b)[41]

---

[40]   'Seine Ausführungen waren leider nicht besonders glücklich, waren in Einzelheiten angreifbar und haben deshalb nicht die Beachtung gefunden, welche die Grundidee seiner Mitteilung hätte finden müssen. Unverständlich ist es, daß Rohleder das Ries vulkanisch, das Steinheimer Becken aber als Meteoriteneinschlag deutet. Gleiche Entstehung muß selbstverständlich für Ries *und* Steinheim angenommen werden, wie es wohl auch sonst von allen Geologen, die sich mit dieser Gegend befaßt haben, geschieht' (Stutzer 1936b).

[41]   'Wie leicht erklärt alles der niederstürzende heiße Meteor, der in den Karsthorizont eindrang, hier durch plötzliche Dampfentwicklung zertrümmertes Gestein aussprengte und etwas tiefer kurz darauf eine schwächere zweite Explosion hervorrief. ... Meine Absicht war es, mit dem Vortrag die Anregung zu geben, das Ries-Problem einmal im Sinne des Meteor-Aufschlags ernstlich durchzudenken und nicht über diese Ansicht mit nichtssagenden Redensarten oder Kopfschütteln hinwegzugehen. Vielen wird es aus Beharrlichkeit schwer werden, den Kryptovulkanismus aufzugeben! Aber es liegt kein Kryptovulkanismus vor,

Stutzer had planned a second talk, which the programme listed as 'The Köfels-widening of the Ötztal-valley (Tirol) as meteor crater'. It was, however, cancelled 'for lack of time' and only an extended abstract was published (Stutzer 1936c; Chapter 5).

The time for this second talk was lacking, because Stutzer's impactist paper provoked an intense and controversial discussion among his audience leading to the public dismissal of Stutzer's arguments and conclusions by the Ries experts.

---

sondern Pseudovulkanismus. Meteor-Krater sind keine Mode, aber es sind Tatsachen. Und das Nördlinger Ries und das Steinheimer Becken sind Meteor-Krater!' (Stutzer 1936b).

Chapter 3
Dismissing Impact I

Both Herbert Rohleder and Otto Stutzer presented their ideas at conferences and they were therefore hard to ignore. Following both presentations there was much discussion and some of it found its way into the literature. Remarkably, it is the contemporary lack of peer review that allows us today to appreciate the form and scope of minority concepts such as the impact hypothesis in the history of geology. There was little to no peer review; the decision to publish rested basically with the editor of the journal, who might be contacted for negotiations well beforehand, while institutional scientists basically had the right to publish in their institutions' journals, and conference papers were published as a matter of course. This is how the papers by Rohleder and Stutzer found their way into the literature, even though they were highly controversial. Instead of rejecting such a conference paper, potential opponents were invited to comment and their comments were then attached to the paper, allowing professional readers to view both sides of the argument and form their own opinions.

Outside the tightly knit circle of Ries experts, there were a few positive and open-minded reactions:

> During the past ten years, the increasingly thorough photographic survey of the sky has led to the conclusion that the ring of asteroids is not contained within the Mars and Jupiter orbits, as had still seemed to be the case around the turn of the century; but that it crosses the orbit of Mars, and recently some asteroids have been discovered that even penetrate within the Earth's orbit and thereby can get alarmingly close to the Earth. Nothing is opposed to the assumption that <u>asteroids have occasionally fallen onto Earth</u>. Because of their small size, the Earth as such is not in danger from such an impact; locally, however, enormous destruction on the Earth's surface can be the outcome. The Meteor Crater in Arizona and the crater field in the Siberian taiga bear witness to this. Additionally, Stutzer (1936[b]) attributes the origin of the *Nördlinger Ries* to the impact of a gigantic meteor of several kilometres diameter. (Nölke 1937: 173)[1]

---

[1] 'Die immer sorgfältiger ausgeführte photographische Überwachung des Himmels hat in den letzten 10 Jahren zu dem Ergebnis geführt, daß der Ring der Planetoiden nicht, wie es noch um die Jahrhundertwende schien, von der Mars- und Jupiterbahn umschlossen wird, sondern über die Marsbahn hinübergreift, und neuerdings sind einige Planetoiden entdeckt worden, die sogar ins Innere der Erdbahn eindringen und dabei der Erde bedrohlich nahe kommen können. Es steht der Annahme nichts im Wege, daß <u>Planetoiden</u>

Friedrich Nölke (1877–1947) had been a teacher of mathematics, physics and geography in Bremen. In 1919 he became a professor at the University of Bremen. His main research was in astronomy, with a focus on the development of our planetary system.[2]

The vast majority of Rohleder's and Stutzer's geological colleagues, however (among them all the relevant Ries experts), quickly and blatantly dismissed the claim of a meteoritic origin for the Steinheim and Ries Basins.

The Austrian Robert Gangolf Schwinner (1878–1953) published a short, polemical comment on Rohleder's paper: 'Meteor craters are one of the latest fashions in geology and are now discovered everywhere in the strangest and wildest regions of the Earth, in Arizona, Arabia, Transvaal and now also in Swabia' (Schwinner 1933).[3]

Schwinner had begun to study engineering at the Technical University in Vienna. Shortly afterwards he switched to the University of Vienna, where he attended lectures in mathematics. He also spent a semester in Jena and Munich to study mathematics and physics. After a break of three years because of ill health, he resumed his studies in Vienna, this time with a focus on meteorology. But again he switched subjects and came to study geology in Vienna. In 1911, already thirty-three years old, he received his doctoral degree in geology. During World War I he worked as a military geologist and whenever home from the front, he used the time to '*habilitate*' himself in geology. In 1919 he became a geological assistant at the Geological Institute of the University of Graz. He was formally promoted to extraordinary professor (professor without a chair)[4] in 1923, but until 1940 he still had to fulfil his former duties as assistant. He retired in 1946 (Fritschl and Hubmann 2003; Metz [n.d.]; see also Chapter 7).

---

gelegentlich auf die Erde herabgestürzt sind. Wegen der Kleinheit ihrer Masse ist die Erde als Weltkörper durch solch einen Aufsturz zwar nicht gefährdet; wohl aber können örtlich auf der Erdoberfläche gewaltige Zerstörungen die Folge sein. Der Meteorkrater in Arizona und das Kraterfeld in der sibirischen Taiga geben Zeugnis davon. Stutzer (1936[b]) führt auch die Entstehung des Nördlinger Rieses auf den Aufsturz eines Riesenmeteors von einigen Kilometern Durchmesser zurück'. (Nölke 1937: 173).

2　<http://de.wikipedia.org/wiki/Friedrich_N%C3%B6lke> accessed 10 July 2013.

3　'Meteor-Krater sind eine der neueren Moden in der Geologie, sie werden nunmehr überall entdeckt in den merkwürdigsten und wildesten Gegenden der Erde, in Arizona, Arabien, Transvaal und nun auch in Schwaben' (Schwinner 1933).

4　A professor with chair or *Professor ordinarius* represented the full academic area in question and often was also head of an institute. Working at such an institute, there may also have been several professors without chair [called a *Professor extraordinarius*] representing a side-area and being subordinated to the professor with chair. Further down the hierarchy would have been the not-yet *habilitated* [scientific] assistants [*Assistenten*] and the freshly *habilitated* lecturers [*Privat-Dozenten*], who had to apply for vacant professorships to continue their academic career.

Schwinner did not believe in an impact in the absence of evidence of actual meteoritic material in its vicinity. He opposed Rohleder with respect to the measurements of magnetic anomalies at the Steinheim Basin and (since both Schwinner and Rohleder thought the Ries to be a volcano), he added that the distance between the Steinheim and Ries Basins was not too great for common volcanism, as such distances were observed in many volcanic provinces.

Schwinner's main argument, however, was the heat balance. He estimated the temperature of the meteorite to have been less than 900°C, because above this threshold the target limestone would decompose and the octahedral structure of meteoritic iron would disappear. This structure, however, was still visible in the Canyon Diablo iron fragments associated with Meteor Crater. If one took the volume of the projectile and its heat capacity into account, Schwinner argued: 'the heat content of the meteorite will be about sufficient to bring the crater lake once to boiling point, but it will by no means be sufficient to power a thermal spring for close to geological times (the thickness of the basin sediments being more than 60 metres)' (Schwinner 1933).[5]

Schwinner concluded with polemic irony: 'It is disappointing that the facts for Steinheim had to be termed 'meagre'. Evidently, there are more abundant facts from other meteor craters (Arizona, Arabia, Ashanti and so on)' (Schwinner 1933).[6]

Stutzer's talk at the annual meeting of the German Geological Society (DGG) held in Kassel three years later (Stutzer 1936b) provoked a similar negative response from his colleagues. By then, some of them seemed to have regarded the impact idea as a nuisance and Edwin Hennig (1882–1977), Richard Hennig's brother, lamented: 'Now, of course, the Meteor Crater of Arizona is not the latest evidence and attempts have not been wanting to explain other somewhat similar dents in the surface of the Earth by means of its example (Rohleder-Steinheim, Kaljuwee [sic]-Ries)' (Hennig 1936).[7]

Edwin Hennig had obtained his PhD from Berlin, where in 1906 he began to work as an assistant. From 1909 to 1911 he partook in the Tendaguru expedition to German East Africa. In 1913 he became a lecturer in Berlin and in 1917 he

---

[5]   'so wird der Wärmeinhalt des Meteoriten etwa langen, den Kratersee einmal zum Aufsieden zu bringen; keineswegs aber genügen, um eine Therme durch einigermaßen geologische Zeit (Mächtigkeit der Beckensedimente über 60 m) zu betreiben' (Schwinner 1933).

[6]   'Es ist sehr betrüblich, daß das Tatsachenmaterial für Steinheim als 'spärlich' bezeichnet werden mußte. Offenbar liegt von den anderen Meteorkratern (Arizona, Arabien, Aschanti, …) reichlicheres vor' (Schwinner 1933).

[7]   'Nun ist freilich der Meteorkrater von Arizona nicht neueste Erfahrung, und an Versuchen, andere irgendwie ähnliche Dellen in der Erdoberfläche nach seinem Vorbilde zu erklären, hat es denn auch bisher schon nicht gefehlt (Rohleder-Steinheim, Kaljuwee[sic]-Ries)' (Hennig 1936).

received the chair of geology and palaeontology in Tübingen (Freyberg 1974: 65), which he held until 1945 when he was dismissed because of his Nazi involvement.

Hennig's arguments against Stutzer were the lack of meteoritic material at Steinheim,[8] which might have been oxidised, but could not have disappeared completely, and the lack of a central hill at Meteor Crater that would match the *Klosterberg* in the middle of the Steinheim Basin. Hennig even doubted that meteorites existed in the geological past, so that there could not be any ancient impact craters, for lack of a causal agent:

> I have repeatedly pointed to the strange fact, that fossil meteorites are unknown. If the finest horny parts and soft-body tissues [of organisms] are preserved, there should also be something recorded over the immeasurable geological times from a constant meteorite shower on earth in marine layers or [preserved] under other favourable conditions. One cannot get round the question, whether at all in former times and always as today, cosmic rock material has been attracted [by the Earth]? A meteorite of Upper Miocene age would in any case be the oldest from which we have any word or at least hints! (Hennig 1936)[9]

Helmut Hölder (1915–2014), then a student at the University of Tübingen, commented on this lack of actualistic thinking:

> Yes, it is astonishing that Hennig at that time formulated the matter like this, but it was connected to the fact that we did not believe meteorites to be capable of anything. They had no role to play at all. ... It was our purpose to explain the Earth 'out of itself' and not to invoke the cosmos. (Hölder interview)[10]

---

[8]   Absence of meteoritic material is no longer considered an argument against impact, as the resulting high temperatures usually cause the projectile to vaporise. Only for craters less than about 1.5 km in diameter, such as Meteor Crater, might fragments of the projectile survive the impact due to spallation during entry into the atmosphere or due to atmospheric drag upon a comparatively low-impact velocity (Koeberl 1998; King 1976: 89).

[9]   'Ich habe wiederholt auf die eigenartige Tatsache hingewiesen, daß fossile Meteoriten unbekannt sind: Wenn sich feinste Hornteile und Weichkörper erhalten, müßte doch von einem dauernd auf die Erde niedergehenden Meteoritenregen durch die unabsehbaren geologischen Zeiten hin irgendwann einmal in marinen Schichten oder sonst wie unter günstigen Umständen etwas überliefert sein. Man wird um die Frage nicht herumkommen, ob denn überhaupt schon früher und stets wie heut kosmisches Gesteinsmaterial eingefangen wurde? Ein obermiocäner Meteorit wäre jedenfalls der weitaus älteste, von dem wir Kunde oder wenigstens Andeutungen hätten!' (Hennig 1936). Meanwhile, there is a project by Pennsylvania State University that recovers fossil meteorites from coal mines (see <http:// www.eurekalert.org/pub_releases/1996–07/PS-PSRP-300796.php> accessed 10 July 2013).

[10]   'Ja das ist überraschend das Hennig damals so formuliert hat, aber das hängt damit zusammen, dass wir den Meteoriten nichts zugetraut haben. Dass die überhaupt eine Rolle

Thus the Earth by definition was regarded as a geologically closed system.

Hennig pointed out that there was more unusual volcanism in southern Germany than just the Ries and Steinheim Basins. The contemporary Swabian 'volcanic embryos' (see Chapter 1) were considered equally enigmatic. Nevertheless, Stutzer (inconsistently, as Hennig saw it) left them out of his 'impactist' considerations. Like other commentators, Hennig mentioned the special locality of the Ries, which had been a place of local uplift at least since the Middle Triassic, and that there had been thermal springs at the Ries and Steinheim Basins:

> Should the meteorite have called them into life like Moses' rod? ... It may be our duty to burden our thoughts with such an additional abundance of questions and the enigmatic but also eternally intriguing [geological] features of southern Germany with new problems. Something *helpful* or even a *solution*, however, I cannot discern in the interpretation we have heard. (Hennig 1936)[11]

Ries experts felt that the geological history of the Ries area before the actual Ries event had been so special that it could not be a mere accident that the Ries structure was situated at this particular spot. It would have been just too improbable! Only with hindsight did some participants (Hölder interview; Helmut Vidal interview; see Chapter 9) in the debate admit that rating the Ries area as a remarkable place was an artefact of the intense research carried out in this region.

Like Hennig, Alfred Bentz (1897–1964) also elaborated on this argument (Bentz 1936). Bentz had studied geology in Munich and Tübingen and obtained his PhD in 1922 from the University of Tübingen with a dissertation on the Mid-Jurassic beds and the tectonics around Bopfingen, a village outside the western Ries rim. Edwin Hennig was his supervisor. From 1922 to 1923, Bentz worked as assistant in Tübingen and in 1923 he became geologist at the Prussian Geological Survey in Berlin. In 1934, the Nazis opened an Institute for Petroleum Geology within the survey and Bentz became its director. In this function – petroleum being a crucial resource for the preparation of what was to become the Second World War – he rose rapidly in the hierarchy, becoming responsible for the petroleum supply of the Third Reich and working directly under the *Generalfeldmarschall* Hermann Göring (1893–1946). Nevertheless, Bentz

---

gespielt haben. ... Wir waren dazu da die Erde aus sich heraus zu erklären und nicht den Kosmos da herbeizuziehen' (Hölder interview).

[11] 'Soll der Meteorit als Mosesstab auch sie ins Leben gerufen haben? ... Wir mögen verpflichtet sein, uns gedanklich solche Mehrbelastung von Fragen aufzuerlegen, die ohnehin Sphinx-artig rätselvollen, aber auch unendlich reizvollen Erscheinungen in Süddeutschland mit neuen Problemstellungen zu erschweren: eine *Erleichterung* oder gar eine *Lösung* vermag ich in der gehörte Deutung nicht zu erblicken' (Hennig 1936).

never became a member of the Nazi party (NSDAP) and it has been argued that he managed to save a number of geologists, prosecuted for political, religious or racist reasons, in eastern Europe from concentration camps by claiming them for his petroleum drilling campaigns. In April 1945, Bentz managed to abscond himself from the final crash of the Nazi regime, while allegedly being off on an official visit to Vienna. In June, he returned to his office in Celle, whereto his institute had been evacuated. There he successfully engaged himself and his colleagues in fruitful collaboration with the British military authorities and in July 1945 he applied for a reestablishment of the *Reichsamt für Bodenforschung*, the national geological survey. A Geological Survey was indeed established in autumn 1945 with Bentz as its director, a position he held until his retirement in 1962 (Seibold and Seibold 2002; Martini 1965; Engelhardt and Hölder 1977: 263; Kockel 2005).

Bentz replied to Stutzer by mentioning the different morphology of the Ries Basin from that of Meteor Crater, as well as the lack of meteoritic material in the Ries polymictic breccia. Bentz (correctly)[12] added that magnetic measurements as well as the negative gravity anomaly in the area were incompatible with the existence of a huge meteoritic mass below the basin floor (Bentz 1936). He concluded:

> I must therefore call it illogical, if one explains the absence of volcanic lava (tuffs are not only very common in the Ries but also in its periphery) by cosmic events of hardly imaginable extent as the cause of the Ries phenomenon – events that cannot be supported by even one milligram of cosmic substance. The supply of magmatic material, which even does not need to consist of lavas alone, is being denied *without the supply of cosmic material having been made in the least probable*. The hypothesis is therefore founded only on morphological similarities and not appropriate for contributing to the clarification of the Ries problem. (Bentz 1936)[13]

---

[12]   Barringer had postulated a large mass of nickel-iron buried underneath his impact crater. Such a mass should lead to a positive gravity anomaly. Meteorite craters, however, usually have a slightly negative gravity anomaly due to the extensive brecciation, which reduces the density of the country rock.

[13]   'Ich muss es demnach als eine innere Unlogik bezeichnen, wenn man das Fehlen eines vulkanischen Schmelzflusses im Ries (Tuffe sind ja nicht nur im Ries, sondern im Vorries sehr verbreitet) damit erklärt, daß kosmische Ereignisse von kaum vorstellbarem Ausmaß die Ursache des Ries-Phänomens waren, kosmische Ereignisse, die durch nicht ein Milligramm kosmischer Substanz belegbar sind. Es wird die Zufuhr magmatischen Materials, das ja nicht nur aus Schmelzflüssen zu bestehen braucht, geleugnet, *ohne daß die Zufuhr kosmischen Materials auch nur im geringsten wahrscheinlich gemacht wird*. Die Hypothese stützt sich demnach nur auf morphologische Ähnlichkeiten und ist nicht geeignet, zur Klärung des Ries-Problems beizutragen' (Bentz 1936).

Wilhelm Ahrens (1894–1968) added arguments against the impact idea that were based on the 'volcanological–petrographical' point of view. He had studied science in Jena and Freiburg. He spent World War I as a military geologist and gunner. After the war, he continued his studies in Hamburg and again in Jena, where he obtained his PhD with a geological/petrographical dissertation on crystalline rocks. In 1922 he began his career at the Prussian Geological Survey, during which time he mapped the volcanic area of the *Laacher See* (East Eifel, Germany) in 1930. In 1945 he became Director of the newly founded Geological Survey in East Berlin. However, he left the German Democratic Republic in 1948, to take up a similar position in Düsseldorf (Federal Republic of Germany). In 1956, he became Director of the Geological Survey of Nordrhein-Westfalen in Krefeld. He retired in 1959 (Schuberth 2007; Wagenbreth 1999: 183).

Suevite, Ahrens claimed, had been erupted in several stages and he noted the presence of volcanic scoria. The presence of aragonite in the lake sediments proved, according to Ahrens, the presence of hot springs and made a 'preceding activity of fumaroles at least probable' (Ahrens 1936).[14]

Stutzer's untimely death from heart failure, only a month later, unfortunately prevented him from carrying the discussions toward a fruitful research programme.

Why?

The historical question is *not* why the impact hypothesis did not succeed some 20 to 30 years earlier than it actually did, soon after the impact nature of Meteor Crater in Arizona had been recognised and accepted by the majority of geoscientists in the 1930s. Compared to Arizona, the situation at the Ries Crater is much more complicated, since much of the evidence is covered by vegetation and lake sediments; the structure is much larger; and volcanology was still making its first uncertain steps toward a real understanding of what goes on in a caldera volcano such as Krakatoa. In fact, volcanologists (as well as mineralogists) were never much interested in the Ries. So it was regional geologists, stratigraphers, many of whom had never seen an active volcano in their lives, who used the icon of a volcano to explain the bizarre geological features of the two German impact craters.

It was also most unhelpful to the impactist cause that Rohleder, for example, proclaimed the Steinheim Basin to be an impact crater but rejected this explanation for the Ries Basin. And Kaljuvee at least was arguing so far out of his depth that he seemed to have lost all reason.

---

[14] '… und eine vorhergehende Fumarolentätigkeit zum mindesten wahrscheinlich gemacht' (Ahrens 1936).

As the previous chapter has shown, the promoters of an impact origin for the German craters did not have a ready-made and convincing theory up their sleeves. So it is no wonder that local experts did not jump to the impact solution.

However, there remains the question of why, in the light of research at Meteor Crater, the impact hypothesis was regarded as completely out of place and was never discussed by Ries experts *alongside* the volcanic hypothesis.

In the 1930s, the meteoritic origin of Meteor Crater, Arizona, had become a 'new paradigm' and the acknowledgement of several other impact craters soon followed. At international conferences, meteorite impacts became a fashionable explanation for all sorts of enigmatic structures. In Germany, however, such discussions quickly died down in the late 1930s and did not lead to a genuine German research effort in impact-crater geology.

Hoyt, in his comprehensive book on the American debate about Meteor Crater, attributed the long-running controversy there, which lasted from the late nineteenth century to well into the 1930s, at least in part, to:

> the unorthodox nature of the idea of impact vis-à-vis the entrenched belief in the volcanic origin of all natural crateriform features, and the tendency on the part of most scientists when faced with unorthodoxy to follow the conventional wisdom on matters on which they have no direct personal knowledge. Acceptance of an impact origin for Coon Mountain by individual scientists, in fact, correlates well with the amount of first-hand data they acquired. (Hoyt 1987: 100)

But at the Ries and Steinheim Basins it was just the other way round. The more people were familiar with the craters, the less they were inclined to think of them as impact craters. For this astonishing fact, as letters in archives and interviews conducted with surviving veterans of Ries research show, there were a number of rather trivial, irrelevant or non-scientific reasons.

Isolation: Politically, Linguistically and Scientifically

In the early twentieth century, German scientists, especially in physics and chemistry, enjoyed a high reputation and German was an internationally important language for scientific exchange. This began to change when, as a consequence of World War I, German scientists were 'held out of international scientific activity and cooperation throughout the 1920s' (Greene 1998: 117). The situation was rapidly aggravated when Nazi Germany started to rebuild the German scientific landscape to its own liking. Numerous scientists had to leave German universities for racist and political reasons and many were murdered or driven into exile (see Chapter 7). Many of those who survived found a new home in English-speaking countries (Snowman 2002). English became the new predominant scientific language and has remained so ever since.

However, it took some time for the remaining German scientists, in our case geologists, to realise that the political isolation due to the Nazi regime was driving them into linguistic and scientific isolation lasting from about 1936 onwards and well into the post-War period. Knowledge of English and consequently of the international literature, was not a matter of course among German geologists until about 1960, when the younger generation, with their post-War education, took up university positions. Before then, geological libraries did not routinely hold international journals.

Johann Thomas Gustav ('Hans') Pichler (born 1931) received a PhD in geology from Ludwig-Maximilian University in Munich in 1957. From 1957 to 1960 he was an assistant at the Bavarian State Collection for General and Applied Geology and at the Geological Institute of the University of Munich. After another assistantship in Catania (Sicily) with the volcanologist Alfred Rittmann and a brief stay at the University of Köln, Pichler became an assistant to Wolf von Engelhardt (see Chapters 9 to 12) at the Mineralogical Institute of the University of Tübingen. In 1974 he became Professor of Volcanology at Tübingen University. He recalled that in the library of the Geological Institute in Munich in 1955 through 1957 there were no English-language journals (Pichler interview):

> Scientific literature was read at that time too. One always searched the literature to know what had already been published about the area [in which one was working] ... and that then was read. But the up-to-date literature was not available to keep oneself informed in a general educational way – to leaf through it and say "Oh, that's interesting, you must read that!" One was restricted. There was the *Geologische Rundschau* and also the *Mitteilungen der Deutschen Geologischen Gesellschaft* [ZDGG] and that was basically all, and I'd say that were great difficulties in the spreading of new ideas in the German-speaking area and probably also elsewhere. (Pichler interview)[15]

Moreover, there were other obstacles. Few Ries geologists were widely travelled or had first-hand experience with active volcanoes, apart from a brief visit to

---

[15] 'Es ist damals auch Literatur gelesen worden. Man hat also auch immer die Literatur gesucht, was ist denn schon erschienen über das Gebiet ... und das hat man dann gelesen. Man hat also nicht ganz allgemein geowissenschaftlich bildend die gerade erschienene Literatur zur Verfügung gehabt und blättert darin und sagt 'Oh das ist eine interessante Sache, das musst du lesen'. Man hat also nur beschränkt, *die Geologische Rundschau* hat man gehabt, dann die *Mitteilungen der Deutschen Geologischen Gesellschaft* und damit war es eigentlich schon zu Ende und das waren große Schwierigkeiten für die Verbreitung neuer Ideen sage ich mal im deutschen Raum und wahrscheinlich auch in anderen Räumen' (Pichler interview).

Mount Vesuvius; and even then the travel itinerary could well have included more Roman remains than geological features (see Bernauer 1939).

In contrast, supporters of the impact idea, Herbert Rohleder as well as Otto Stutzer, had a rather more cosmopolitan background and experience. Both were widely travelled and fluent in English (Rohleder as a matter of course).

Artificial Boundaries and their Influence on Geological Research

A quick look at a political map shows that the two German impact craters are situated in two political jurisdictions. The Steinheim Basin belonged to Württemberg (since 1948 Baden-Württemberg), whereas most of the Ries Basin is situated in Bavaria – different states over historical time and even now served by two separate geological surveys. Whereas Nature does not care about such artificial boundaries, they can influence research programmes, and field geologists tended not to cross this particular political boundary even though the close genetic relationship between the two craters had been recognised and accepted by German geologists for quite some time (interviews Hüttner, Pichler).

'I must unfortunately say that Professor Dehm very strongly saw to it that we geologists from Württemberg did not cross the Bavarian frontier' (Hölder interview).[16] This led to a sort of mock ceremony when field trips with students from Tübingen crossed into Bavarian territory: '"Be careful!" it was said: "The Bavarian Lion is roaring!"' (Hüttner interview).[17]

Georg Wagner, then Professor Emeritus at Tübingen, also recalled with some resentment: 'But the area was far from Munich; there was neither beer nor wine. The Bavarian said: I rest and possess, let me sleep! The Bavarians jealously regarded the Ries as their domain. And despite the numerous doctoral theses [about it, they] have achieved little in summing up [data] and leading on [to solutions of scientific questions]' (letter from Georg Wagner to Leonard Rückert, 29 September 1969, UAT 605/343).[18]

Thus, the Steinheim and Ries Basins had never been treated as one structural unit. Ries research until the late 1950s was conducted with a stratigraphic rather

---

[16]   '… da muss ich leider sagen, dass Professor Dehm in der Richtung sehr streng darauf achtete, dass wir württembergischen Geologen nicht über die bayerische Grenze gingen' (Hölder interview).

[17]   'Vorsicht, der bayerische Löwe brüllt!' (Hüttner interview). The Bavarian coat of arms is flanked by two lions, while the Württemberg coat of arms was supported by stags.

[18]   'Aber das Gebiet war weit von München; es gab dort weder Bier noch Wein. Der Bayer sagte: Ich lieg und besitze, laßt mich schlafen! Das Ries haben die Bayern eifersüchtig als ihre Domäne betrachtet. Und trotz der vielen Doktorarbeiten wenig zusammenfassend und weiterführend geleistet' (letter from Georg Wagner to Leonard Rückert, 29 September 1969, UAT 605/343).

than a structural focus; it had classical stratigraphy and geological mapping in mind. For this purpose, the area was divided into square topographic map-sheets, each of which had to be marked up by a PhD student. And for these students, the Ries rocks were part of the stratigraphical succession and not members of an independent structure. Consequently, nobody had an overview of the entire Ries Crater (Hüttner interview).

What was the custom for the large political units also held true for the individual map-sheets. Crossing boundaries was seen as an act of academic aggression rather than one of natural curiosity and a chance for a friendly chat and scientific exchange:

> In former times that was very deeply entrenched. Thank God it is abolished now. It was simply, how should I say, a sort of tradition; it was not done that you simply went into neighbouring areas where a colleague was working just to see what was going on there. The other person might take offence and you wouldn't like it yourself if someone came [into your area] and so on. ... In former times, it was much more common that one simply spared the other; as I said, one kept to one's own hunting-ground and didn't break into the geological hunting-ground of colleagues. (Pichler interview)[19]

Likewise, Rudolf Hüttner, who did his PhD by mapping on the Württemberg side of the Ries (Engelhardt and Hölder 1977: 255) and later worked for the geological survey of Baden-Württemberg in Freiburg (Chapter 11), recalled:

> Yes, I have stopped mapping at the boundary [with Bavaria]. I would have continued mapping across the frontier, but the Bavarian maps at that time were still so bad. It was nearly impossible to orient oneself there; ancient stuff – There had been new [topographic] mapping from Württemberg and in the Bavarian part there was the old mapping and when linked up it did not fit.[20] So I was glad that I didn't need to venture into [Bavaria]. (Hüttner interview)[21]

---

[19] 'Das war früher ganz stark ausgeprägt, das ist heute Gott sei gedankt abgeschafft, nicht, das man ganz einfach in Nachbargebiete geht wo noch Kollegen arbeiten und man da nach dem Rechten sieht, das war ganz einfach eine, na wie soll ich sagen, eine Art von Tradition, tut man nicht, der andere könnte es krumm nehmen und du hast es auch nicht gerne wenn da jemand kommt und so weiter. ... Also das war früher sehr viel weiter ausgeprägt, das man das einfach geschont hat oder wie gesagt sein Jagdrevier eingehalten hat und nicht in die geologischen Jagdreviere der Kollegen eingebrochen ist' (Pichler interview).

[20] See Hüttner et al. (1969).

[21] 'Doch, ich hab meine Kartierung an der Grenze aufgehört, ich hätte zwar auch noch über die Grenze hinüber kartiert aber die bayerischen Karten waren damals noch so schlecht, man konnte sich dort kaum noch orientieren; uraltes Sach – Es war eben noch aus einer neuen kartographischen Darstellung aus Württemberg und im bayrischen Teil die alte

It was no better on the Bavarian side. The recollections of Jean Pohl (Munich Institute of Geophysics) show that the issue was not only territorial boundaries but also about the boundaries of fields of investigations and their perceived responsibilities:

> [B]ut we didn't bother ourselves about Steinheim. There were the people in Baden-Württemberg there. At that time, the Ries was such a huge object; there was so much to research that Steinheim was initially neglected. Yes, of course, the two were named together ... As to the others in Munich, I don't know whether the astronomers had an opinion about it, I can't remember, or whether they were involved or consulted in any way. Other research? It seems possible that Angenheister[22] tried to take part in the discussion, but I can't remember exactly. And the other geologists? I believe that the 'proper geologists'[23] did not participate. With respect to the geological mapping, the Ries was a territory for the palaeontologists. (Pohl interview)[24]

Such artificial boundaries were only finally broken down in 1960 under external pressure, which by some was seen as a catastrophic intrusion of new ideas into the usual routine of Ries research:

> That one side might say, "You're doing the neighbouring sheet, why don't we talk about the fit at the boundary?" was not done. Do you know when it was done? When under the pressure of the new theory, when Chao and Shoemaker [in 1960–61, see Chapter 9] came. And then suddenly they all came together and looked for mutual support. Not only the people from Munich and Württemberg

---

Darstellung und das hat man dann zusammengefügt und das hat nicht zusammengepasst und da war ich froh, dass ich nicht hinein gehen musste' (Hüttner interview).

[22]   See Chapter 9.

[23]   From the Munich Institute for General and Applied Geology, situated directly adjacent to the palaeontologists and stratigraphers.

[24]   ' ... aber wir haben uns um Steinheim nicht gekümmert, da waren die Leute in Baden-Württemberg. Das Ries war damals ein so großes Objekt, da gab es so viel zu erforschen, so dass Steinheim zunächst einmal vernachlässigt wurde. Ja sicher, die beiden sind dann schon zusammen genannt worden. ... Und ansonsten in München ich weiß nicht, ob die Astronomen da eine Meinung dazu hatten, ich kann mich nicht so erinnern. Oder ob die überhaupt beteiligt oder herangezogen wurden. Andere Erforschung? Es kann schon sein, dass Angenheister da versucht hat, sich auch an der Diskussion zu beteiligen. Ich kann mich explizit nicht erinnern und die anderen Geologen, ich glaube, da haben sich die richtigen Geologen hier in der Luisenstraße haben sich da nicht beteiligt. Das Ries war ein Feld für die Paläontologen, was die geologische Kartierung betrifft' (Pohl interview).

wanted to support each other and probably did so. They only came together under the intrusion of this catastrophic development. (Pichler interview)[25]

The rivalry between Württemberg and Bavaria was not only a territorial one with a long general history in the German past; it was also because the two political entities stood for two scientific theories or rather 'schools' of Ries genesis (see Chapter 1). Württemberg, that is the University of Tübingen and the Württemberg Geological Survey stood for the Kranz–Löffler model of a single huge central explosion that blasted the Steinheim and Ries Basins respectively, which was founded mainly on field evidence especially at the Steinheim Basin, which lay completely within Württemberg and was supported by the observation of similar structures in a small experimental explosion crater produced in an experiment by Walter Kranz. In Munich, however, the people at Ludwig-Maximilian University and the Geological Survey of Bavaria for theoretical reasons supported an only slightly modified version of Branca's and Fraas' theory (see Chapter 1), voiced by Joachim Schröder (in 1934) and by Dorn (in 1942) (see Dorn 1950). They postulated several smaller explosions at various locations within the Ries and in the periphery outside the morphological basin to account for the abundance and distribution of explosion debris:

> But it had been also a scientific conflict, because on the Württemberg side, one had the explosion theory in view and had defended it through Kranz, Löffler and of course Schorsch[26] Wagner as the great propagandist of this theory, while on the Bavarian side one had always operated on the assumption of local outbursts, but nobody was able to show, that is prove, these local outbursts properly – and something like that annoyed Schorsch. Schorsch always said that it must be possible to draw something; they [the Bavarian researchers] should make a drawing. But nothing ever developed about how one might explain such a local outburst. And there was something else: when the new impact theory arrived [in 1960–61], Gerold Wagner [Georg Wagner's son, a young geologist and a close friend of the interviewee] and I said: now the Bavarians must go to it – they must finally show their local outbursts – if there really are local outbursts outside the Basin then there was probably no impact. But then nothing came

---

25 '… dass die eine Seite gesagt hat, ihr macht das Nachbarblatt wollen wir nicht die Grenze abstimmen? Sowas hat man nicht gemacht, wissen Sie, wann man es gemacht hat als unter dem Zwang der neuen Theorie als dann Chao und Shoemaker kamen und dann plötzlich sind sie alle zusammengekommen, haben sich Unterstützung gesucht, nicht nur die Münchner und die Württemberger haben sich gegenseitig stützen wollen; vielleicht auch gestützt. Erst unter dem Einbruch dieser katastrophalen Entwicklung ist das so gelaufen, sind die zusammengekommen' (Pichler interview).

26 Swabian colloquial for Georg, pronounced similar to the French 'George', from which it was probably derived during the Napoleonic occupation of Württemberg.

from the Bavarian side, because these local outbursts had only been hypothetical constructs. They found themselves forced to assume such a thing because – I assume they were more theoretically minded [the interviewee stressed this point emphatically several times] – a volcanic explosion of such an enormous extent is hard to imagine, so we prefer to imagine there having been additional smaller explosions in the surrounding region. You see, that was the idea. While on the Württemberg side, we reasoned from what we'd seen in the field – how is the bedding of the stuff arranged? – everything indicates that it came from inside the crater. There were two different approaches. (Hüttner Interview)[27]

As Hüttner rightly noted, the promotion of the idea of 'local outbursts' had been basically a means to circumvent one of the major problems at the Ries Basin, namely the sheer size of the single explosion that created it, which otherwise had to be assumed. By 'local outbursts' the force was divided into numerous smaller explosions and thus came down to the 'normal' and imaginable range of volcanic forces. Consequently, there was no need to look for strange and alien mechanisms for the explosion(s).

In a short review paper, Bentz (1929) had tried to reconcile the Kranz–Löffler model of Ries genesis with the ideas of Branca and Fraas. Geological mapping, profiles and descriptions of outcrops in accord with Bentz's conviction had definitely shown that the breccias around the Ries Crater were allochthonous at all the disputed outcrops, while new gravimetric measurements had proven that, contrary to Kranz's initial assumption, only the innermost zone of some 12 km

---

[27] 'Das war aber auch eben ein wissenschaftlicher Konflikt, weil man eben auf Württembergischer Seite durch Kranz durch Löffler und natürlich Schorsch Wagner als dem großen Propagandisten dieser Theorie die reine Sprengtheorie im Auge hatte und verteidigt hatte, während man auf bayrischer Seite immer mit den örtlichen Aufbrüchen da immer operiert hat und niemand konnte die örtlichen Aufbrüche richtig zeigen – nachweisen und so was hat den Schorsch geärgert – der Schorsch hat immer gesagt: man muss etwas zeichnen können – die sollen doch mal eine Zeichnung machen und da kam eben nichts wie man so einen örtlichen Aufbruch erklären kann und eines war noch damals, dass die neue die Impakttheorie dann eben da war und da haben wir eben gesagt, Gerold Wagner und ich: jetzt müssen die Bayern noch ran – jetzt müssen sie endlich ihre örtlichen Aufbrüche noch zeigen – wenn es da wirklich örtliche Ausbrüche außerhalb des Kessels gibt ja dann ist es wohl kein Impakt. Aber dann kam eben von bayerischer Seite nichts, weil sie eben – ja es war nur eine gedankliche Konstruktion die örtlichen Aufbrüche – sie sahen sich gezwungen so etwas anzunehmen, weil sie – sie gingen mehr vom Theoretischen aus – von der theoretischen Vorstellung – so denke ich eine Sprengung vulkanischer Art in so großem Ausmaß, das ist so schwer vorstellbar also stellen wir uns lieber vor, dass es da noch kleinere Sprengungen, da im Umkreis gegeben hat, nicht, so war die Vorstellung. Während man auf württembergischer Seite davon ausging – was hab ich jetzt im Gelände gesehen – wie lagert das Zeug – es spricht alles dafür, dass es aus dem Krater gekommen ist – es sind zwei verschiedene Auffassungen'. (Hüttner interview).

Figure 3.1   Richard Dehm denied a central explosion at the Ries Basin and
              opted alternatively for smaller, local volcanic explosions to explain
              the widespread brecciation. (Photo from Fahlbusch 1996: 297).

diameter, and not the whole of the Ries Basin, represented an explosion crater.
The periphery seemed to consist of downfaulted and slumped blocks and slivers.
Consistent with this was the fact that at the Ries rim, the Jurassic country rock
did not descend in a smooth plain under the basin fill, but broke off steeply near
the Basin. From this Bentz concluded:

that today's escarpment around the basin is caused by young faults or subsidence phenomena. ... Thus, we must consider today's Ries Basin as a caldera, similar to the opinion of Branca and Fraas, but with the important difference that the cause of the subsidence is not the sinking of a hypothetical laccolith; it must be attributed to phenomena of subsidence in the vicinity of an explosion crater. (Bentz 1929)[28]

The paper, however, though containing some valuable thoughts, had little effect in reconciling the two schools of Ries geology.

It was mainly Richard Dehm (1907–96), Professor of Palaeontology and Stratigraphy in Munich (Figure 3.1) and his PhD supervisor, friend and colleague Joachim Schröder (born 1891), from 1920 to 1925 an assistant at the Palaeontological Institute at Munich and afterwards curator at the Bavarian State Collection for Palaeontology and Stratigraphy (Freyberg 1974: 142), who tended to ignore the field data, which pointed to a central explosion and postulated small and local volcanic eruptions of suevite instead.

These were then duly found by their students in the field, but were imperceptible to anybody else. One of these students recalled that Dehm taught 'some sort of volcanism, even though nobody really knew what this should be' (Volker Fahlbusch, 1934–2008, personal communication by phone, 17 September 2001):[29]

> If one reads the Bavarian PhD theses, done in the 1950s, at the end they always say ... that proofs or [even] hints of local outbursts could not be found, but that they could also not be excluded. That was always the bowing to the PhD supervisor. (Hüttner interview)[30]

The rivalry between Württemberg and Bavaria has had a tradition over many centuries and no one can tell when it began. But it seems that the usual mild

---

[28] ' ... daß der heutige Absturz zum Kessel durch junge Verwerfungen oder Sackungserscheinungen bedingt ist. ... Wir müssen daher den heutigen Rieskessel als ein Einsturzbecken auffassen, ähnlich wie dies auch Branca und Fraas getan haben, allerdings mit dem wichtigen Unterschied, daß die Ursache des Einsturzes nicht im Zurücksinken eines hypothetischen Lakkolithen besteht, sondern auf Senkungserscheinungen im Umkreis eines Sprengtrichters zurückgeführt werden muß' (Bentz 1929).

[29] ' ... irgendeine Art von Vulkanismus, auch wenn man nicht so genau wusste, wie das gehen sollte' (Volker Fahlbusch about Dehm, personal communication to the author by phone, 17 September 2001).

[30] 'Wenn man die bayerischen Doktorarbeiten, die damals in den fünfziger Jahren gemacht worden sind, liest, dann heißt das dann zum Schluss immer, ... Beweise oder Hinweise auf örtliche Aufbrüche konnten nicht gefunden werden, aber man kann sie auch nicht ausschließen. Das war immer die Referenz vor dem Doktorvater' (Hüttner interview).

animosity was magnified by a personal component, derived from Dehm's biography:

> Probably, there was a certain jealousy, ... because Dehm had been Professor[31] at Tübingen and was among those who gave the most inspiring lectures at that time, ... Then, when he was in Munich, he was a 'Municher'. But really he was always a 'Municher', since Tübingen had only been some sort of transit station from Strasbourg – he had been, so to speak, on sufferance in Tübingen and had been more than glad to regain his stable position in Munich. After the War, he had had some difficulties, because he had been a professor in Strasbourg. – And so there had been some tension, as usual – "this is now *my* map boundary" and similar things. (Seibold interview)[32]

Richard Dehm had studied science at the universities of Erlangen and Munich. In 1930 he did his PhD with a mapping project in the Ries Basin. Afterwards, he continued with a practical teacher education at schools in Nördlingen and Munich. He passed his teaching examination in 1932 and afterwards became an assistant at the Institute for Palaeontology and Stratigraphy at the University of Munich. There he did scientific work again and was '*habilitated*' in June 1935 with a dissertation on Tertiary fissure fillings.

In 1939–40 Dehm undertook a research journey to India and Australia together with Joachim Schröder. In 1941 they were interned but somehow managed to return to Germany via Siberia. In October 1941 Dehm became a professor at the *Reichsuniversität Straßburg* and Director of its Palaeontological Institute, a position which he kept till 1945. This university had been founded in the occupied territory of Strasbourg in 1941 by Nazi officials. With the withdrawal of the German army, it was moved to Tübingen and disbanded at the end of World War II. '[Dehm] avoided internment, because he had been travelling on behalf of the transfer of precious scientific goods to universities on this side [the eastern bank] of the River Rhine, when on the day of his intended

---

31 *Lecturer* would be more correct.

32 'Wahrscheinlich war eine gewisse Eifersucht da, ... denn Dehm war ja Professor in Tübingen und hat mit die aufregendsten Vorlesungen damals gehalten, ... als er dann in München war, war er Münchner, er war immer Münchner, denn Tübingen war ja nur so eine Art Übergangsstation von Straßburg – er war ja sozusagen geduldet in Tübingen und war dann heilfroh, als er wieder seine feste Stellung in München hatte. Er hatte ja nach dem Krieg, weil er in Straßburg Professor war, einige Schwierigkeiten. – Und so war da halt Spannung da, wie üblich: 'und das ist eben nun meine Blattgrenze' und so was Ähnliches' (Seibold interview).

return he heard of the taking of Strasbourg [on 23 November 1944]' (Engelhardt and Hölder 1977: 155).[33]

Dehm began to lecture at the University of Tübingen until it was closed for the summer semester of 1945. Denazification at first prevented Dehm from resuming his lectureship in the autumn of 1945, but in May 1946 'three former professors of the *Reichsuniversität Straßburg* [among them Dehm], who, late in 1944 upon taking refuge from the French troops that had marched into Alsace, had managed to get away to Tübingen, received the permission of the Governor [of the occupying force] to continue their lectureships at the *Eberhard-Karls-*University [of Tübingen]' (Zauner 2010: 963).[34]

From 1948 to 1950 Dehm was appointed Curator of the palaeontological collections at the Tübingen Institute of Geology and Palaeontology. From 1950 to 1976, until his retirement, he was Professor for Palaeontology and Stratigraphy at the University of Munich and again director of a palaeontological institute. His main research focus was the Ries, the Bavarian part of the Alps and the palaeontology of mammals.[35]

During his time in Munich, Dehm and Schröder sent more than thirty students to map the Bavarian side of the Ries area for their diplomas or PhD theses on a scale of 1:25,000, possibly 'because Dehm had also done his PhD thesis in the Ries and had also mapped one sheet; and Schröder likewise' (Pichler interview).[36]

Dehm and Schröder concentrated on:

> hitherto unmapped Ries sheets, especially within the Ries and some on previously mapped sheets, a total of some fifteen large-scale maps. Most of these mapping projects were undertaken when the volcanic origin of the Ries was still regarded as certain. With detailed observations concerning the pre-Riesian biostratigraphy and palaeontology within the scope of the specially prescribed mapping work

---

[33]   'Er entging der Internierung dadurch, daß er zum Zweck der Verlagerung wertvollen Wissenschaftsguts in Universitäten diesseits des Rheins unterwegs war, als er am Tag seiner beabsichtigten Rückkehr von der Besetzung Straßburgs erfuhr' (Engelhardt und Hölder 1977: 155).

[34]   '... erhielten drei ehemalige Professoren der 'Reichsuniversität' Straßburg, die sich Ende 1944 auf der Flucht vor den im Elsass einmarschierenden französischen Truppen nach Tübingen abgesetzt hatten, die Genehmigung des Gouverneurs, jetzt ihre damals begonnene Tätigkeit als Lehrbeauftragte an der Eberhard-Karls-Universität fortzusetzen' (Zauner 2010: 963).

[35]   Biographical information from Vidal in Barsig and Kavasch 1983: 24–8; Engelhardt and Hölder 1977: 155; Zauner 2010: 963; <http://de.wikipedia.org/wiki/Reichsuniversit%C3%A4t_Stra%C3%9Fburg> accessed 10 July 2013.

[36]   'Ich glaube das lag daran, dass der Dehm auch seine Doktorarbeit im Ries gemacht hat und auch ein Blatt kartiert hat und Schröder auch' (Pichler interview).

within the Ries itself and beyond, Dehm hoped to trace the still unresolved riddle of the geological prehistory of the Ries. (Barsig and Kavasch 1983: 27)[37]

Ironically, of course, this was useless for the explicit investigation of an impact crater, which bears no genetic relationship to the geological history of 'ground zero' (only the chemical and rock-mechanical properties of the target area being of interest for the cratering mechanism).

The theoretical underpinning of the Munich School did not seem to have been discussed much, at least not on field trips with students:

> Back in the 1950s, things were structured differently. Approaching today's professors: you simply go there and say, can I talk to you about this or that? This possibility did not exist then. At that time, we were led. I remember we went on a Ries excursion one day. At first the topic was palaeontology, but not exclusively so – the freshwater limestones with the freshwater snails were shown – and it was simply stated that there was this theory, that there was a structure made by explosions or by subsequent subsidence, just a rehash of the old models. (Pichler interview)[38]

If we follow this recollection of Hans Pichler, we must note that the custom at least on Dehm's field trips was to extemporise at length on the local palaeontology while the interpretation of Ries genesis was not shown to be reasonable by field evidence, but was only stated to students as an unquestioned narrative to make sense of the whole structure. It was a 'story', not a theoretical framework, or at least a well-founded hypothesis to integrate a comprehensive dataset, as one would expect of proper procedures for scientific research.

---

[37] 'Er ließ ... bisher noch nicht kartierten Riesblätter, insbesondere im Riesinneren, und einige bereits früher kartierte Blätter, insgesamt etwa 15, großmaßstäblich neu kartieren. Die meisten dieser Aufnahmen sind zu einer Zeit durchgeführt worden, als die vulkanische Entstehung des Rieses allgemein noch als gesichert galt. Mit detaillierten Beobachtungen zur präriesischen Biostratigraphie und Paläontologie hoffte Richard Dehm im Rahmen speziell darauf angesetzter Kartierungsarbeiten im Ries selbst und darüber hinaus dem Rätsel der noch immer ungeklärten geologischen Vorgeschichte des Rieses auf die Spur zu kommen' (Barsig and Kavasch 1983: 27).

[38] '[D]as war in der damaligen Zeit in den 50er Jahren einfach, das andere das war einfach ganz anders strukturiert. Der Zugang zu den heutigen Professoren, da geht man hin und sagt, kann ich das und das mit ihnen besprechen, das war damals nicht, da wurden wir hingeführt. Ich weiß, wir waren einmal auf einer Riesexkursion, da ging es zunächst um die Paläontologie, aber nicht nur um die, diese Süßwasserkalke mit den Süßwasserschnecken, die wurden vorgeführt und ganz einfach gesagt, es wäre eine Theorie, eine Struktur, die durch Aussprengungen entstanden ist oder durch Nachsacken, eben das Wiederkäuen des alten Modells' (Pichler interview).

But not everybody in Munich was happy about the way the Bavarian side handled the mapping at the Ries Basin. (Georg) Hermann Reich (1891–1976) had studied geology in Heidelberg, Erlangen and Freiburg, where he received his doctorate degree with a dissertation on the stratigraphy and tectonics of the area encompassed by the diatremes of the Swabian Volcano. From 1919 to 1921, after military service during World War I, Reich became assistant to the seismologist Emil Wiechert (1861–1928). In 1921 he became a geologist at the Prussian Geological Survey. After his *Habilitation* in 1925, Reich began lecturing at the Technical University in Berlin-Charlottenburg. In addition to his survey job, he was appointed professor without chair at this same technical university in 1931. In 1945, Reich switched to the University of Göttingen and in 1948 he became the Professor of Geophysics in Munich, where he developed the new seismological observatory in Fürstenfeldbruck. Reich retired in 1957 (Porstendorfer 2003).

As a still fairly new Munich professor of geophysics, Hermann Reich would have preferred a comprehensive mapping programme handled by geologists rather than palaeontologists:

> I cannot understand why the Geological Survey in Munich has not mapped this undivided structure and as objectively as possible; instead it has left the area lying around as a hunting-place for doctoral theses of young [inexperienced] people. Unfortunately an opportunity has been missed there, by which the Munich Survey has dissociated itself from something that might well have brought it international approbation. (Letter from Reich to Walter Kranz, 5 January 1950, GA 19411)[39]

The reason for not mapping the Ries Basin with staff from the survey lay in the lack of proper topographic maps with contour lines for that part of Bavaria; instead, the maps depicted topography only by different shading. Thus, orientation in the more or less level Ries Basin was rather difficult and the survey therefore precluded the area from geological mapping (Dehm 1977: 8).

In the previous autumn, Reich had planned a seismic survey of the Ries. Despite his feelings concerning the mapping efforts of the Dehm school, he nevertheless loyally supported the Munich interpretation of 'local outbursts' when the exasperated Kranz tried to exert an influence on the Munich school by way of a letter to the Rector of Munich University. This letter had been written

---

[39]   'Ich kann es nicht verstehen, daß das geologische Landesamt in München dieses einzigartige Gebilde nicht einheitlich und möglichst objektiv aufgenommen hat und statt dessen diesen Bereich als Jagdgrund für Doktorarbeiten junger Leute liegen läßt. Da ist leider eine Gelegenheit verpaßt worden, durch die sich das Münchener Amt einer Sache entzogen hat, die ihm einen internationalen Ruf hätte bringen können' (letter from Reich to Walter Kranz, 5 January 1950, GA 19411).

on 5 September 1949. The Rector, however, had forwarded it to Hermann Reich, who replied to Kranz:

> Then the time will have arrived to discuss the various Ries theories. My personal view is that the so-called tectonic theories, like the meteorite theories – Quiring[40] of all people raises it again in his latest publication – are completely untenable. To my mind an explosion theory is undoubtedly called for – and it is your merit to have stressed that categorically again and again, after you were actually the first to present this thought in a form that is technically and geologically convincing and that can also partly serve as proof. Now I only add that it was not a single explosion process from one crater, but that it was a much more complicated event, in which explosion sites were situated on the periphery, along the rims of the Ries Basin. Moreover, the seismology has shown plainly that the crystalline basement in the Ries area has experienced a very considerable uplift. This, too must be somehow included in the theory. (Letter from Reich to Walter Kranz, 8 October 1949, GA 19409)[41]

While Reich in his letter was prepared to acknowledge Kranz's merits, Reinhold Seemann, who favoured a tectonic, non-explosive Ries origin (see below), noted with considerable glee and triumph that Reich's seismic investigations would preclude Kranz's explosion model:

> I heard that Reich has stressed that there "can be no talk about a funnel, not even a wide one and the plain above which the rocks are brecciated is a flat plain". That is, a central but 300 m deep explosion bowl, filled with breccias that allegedly originated above an extended explosion surface, has taken the place of the now happily discarded explosion funnel. Thus the lateral blasting and scattering effect is restricted considerably and Kranz's theory at least is made

---

[40] See Chapters 4 and 5.

[41] 'Es wird dann die Zeit gekommen sein, die verschiedenen Ries-Theorie zu diskutieren. Meine persönliche Einstellung geht dahin, dass die sog. tektonischen Theorien, ebenso wie die Meteoritentheorien – ausgerechnet Quiring bringt sie wieder in seiner neusten Veröffentlichung – völlig unhaltbar sind. Eine Sprengtheorie ist m. E. unbedingt zu fordern – und es ist Ihr Verdienst, das immer wieder nachhaltig betont zu haben, nachdem Sie überhaupt als erster diesen Gedanken in einer Form dargestellt haben, die technisch und geologisch einleuchtend sind z.T. auch beweisend ist. Ich zeige nur jetzt dazu, dass es sich nicht um einen einmaligen Sprengvorgang aus einem Krater gehandelt hat, sondern um einen wesentlich komplizierteren Vorgang, bei dem auch periphere Sprengherde an den Rändern des Rieskessels auftraten. Ferner ist durch die Seismik ganz einwandfrei erwiesen, dass das Kristallin im Riesbereich eine sehr erhebliche Hebung erfahren hat. Auch das muss irgendwie in die Theorie aufgenommen werden' (letter from Reich to Walter Kranz, 8 October 1949, GA 19409).

impossible. One can still talk about the laccolith theory related to the explosion concept of the Munich School, although the seismic data are not unequivocal. A tectonic and morphologic interpretation of the data (subsidence of Triassic and Jurassic sedimentary blocks) is also possible. ... Perhaps G[eor]g Wagner secretly regrets not having followed my advice, to refrain from giving so much room to Kranz's explosion theory in his text book.[42] He and his friends will hardly acknowledge their error openly and the fight about the interpretation of the 'Ries bowl beds' will continue. I at least have the satisfaction that the path to a better solution is free again, after it has been obstructed for so long by the Swabian geologist[s] (excluding you!); by blasting debris. (Letter from Seemann to Helmut Hölder, 22 September 1950, GA 3/323)[43]

But true to his geological upbringing in Tübingen, Kranz was not convinced and, indeed, the issue was still unresolved some two years later:

This spring, Professor Reich has again made seismic investigations with money from our Survey in the region of Fünfstetten [some 5 km east of the Ries rim] in order to check, for once, one of the alleged local outbursts of Schröder and Dehm in detail. The result of the preliminary analysis by Reich is unequivocally that at that place the polymictic breccias with the large blocks of Lias [Lower Jurassic], Keuper [Upper Triassic] and crystalline [basement] rest <u>on top</u> of the Malm [Upper Jurassic]. The result, however, did not make the slightest impression upon the 'local outburst' enthusiasts in Munich; they now excuse themselves by saying that this mysterious thing then must be somewhere close by. Preformed notions

---

 42 Wagner (1931: 540–46); 2nd edition in 1950.

 43 '[Ich] erfuhr aber, dass Reich betont habe "von <u>keinem</u> Trichter, auch nicht von einem flachen könne die Rede sein u. die Fläche, worüber die Gesteine zertrümmert sind, sei eine <u>ebene</u> Fläche." An Stelle des jetzt glücklich erledigten Spreng<u>trichters</u> ist also ein zentraler, aber 300m tief mit Trümmerschichten erfüllter Spreng<u>kessel</u> getreten, der über einem flächenhaften Sprengherd entstanden sein soll. Dadurch wird aber doch die seitliche Spreng- u. Streuwirkung wesentlich eingeschränkt u. zum mindesten die Kranz'sche Theorie unmöglich. Über die mit Sprengung verbundene Lakkoliththeorie der Münchener Schule lässt sich noch reden, obwohl da die seismischen Befunde nicht eindeutig sind. Eine tektonische und morphologische Deutung des Befundes (Versenkung triassischer u. jurassischer Deckgebirgsschollen) ist ebenso gut möglich. ... Vielleicht bedauert es Gg. Wagner doch im Stillen, meinem Rat, die Darstellung der Kranz'schen Sprengtheorie in seinem Lehrbuch nicht gar so ausführlich zu machen, nicht befolgt zu haben. Offen ihren Irrtum zugeben werden er u. seine Freunde wohl kaum u. der Kampf wird um die Deutung der 'Rieskesselschichten' weitergehen. Ich habe wenigstens die Genugtuung, dass der Weg zu einer besseren Lösung wieder frei ist, nachdem er so lange durch [?] den [sic] schwäbischen Geologen (Dich ausgenommen!) durch Sprengschutt verbaut war' (letter from Seemann to Hölder, 22 September 1950, GA 3/323).

are hard to counteract! (Letter from Alfred Bentz to Walter Kranz, 4 July 1952, GA 19347)[44]

It was not until 1963 that Dehm finally admitted that 'proving the 'local outburst', which have been postulated several times (W. Branca, A. Moos,[45] J. Schröder and R. Dehm), has not yet been successful' (Dehm in AG Ries 1963: 9),[46] and that Reich's seismological investigations actually disproved some of the specific locations assumed to represent 'local outbursts' (Dehm in AG Ries 1963: 9).

Actualism – Method Confused as Fact

Until well into the 1960s, the main argument of traditional Ries experts against the impact hypothesis was that it violated the most important principle of modern geology, the principle of 'actualism',[47] which because of its successful application had tended to become a rigid dogma not just in Germany (Hooykaas 1959). As the principle was commonly understood, it not only demanded that the phenomena of the past may only be explained by processes observable in the present, but also implied a strictly gradual, non-catastrophic nature for geological processes and that the earth should be regarded as a closed system. Consequently, the impact hypothesis was seen as an unscientific throwback to former catastrophic explanations, since the violent influence of impacting or closely passing celestial bodies had been suggested as a causal agent in 'theories

---

[44] 'Dieses Frühjahr hat Professor Reich mit Mitteln von unserem Amt in der Gegend von Fünfstetten wieder Seismik gemacht, um einmal einen der angeblichen örtlichen Aufbrüche von Schröder & Dehm im detail [sic] nachzuprüfen. Ergebnis nach der vorläufigen Auswertung von Reich eindeutig dies, dass die dortige bunte Breccie mit den grossen [sic] Schollen von Lias, Keuper und Kristallin auf dem Malm liegt. Das Ergebnis machte aber auf die örtlichen Aufbruchs-Enthusiasten in München nicht den geringsten Eindruck; sie reden sich jetzt damit heraus, dass dieses geheimnisvolle Ding dann eben irgendwo dicht dabei sein müsse. Gegen vorgefasste Meinungen ist schwer aufzukommen!' (Letter from Alfred Bentz to Walter Kranz, 4 July 1952, GA 19347).

[45] August Moos (see Chapter 7).

[46] 'Der Nachweis der mehrmals postulierten 'örtlichen Aufbrüche' (W. Branca, A. Moos, J. Schröder und R. Dehm) ist noch nicht gelungen' (Dehm in AG Ries 1963: 9).

[47] The German term *Aktualismus* [actualism] is derived from the French *actualisme* referring to *causes actuelles*, that is present-day causes. Actualism for most purposes was and is fairly close to uniformitarianism, although – like the modern Anglo–Saxon view of uniformitarianism – it does not require a cyclical world-view (see also Shea 1982 for a detailed discussion of its meaning). During the Nazi period, however, 'actualism' was redefined under the heading of 'German Geology' (see Chapter 7) although by no means all German geologists were prepared to follow this ideological redefinition.

of the Earth' of the eighteenth and up to the early nineteenth century as explanations for the Deluge:

> I shall conclude by remarking, that if the hypothesis of a shock derived from the passage either of a Comet or one of those numerous, important, and long neglected bodies, often of great magnitude and velocity, which occasion meteors, and shower down stones upon the earth, would explain the phenomena of the deluge ... we need not be deterred from embracing that hypothesis under an apprehension that there is in it anything extravagant or absurd. In the limited period of a few centuries, there is little probability of the interference of two bodies so small in comparison with the immensity of space; but the number of these bodies is extremely great, and it is therefore by no means improbable, says La Place, that such interference should take place in a vast number of years. (Greenough 1819: 198–9)

Such deluge theories were now regarded as obsolete, and this was extended without further thought to the general possibility of an interaction of the Earth with celestial bodies bigger than the common meteorite.

Georg Wagner, Professor of Geology at Tübingen (see Chapters 9 and 10), and one of the most ardent critics of the impact theory after 1960, reacted by denigrating the impact idea as a 'Fizz-Bang-Hypothesis' (Hüttner interview).[48]

The principle of actualism 'was an absolute dogma; it was so for me too, and only afterwards have 'events' been recognized in geology that, when you think more closely about them, can be reconciled with actualism too' (Seibold interview).[49]

In following this dogma, people forgot that actualism or uniformitarianism, however defined, cannot be more than a tool of geological investigation, a method or approach, rather than an unquestioned fact. And on closer inspection, the notion of actualism, from the early 1930s to the 1960s as it was pitted against the impact hypothesis, was really nothing more than a catch-word, without depth or substance. The promoters of an impact hypothesis were asked by their critics to obey the rules of geology – that is to restrict themselves to actualistic processes – while the defenders of a volcanic origin of the Ries Crater did not hesitate to violate the principle of uniformity for their own purposes. They claimed that the Ries must have been formed by a quite *singular* volcanic process for which there was no present-day analogue[50] (Wagner 1931: 540; Wagner 1960: 596).

---

[48]   'Ratsch-Bum-Hypothese' (Hüttner interview).

[49]   'Das war eine absolute Glaubensregel – das war's auch für mich – und danach kamen dann erst die Eventgeschichten, die im Grunde durchaus mit dem Aktualismus zu erklären sind' (Seibold interview).

[50]   This, however, was perfectly reconcilable with 'actualism', as perverted by Nazi ideology in the theoretical framework of 'German Geology' (see Chapter 7).

Figure 3.2   Reinhold Seemann was the main promoter of the idea that the features of the Ries Basin evolved through a long history of gradual tectonic processes. (Photo from Hölder 1976: 204).

A tiny minority favoured an equally non-uniformitarian tectonic origin for the Ries, involving a hypothetical subterranean 'spike' of rock moving north – triggered by the building of the Alpine mountains – which supposedly surfaced at the Ries.

For Reinhold Seemann (1888–1975), the geological curator at the Natural History Museum in Stuttgart (Figure 3.2), the main promoter of the idea that the features of the Ries Basin evolved from a long history of gradual tectonic processes, both 'impactists' and 'volcanists' were simply not diligent and patient enough to figure out the gradual tectonic evolution of the Ries Crater (Hölder 1962: 27; see Chapter 9).

Although, as we shall see, Seemann felt otherwise, it was already recognised by contemporaries that his idea itself violated the actualistic principle:

> In recent years I have continued to read a least the Ries literature, even though I no longer have the time to engage myself with the problem actively. But I must say that I am pretty much disappointed about the development that is reflected in it [in the literature]. Nearly always, people try to 'doctor around at the symptoms', and to explain the phenomena locally, while neglecting that the biggest riddle remains completely unexplained, now as before. The so-called tectonic explanation by Seemann is also no explanation at all, but an attempt using unsuitable means to attribute things to a vaguely defined 'tectonics', for which it was never responsible anywhere else. (Letter from Alfred Bentz to Walter Kranz, 11 January 1947, GA 19340)[51]

Reinhold Seemann studied science in Tübingen, Freiburg, Berlin and Stuttgart, with an emphasis on geology and mineralogy. During World War I he was an officer at the front. During a leave of absence for recovery from a wound, he was awarded a PhD in Tübingen with a dissertation on the petrology of Triassic sediments near Freudenstadt in the Black Forest. After returning to the battlefields, he was again severely wounded and lost one of his eyes. After World War I, he worked first as a geologist for the *Deutsch-Luxemburgische Bergwerks- und Hütten-A.G.* (German-Luxembourg Mining and Smelting Corporation) in

---

[51] 'Ich habe in den letzten Jahren wenigstens die Riesliteratur weiter verfolgt, wenn ich auch selbst nicht mehr die Zeit hatte, mich aktiv mit den Problemen zu beschäftigen. Ich muß aber sagen, daß ich über die Entwicklung, die sich darin abspiegelt, recht enttäuscht bin. Man versucht fast auf der ganzen Linie 'an den Symptomen zu kurieren' und die Erscheinungen lokal zu erklären, wobei man dann nur übersieht, daß dann das größte Rätsel nach wie vor völlig unerklärt bleibt. Auch die sogenannte tektonische Erklärung von Seemann ist überhaupt keine Erklärung sondern ein mit untauglichen Mitteln unternommener Versuch, einer nicht näher definierten 'Tektonik' Dinge in die Schuhe zu schieben, für die sie nirgends je verantwortlich war' (letter from Alfred Bentz to Walter Kranz, 11 January 1947, GA 19340).

Hersbruck (Bavaria) and from 1925 he was Curator at the Württemberg natural history collections in Stuttgart.

Between 1940 and 1946, in addition to his curatorship, he lectured on geology and soil sciences at the Agricultural College in Hohenheim near Stuttgart. From 1946 to 1949 he was Acting Director of the Württemberg natural history collections. Seemann retired in 1950 (Hölder 1976).

Seemann's tectonic alternative (Seemann 1939) to the explosion theories concerning the Ries (whether volcanist or impactist in nature) was based on a thorough reconstruction of the events at the Ries Crater and on the assumption that the temporal, historical sequence that he sought to elucidate encompassed considerable geological time:

> I know that my work is full of errors too, but I think it is so only in the interpretation of individual single cases. My conviction today is, as before, that in the Ries, too, everything went on naturally without the miracle of the Ries explosion, and that only our restricted knowledge seems to make it necessary to fall back on this *deus ex machina*. I think it is best and easiest to characterize ~~my view of~~ the Ries with the expression that I have told you before: it is a 'nerve point' in the geology of South Germany, which was formed by epirogenic and syn-orogenic movements in the course of a <u>long</u> geological history, and was repeatedly changed, and especially strongly by morphological forces and once (or several times? for a longer duration of time?) was afflicted by volcanic forces. Through the mingling and succession of these factors, the enigmatic chaos that is so difficult to disentangle was created. For its unravelling much geological knowledge and experience is necessary, as well as a clear eye unclouded by the explosion theory. (Letter from Reinhold Seemann to Helmut Hölder, 6 January 1956, GA 3/315)[52]

---

[52] 'Ich weiß, daß meine Arbeiten auch voll Irrtümern ist [sic], aber m.E. nur in der Deutung einzelner, örtlicher Fälle. Meine Überzeugung ist heute nach wie vor die, daß auch im Ries alles normal zugegangen ist, ohne das Wunder der Riessprengung, u. daß nur unsere beschränkte Erkenntnis zu diesem deus ex machina greifen zu müssen glaubt. Am besten ~~ist mein Standpunkt~~ u. kürzesten ist das Ries wohl mit dem Dir gegenüber schon einmal gebrauchten Ausdruck neuralgischer Punkt der Geologie Süddeutschlands gekennzeichnet, der im Laufe einer <u>langen</u> geol. Geschichte durch epirogene u. synorogene Bewegungen entstand u. immer wieder u. besonders stark von morphologischen Kräften umgeformt u. auch einmal (öfter? länger?) von vulkanischen Kräften heimgesucht wurde. Im Durch- u. Nacheinander dieser Faktoren entstand das rätselhafte, schwer entwirrbare Chaos. Seine Entwirrung benötigt viel geologische Kenntnis u. Erfahrung u. ein klares, von der Sprengtheorie nicht getrübtes Auge' (letter from Reinhold Seemann to Helmut Hölder, 6 January 1956, GA 3/315).

Seemann was convinced that traditional geological expertise was the key to the mysterious Ries Basin. For him, only the faithful upholding of tradition could throw light on the problem, not fancy new ideas:

> I believe that no geological tyro can see through the confusing possibilities of various interdependent factors, disentangle them, and sort them into an historical succession. Indeed, it is questionable whether this is today possible even for an experienced geologist. The explosion theory, at any rate, is the most unsatisfactory solution and should be banned as a 'loophole'. (Letter from Reinhold Seemann to Helmut Hölder, 30 March 1956, GA 3/314)[53]

His doctoral student Helmut Hölder, in an interview some decades after the events, characterised Seemann's extremely conservative attitude in an interview:

> Seemann was under a sort of spell. He was not yet able to conceive that a meteorite impact could have caused such a thing on Earth. This had deeper reasons. He was a man of orderliness, almost in the sense of Goethe. Things in nature are orderly; they are not created by chaos. Even volcanism for Seemann was something too chaotic, especially such a gigantic water-vapour explosion, which does not occur in Europe. ... It was horrible to him. Intuitively – feelings always come into play – intuitively he rejected it. ... Goethe, for example, rejected all chaotic processes. [Hölder continued to quote from Goethe: *Faust II*:] 'Never was Nature and her lively power restricted to day and night and hour. She forms in order every pose and even when great it's not by force'. ... In any case, to this view that Nature works in an orderly and gentle way, which is directly in Goethe's tradition, with everything disordered being explained away if possible, was Seemann's mind – I don't want to say addicted – but he was bound to it. Because of that, he wanted to explain Nature as orderly: hence his tectonic explanation of the thrusts in the Ries. [Tectonic] brecciation too was closer to Seemann's nature than an instantaneous event, whether it was one from below (volcanic) or above (meteoritic). This, for Seemann, was akin to an act of terrorism; and he didn't want to see the Earth as subject to such an act of terrorism, whether from below or (especially not) from above. From above! That for him was a completely incomprehensible idea. (Hölder interview)[54]

---

[53]   'Ich glaube, kein junger Geologe kann die verwirrenden Möglichkeiten verschiedener ineinander arbeitender Faktoren durchschauen, sie entwirren u. in einen geschichtlichen Ablauf ordnen. Ja, es ist die Frage, ob dies heute noch selbst einem erfahrenen Geologen möglich ist. Die Sprengtheorie jedenfalls ist dazu das untauglichste Mittel u. sollte auch aus 'Ausweg' verboten werden' (letter from Reinhold Seemann to Helmut Hölder, 30 March 1956, GA 3/314).

[54]   'Seemann, war da in einem gewissen Bann, er konnte sich noch nicht vorstellen, dass ein Meteoriteneinschlag ja auf der Erde so etwas bewirkt haben sollte. Das hat nun

Seemann was basically in a minority of one, as even his PhD student Helmut Hölder would not follow him to the extremes and so Seemann showed himself disappointed when a scientific generation later Hölder's own PhD student, Rudolf Hüttner, wasn't prepared to follow Seemann's theory (see letter from Seemann to Hölder, 14 January 1954, GA3/320).

Late in his career, Seemann saw himself as a lone voice in the wilderness. He never wavered in his opinion, but gave up any idea of being able to convince others of the truth as he saw it. When his retirement was imminent, he planned to move away from Stuttgart, where he had spent the last years of his geological career. There was some bitterness in his resignation:

> Even though I walk alone through the Ries, I nevertheless believe I see more than all the rest, who move in a big crowd with loud exclamations. Maybe my work will still help to open the eyes of others. If not, then I at least have learned a lot for myself, not only in the interpretation and assessment of landscape but also of people; especially those who concern themselves with 'science' and are leaders in this endeavour. That here too things should be conducted in a human way was only to be expected, but that the proceedings have been so very human has been a certain disappointment [to me]. It will cure [my illusions] and will render my farewell from the scientific hustle and bustle, and from here, easier. (Letter from Seemann to Helmut Hölder, 22 September 1950, GA 3/323)[55]

---

tiefere Gründe, er war ein Mann der Ordnung, beinahe im Goethe'schen Sinn, die Dinge der Natur sind geordnet, sie sind nicht durch Chaos entstanden. Schon der Vulkanismus war Seemann etwas zu Chaotisches, vor allem solch eine Riesen-Wasserdampfexplosion, die es im europäischen Bereich nicht gibt. ... Das war ihm ein Horror eigentlich. Er hat sie gefühlsmäßig, das spielt ja auch mit, das hat er gefühlsmäßig abgelegt. ... z. B. Goethe [hat] alle turbulenten Erscheinungen abgelehnt. ... : "Nie war Natur und ihr lebendig Fließen, auf Tag und Nacht und Stunden angewiesen, sie [bildet regelnd] jegliche Gestalt und selbst [im Großen] ist es nicht Gewalt". ... Also jedenfalls, diese Auffassung dass Natur geordnet, ruhig arbeitend, das steht ganz im Goethe'schen Sinn, alles Turbulente möglichst wegdeutend, dieser Einstellung war der Geist von Reinhold Seemann an sich, verfallen möchte ich nicht sagen, sondern er war daran gebunden. Deshalb wollte er die Natur geordnet erklären, deshalb seine tektonische Erklärung der Überschiebungen im Ries. Auch die Zertrümmerung lag für Seemann einfach viel näher, als ein Augenblicksereignis, sei es von unten her vulkanisch, oder komme es von oben her meteoritisch, das ist für Seemann so etwas gewesen, wie ein Terroranschlag und diesem Terroranschlag wollte er die Erde nicht ausgesetzt sehen, weder von unten und gar nicht von oben her. Von oben her, das war eine für ihn völlig nicht nachvollziehbare Vorstellung' (Hölder interview).

[55] 'Auch wenn ich allein durchs Ries wandere, so glaube ich doch mehr zu sehen, als alle, die dies in grossen Haufen u. mit lautem Geschrei tun. Vielleicht hilft meine Arbeit doch noch, auch anderen die Augen zu öffnen. Wenn nicht, dann habe ich wenigstens für mich viel gelernt, nicht nur in der Deutung u. Wertung der Landschaft, sondern auch in der der Menschen, besonders derer, die sich mit 'Wissenschaft' abgeben u. darin führend sind.

Mineralogy and Volcanology

Although there had been much effort to map the Ries area, especially on the much larger Bavarian side by the working group of Dehm and Schröder, there was a most surprising lack of interest in the properties of the actual rocks, which even from their appearance in hand-specimens are most remarkable. This seems to have been typical for the geologists of that time. Otto Pfarrherr, a microscope salesman, who served many a scientific generation and is now in his eighties, loves to tell the following anecdote: When he tried to sell a microscope to the Erlangen Geological Institute in 1957/58, Professor Bruno von Freyberg (born 1894) told him: 'A geologist needs a hammer, not a microscope!' (Otto Pfarrherr, pers. comm., 20 July 2011).[56]

Thin-sections and chemical analyses belonged in the mineralogists' toolbox. Between 1866 and 1956 there were thirty-eight mineralogical/petrological doctoral theses at the University of Tübingen, which – as far as can be discerned from the titles only – were not related to any larger research programmes, as one would expect from successive working groups consisting of a professor, assistant and several PhD students working closely together. Instead, the topics of the dissertations reveal individuals working separately on all sorts of disparate issues. Not every year saw a PhD, and particularly after 1914 there were considerable gaps between PhD students (Engelhardt and Hölder 1977: 83–4).

Wilhelm Branca, Professor of Mineralogy at Tübingen until 1894, worked on the Ries Basin together with Eberhard Fraas from 1901 onwards. But as this was only after his time in Tübingen, he did not initiate any mineralogical Ries tradition there. Thus, there have been only two mineralogical Ries dissertations, both in the first decade of the twentieth century: one by Richard Oberdorfer from Ludwigsburg in 1904, titled *Die vulkanischen Tuffe des Rieses bei Nördlingen* [The Volcanic Tuffs of the Ries near Nördlingen], the second by Richard Löffler from Stuttgart in 1910, on *Die Zusammensetzung des Grundgebirges im Ries* [The Composition of the Crystalline Basement at the Ries].

Löffler found nothing disquieting, even though he investigated thin-sections, with which he had some help from Adolf Sauer (1852–1932), chair at the *Königliche Technologische Hochschule Stuttgart* [Royal Technical University in Stuttgart]. Löffler seems to have only been interested in petrography and rock classification (Löffler 1912).

Dass es auch hier menschlich zugeht, war zu erwarten, dass es aber <u>so</u> menschlich zugeht, war nun doch eine gewisse Enttäuschung. Sie ist mir heilsam u. wird mir den Abschied von dem wissenschaftlichen Betrieb u. von hier erleichtern' (letter from Seemann to Helmut Hölder, 22 September 1950, GA 3/323).

[56] 'Ein Geologe braucht einen Hammer; kein Mikroskop!' (Otto Pfarrherr, pers. comm., 20 July 2011).

Oberdorfer's dissertation (Chapter 1) may have been provoked by the interest of Adolf Sauer. From 1903 onwards, Sauer was head of the geology department at the *kgl. Württembergisches Statistisches Landesamt* [Royal Bureau of Statistics in Württemberg], where he worked on the petrography of the 'volcanic rocks' of the Swabian Alb and the Ries Basin, assuming that phonolitic magma[57] had resorbed large amounts of granite (see Ahrens 1929). The Technical University offered a prize on this topic, for which Oberdorfer submitted his thesis. He investigated some 120 thin-sections and performed several chemical analyses of suevite (Sachs 2011). Oberdorfer noted and described unusual and strange mineralogical features, such as isotropic feldspars, which must have been transformed into glass apparently while remaining in the solid state, that is without having been melted. Quartz too showed strange and unusual planar features (Figure 12.1) and sometimes was isotropic as well. The suevite chemistry of the Ries Basin proved to be extremely heterogeneous and quite abnormal for a volcanic melt. The content of MgO, which normally correlates inversely with SiO_2, varied considerably, while the SiO_2 concentration was relatively high. Oberdorfer assumed a strongly mafic original magma, which had assimilated large amounts of granite and gneiss. He also talked of a mechanical mixture (Oberdorfer 1904).

While during this early period mineralogical research in Tübingen was conducted under the same roof as geology, the two branches of geoscience went separate ways in the second decade of the twentieth century. In 1911 Ernst Koken applied for the establishment of an independent chair of mineralogy because '[m]ineralogy, geology and palaeontology have diverged so far in their aims and methods, that the full coverage of the subjects surpasses by far the power of a single person' (Koken quoted after Engelhardt and Hölder 1977: 40).[58] He envisaged a closer connection of mineralogy with physics and chemistry. His wish was granted and in 1913 mineralogy moved to separate quarters. But the timing was ill chosen for this new chair of mineralogy in Tübingen because, before everything was settled properly, World War I began and in consequence the financial backing of the mineralogical chair was bad and the lack of scientific instruments was deplorable. This remained so until after World War II and consequently none of the appointed professors remained long, but simply used the position as a 'spring-board' to more attractive mineralogical chairs (Engelhardt and Hölder 1977: 41–2). Under such conditions, the formation of proper scientific working groups with joint research programmes remained

---

[57] i.e. especially rich in potassium but also sodium and with intermediate silica content.

[58] 'Mineralogie, Geologie und Paläontologie haben sich in Zielen und Methoden so weit voneinander entfernt, daß die volle Vertretung der Fächer die Kraft eines Einzelnen weit übersteigt' (Koken quoted after Engelhardt and Hölder 1977: 40).

a wishful dream and as regards research on the Ries Basin: 'The mineralogists formerly had nothing to do with the Ries' (Hölder interview).[59]

In Munich as well, there seemed not much encouragement for mineralogical work at the Ries Basin:

> Professor [Georg] Fischer [born 1899] of the petrographical institute has also concerned himself with the Ries and already in the 1950s had assigned a (sadly not much regarded) dissertation there: His PhD student, Willi [Wilhelm] Ackermann ..., long before Shoemaker and Chao, found spectacular anomalies as to mineral metamorphosis, which could hardly be explained by normal PT-conditions in volcanism. Rumour has it that Ackermann could have contributed even more observations but was prevented from publishing them because what must not be, cannot be. (Letter from Wolf-Dieter Grimm to the author, Munich, 9 September 2001)[60]

Ackermann did among other things report that he too had found strange feldspars and quartz grains which had turned into glass, something that Oberdorfer already noticed. Ackermann thought they might be eutectic melts, generated where quartz and feldspar came into contact:

> The material thus should be understood as being in the act of migmatising.[61] The preservation of this state could be caused by the sudden quenching, which can be equated with the Ries explosion. (Ackermann 1958: 159)[62]

Ackermann thus interpreted the strange features within the volcanist paradigm: Melting of feldspars and quartz was supposed to have been caused by rising temperatures from below, but further melting was suddenly arrested by the

---

59   'Die Mineralogen haben früher nichts mit dem Ries zu tun gehabt' (Hölder interview).

60   'Auch Prof. Fischer, Petrographisches Institut, hat sich mit dem Ries beschäftigt und dort schon in den Fünfzigerjahren eine (leider wenig beachtete) Dissertation angesetzt: Sei Doktorand Willi Ackermann ... fand lange vor Shoemaker und Chao spektakuläre Anomalien in Form von Mineralumwandlungen, die durch normale PT-Bedingungen beim Vulkanismus kaum erklärbar waren. Es wird kolportiert, dass Ackermann noch weitere Beobachtungen hätte beitragen können, diese aber nicht publizieren sollte, da nicht sein kann, was nicht sein darf' (letter from Wolf-Dieter Grimm to the author, Munich, 9 September 2001).

61   The minerals with the lowest melting temperatures begin to melt; according to Ackermann at a temperature between 500 and 600°C.

62   'Das Material wäre dann so aufzufassen, als im Begriff der Migmatisierung. Die Erhaltung dieses Zustandes könnte die Folge plötzlicher Abschreckung sein, die mit der Riesexplosion gleichgesetzt werden müßte' (Ackermann 1958: 159).

(volcanogenic) explosion, which caused rapid cooling of the formerly heated rock. Additionally, according to Ackermann, there were other changes in minerals. The whole basement seemed hydrothermally altered as a consequence of the Tertiary 'volcanism' (Ackermann 1958).

In addition to the neglect of mineralogical and petrological data, many Ries investigators had hardly any geological experience outside Central Europe, which because of its dense cover of vegetation restricts traditional geological work largely to artificial outcrops, and most investigators only knew active volcanoes from the literature.

Richard Dehm's views of the Ries structure's genesis were, according to Pichler:

> totally conservative. But here once again comes into play the fact that he himself had simply picked up this theory [from others]. He himself – he did have a good geological education, mostly palaeontologically oriented – but was completely lacking in knowledge of volcanism, and since he did not have that, he was unable to cultivate a critical attitude in that regard and fell back on ideas he had picked up and what had been said on this topic before. Here comes into play ... a lack of knowledge of a most fundamental aspect: volcanism – the phenomena of volcanism. And you can get that only if you have in fact worked on volcanoes and have been occupied with volcanoes. This was completely missing in the case of the two [Richard Dehm and Joachim Schröder] and of course for the students too; they didn't have it either. (Pichler interview)[63]

> In former times, geologists at best went to Italy, to Vesuvius and maybe also Mount Etna; and if somebody had travelled to Iceland, that was a very great thing, but for the most part they didn't venture beyond and thus the experience and comparisons were missing. [Otto] Stutzer, [the impactist,] however, had also worked in South America, for example, I think, in Venezuela and Colombia. He ventured far abroad, had worked for the petroleum [industry] and had also seen the meteorite crater in Arizona; and from his travel experiences he drew the

---

[63] 'Die Einstellung von Dehm und Schröder ..., die sind ja ganz konservativ gewesen, aber da kommt wieder mal die Sache rein, dass er auch ganz einfach diese Theorie übernommen hat. Er selber war, er hatte zwar eine gute geologische Vorbildung, er war vor allen Dingen paläontologisch orientiert, aber er war völlig ohne, er hatte keine Kenntnisse über den Vulkanismus und da er das nicht hatte, konnte er dem auch nicht entsprechend kritisch gegenüber treten und hat sich auf das gestützt und das übernommen was eben da bisher gesagt wurde, das kommt eben da rein, ... ganz einfach auch ein Mangel an Wissen über einen ganz wesentlichen Aspekt, den Vulkanismus, Phänomene des Vulkanismus, nicht? Und die gewinnt man nur, wenn man tatsächlich auch an Vulkanen gearbeitet hat und sich mit Vulkanen beschäftigt hat. Das hat den beiden völlig gefehlt und auch den Schülern natürlich, die hatten es auch nicht' (Pichler interview).

conclusion that [the Ries] was formed by an impact event .... The 'local geologists' [on the other hand] ... quite simply had been unable to draw on the experience that you get from travelling in the outside world. (Pichler interview)[64]

Pichler himself, as he recalled, achieved this kind of experience rather late and it was only then that he started to wonder:

> I have thought to myself, 'funny, such a gigantic crater, if you look at the caldera structure'. In 1960, I went to Santorini for the first time and got to know Santorini – the large caldera in Santorini; and then when I came to Tübingen, I wondered why such a gigantic structure [as the Ries] didn't have lavas too. There definitely should be lavas and not only this suevite. Moreover, the quantity of suevite didn't match the size of this big structure. (Pichler Interview)[65]

For their want of personal experience with volcanoes and their associated phenomena, people tended to fall back on what one might call geological myths, invoking large volcanic eruptions such as Krakatoa (Indonesia, 1883) and Bandai-San (Japan, 1888). But there was hardly anything known about these eruptions at the Tübingen Institute – neither about the eruption mechanisms nor their time frames; however, 'volcanism' was a convenient term to account for anything and everything and was most useful for putting a stop to further questions.

The seven-kilometre-wide caldera of Krakatoa in the Sunda Strait between Java and Sumatra experienced its latest caldera collapse in August 1883. The eruption had begun in May with mild steam and ash eruptions, sometimes from

---

[64] 'Stutzer ... hat natürlich die Erfahrung gehabt und zwar die Erfahrung eines weitgereisten Geologen, es war ja früher so, dass die Geologen allenfalls nach Italien gekommen sind zum Vesuv und vielleicht auch zum Ätna und wenn einer nach Island gefahren ist, das war eine ganz tolle Sache, aber viel weiter sind die nicht rausgekommen und damit hat die Erfahrung, der Vergleich gefehlt und der Stutzer, der ja auch in Südamerika gearbeitet hat, zum Beispiel in, ich meine in Venezuela und in Kolumbien, der ist weit herumgekommen, hat fürs Erdöl gearbeitet, hat auch den Meteoritenkrater in Arizona gesehen und von dieser Reiseerfahrung hat er den Schluss abgeleitet, das ist ein Impaktereignis ... . Die 'Heimatgeologen' ... [waren] eben ganz einfach nicht in der Lage, von ihrer Erfahrung, die man durch Reisen in die Welt erwirbt – ; dass man davon zehrt' (Pichler interview).

[65] 'Ich habe mir schon gedacht, komisch, so ein Riesenkrater, wenn man die Calderastruktur ansieht. Ich war ja 1960 das erste Mal in Santorin und habe Santorin kennen gelernt – große Caldera in Santorin – und habe mir dann doch gedacht als ich nach Tübingen kam, warum hat denn so eine Riesenstruktur nicht auch Laven, da müssen doch auch Laven vorkommen nicht nur dieser Suevit, und überdies war das Vorkommen von Suevit auch in einer Quantität in einem Missverhältnis zu dieser großen Struktur' (Pichler interview).

several vents simultaneously. The eruptions intensified from 24 August onwards, producing large amounts of pumice and culminating on 27 August in a caldera collapse. The vast amount of hot ash and pumice entering the sea, together with the displacement of the caldera floor, generated a devastating tsunami, which reached heights of up to 40 m and killed more than 34,000 people.[66]

The eruption of Bandai-San in 1888 had been the worst volcanic disaster in recent Japanese history. The stratovolcano lay dormant for some thousand years but then awoke, producing landslides, lahars and pyroclastic flows. Some 1.5 km^3 of the summit collapsed, resembling what happened in the Mount St Helens eruption of 1980.[67]

In the 1930s, however, the eruption mechanisms of these two volcanoes were unknown. The volcanoes were chosen because of their size, which was similar to the order of magnitude of the Ries and Steinheim basins. Bandai-San was additionally suitable for comparison, in that the eruption, like the event that created the Steinheim Basin, produced no juvenile volcanics. Bandai-San came under the heading of 'phreatic explosive eruptions', which were interpreted as:

> a special type of eruption, which begin like the initialising eruptions in long-dormant volcanoes with a blocked-up vent. ... They only erupt old rock material and no fresh magmatic matter apart from gases; and even among these, it seems that resurgent water vapour plays the main role. ... The best-known example is the eruption of Bandai-San in Japan on 15 July, 1888. At 7 o'clock [a.m.] rumbling was heard from the mountain. Half an hour later earthquakes occurred and shortly afterwards a column of vapour and dust rose 1,300 m high. Then followed fifteen to twenty explosions, the last of which hurled out old rock material nearly horizontally. The dust-laden eruption cloud rose more than 6,000 m, hot ash from pulverized old rock material caused complete darkness. A horseshoe-shaped crater, open to the north, with a diameter of two kilometres, was blasted open. The material, which was hurled laterally, descended like a glowing avalanche, although the temperature of the dust-dry emulsion was certainly much lower than in true glowing avalanches. A battering wind [a pressure wave] with a velocity of forty metres per second preceded the oncoming mass. Seventy-one square kilometres of land were covered by debris [and] 461 people were killed. The whole eruption lasted only about two hours. (Rittmann 1936: 40–41)[68]

[66]   <http://vulcan.wr.usgs.gov/Volcanoes/Indonesia/description_krakatau_1883_eruption.html> accessed 21 June 2010.

[67]   <http://en.wikipedia.org/wiki/1888_Eruption_of_Mount_Bandai> accessed 21 June 2010.

[68]   'Phreatische Explosivausbrüche // Eine besondere Art von Ausbrüchen die wie die Initialausbrüche bei verschlossenem Schlot lang ruhender Vulkane einsetzen, wurden als halbvulkanische (Dana), ultravulkanische (Mercalli) oder indirekte (v. Wolff) bezeichnet. Sie fördern nur altes Gesteinsmaterial aber keine frischen magmatischen Stoffe außer Gasen,

The eruption of Mount Bandai was explained by the action of groundwater, which:

> percolated into the depth and collected in fissures and clefts. Perhaps it was partly also condensed fumarole vapours. This water came into contact with hot gases, which rose from the depth, was vaporized and caused phreatic explosions that blew off the summit of the mountain. The process can only be understood if one assumes a rise of temperature in the blocked conduit prior to eruption, caused by the abundant production of hot juvenile gases. (Rittmann 1936: 142)[69]

When questioned, many of the Ries geologists evoked these natural disasters as witnesses to the extraordinary forces that volcanoes could generate. 'Meanwhile we are in the realm of geological myths! ... They were the model to show that something like that is possible' (Pichler interview).[70]

Myth-making was furthered by the fact that volcanology was still in its infancy and nothing much was known about large caldera-volcanoes – the only suitable volcanic edifices that might have served as analogues to the 24-km-wide Ries Basin.

Volcanology in the early twentieth century was an esoteric subject in German geology. Hardly anybody was an expert on it or had first-hand scientific experience with active volcanoes. Richard Löffler, for example, was regarded as a

---

und auch unter diesen spielt anscheinend resurgenter Wasserdampf die Hauptrolle. ... Das bekannteste Beispiel ist der Ausbruch des Bandai-San in Japan am 15. Juli 1888. Um 7 Uhr war Getöse vom Berg her zu hören, eine halbe Stunde später ereigneten sich Erdstöße, und bald danach stieg eine Dampf- und Staubsäule 1300m hoch empor. Darauf folgten 15 bis 20 Explosionen, deren letzte alte Gesteinsmassen fast horizontal ausschleuderte. Die staubbeladene Ausbruchswolke erhob sich über 6000m hoch, heiße Asche aus zerriebenem altem Gesteinsmaterial verbreitete völlige Finsternis. Ein nach Norden offener, hufeisenförmiger Krater von 2 km Durchmesser wurde ausgesprengt. Das seitwärts fortgeschleuderte Material ging nach Art einer Glutlawine nieder, obschon die Temperatur der staubtrockenen Emulsion sicher viel tiefer war, als bei den echten Glutlawinen. Ein Schlagwind von 40 m/sec Geschwindigkeit eilte der herabsausenden Masse voraus. 71 qkm Land wurde mit Schutt bedeckt; 461 Menschen kamen ums Leben. Der ganze Ausbruch dauerte nur etwa 2 Stunden' (Rittmann 1936: 40–41).

   69 'Das Grundwasser sickerte in die Tiefe und sammelte sich in Spalten und Klüften an. Vielleicht handelte es sich z.T. Auch um kondensierte Fumarolendämpfe. Dieses Wasser kann mit heißen Gasen, die aus der Tiefe emporstiegen, in Berührung, wurde verdampft und verursachte phreatische Explosionen, die den Berggipfel wegsprengten. Der Vorgang lässt sich nur verstehen, wenn man eine, der Eruption vorhergehende Temperatursteigerung im verstopften Schlot annimmt, die durch reichlichere Förderung heißer juveniler Gase verursacht wurde' (Rittmann 1936: 142).

   70 'Da sind wir mittlerweile schon in der Sparte geologische Mythen! ... Das ist eben das Vorbild, dass so was möglich ist' (Pichler interview).

volcanologist (Hölder interview; cf. Chapter 9), but this was because he was an expert on the Ries crater, the only 'volcano' he knew anything about.

Rudolf Hüttner reported an effort to encourage a German volcanologist to work at the Ries Basin. He remembered:

> a letter by Mr Löffler, which he had written to Mr Reck. Reck was a volcanologist living in Berlin, but who had been investigating the Hegau volcanoes[71] in the 1920s. [Because] he [Löffler] had seen his own inefficiency – as he had no institute where he would have been able to employ people, where he had devices and so on – he was keen to involve a volcanologist who could draw on modern equipment, who should occupy himself with the Ries. He wrote a letter to him [Reck]: "The secret of the Ries lies in the suevite". ... Reck has declined – several times – to bother himself with the Ries. (Hüttner interview)[72]

The geologist and palaeontologist Hans Reck (1886–1937) had studied science in Munich, Würzburg, Berlin (under Branca) and London. At Branca's instigation he conducted an expedition to Iceland in 1908. In 1909 Reck received his PhD from Munich with a description of mass eruptions in Iceland. In 1910 he became an assistant to Branca in Berlin and the following year was appointed to supervise the dinosaur excavations at Tendaguru. He also led an expedition to the Olduvai Gorge. Beginning in 1914, Reck worked for the geological survey in German East Africa and there also participated in World War I. From 1915 to 1919 he was detained as a prisoner of war by British troops. After his release, he became a professor without chair at the Geological and Palaeontological Institute in Berlin. In the years 1925 to 1928, together with Greek colleagues, he investigated the volcanic island of Santorini. In 1931 he accompanied Louis Leakey (1903–72) on the second Olduvai Expedition. Reck died during another expedition in 1937. During his career, Reck published numerous papers and reports on geomorphologic, stratigraphic, volcanological and palaeontological topics (Mayr 2003; Kohring and Schlüter 2009; Mohr 2010: 303–6).

Even though Reck refused to follow Löffler's invitation, he had a decidedly volcanist opinion of the Ries, which was close to the Munich model of Ries genesis and was quoted accordingly by Kranz:

---

[71] Heavily eroded Tertiary volcanoes near Lake Constance

[72] ' ... dann habe ich mich erinnert an einen Brief des Herrn Löffler, den er an den Herrn Reck geschrieben hatte, Reck war ein Vulkanologe, der in Berlin lebte der aber die Hegau-Vulkane in den Zwanziger Jahren untersucht hat und dem Löffler lag es dran – er hat sein eigenes Ungenügen gesehen er hat ja kein Institut gehabt, wo er da Leute anstellen kann, wo er Apparate hat und so – da lag's ihm dran einen Vulkanologen heran zuholen, der moderne Einrichtungen zur Verfügung hat, der sich mit dem Ries beschäftigen sollte und dem schrieb er in einem Brief 'das Geheimnis des Rieses liegt im Suevit' ... der Reck hat es abgelehnt – mehrfach sich mit dem Ries zu beschäftigen' (Hüttner interview).

To my mind Ries and likewise Köfels,[73] represent weakest utterances of volcanic eruptions to the surface; volcanoes that have been suffocated during their genesis underneath the load of the roof; products of the utmost effort of a rising melt, which, however, was insufficient for full liberation, but let only insignificant descendants reach daylight along points of weakness. The projection of the forceful rising is the brecciation of the roof; the projection of the following exhaustion, however, is the caldera formation ... rim thrusts as well as local fine fragmentation; breaking of rocks and disintegration can just as well be explained by a central 'Ries mountain' – as long as they are not landslide debris – or by finger-fashioned, branching local extrusions and formation of blocks in the roof over a deep seated magmatic intrusion (Reck, quoted after Kranz 1928: 295–6; original reference missing)[74]

Kranz rejected this scenario as 'a personal credo ... without factual arguments for it and contrary to all hitherto actual geological observations and explosion-technical experience'.[75] The Ries breccias, Kranz affirmed, are irreconcilable with a volcanism that remained underground (Kranz 1928: 295):

The crushing of the fine debris to such an extent and the crunching together of most varied formations in the polymictic breccia, can only be explained by volcanic action at the surface, by fullest 'liberation' through a most forceful explosion and never by cryptovolcanic 'brecciation of the roof' with or without a 'Ries mountain'; in the literature on landslides too, I have so far not met with such an intensive blasting and crunching together of most varied rock formations. Friction breccias can also originate tectonically, but hardly such masses of fine

---

[73]   Köfels is a large landslide in the Ötztal, Austrian Alps, where pumice-like melts had been found (see Chapter 5).

[74]   'Mir verkörpern Ries wie Köfels schwächste Äußerungen vulkanischer Durchbrüche auf der Oberfläche; im Werden unter der Last des Daches erstickte Vulkane, Produkte einer äußersten Kraftanstrengung empordrängenden Schmelzflusses, die jedoch nicht zur vollen Befreiung genügte, sondern höchstens unbedeutende Nachläufer auf Schwächepunkte bis zum Tage vordringen ließ. Die Projektion des Kraftaufstiegs ist die Dachzerrüttung, die Projektion der folgenden Erschöpfung aber die Kalderabildung ... Randüberschiebungen wie lokale Griesbildungen, Gesteinszerschlagung und Zerrüttung aber lassen sich ebensowohl wie durch einen zentralen ‚Riesberg' – soweit es nicht Bergrutschmassen sind – auch durch fingerförmig verzweigte lokale Auftreibungen und Schollenbidungen [sic] in Dach über einer Tiefenintrusion verstehen' (Reck, quoted after Kranz 1928: 295–6; original reference missing).

[75]   'ein persönliches Glaubensbekenntnis ... ohne tatsächliche Unterlagen dafür und entgegen allen bisherigen tatsächlichen geologischen Beobachtungen und sprengtechnischen Erfahrungen' (Kranz 1928: 295).

debris, extended over a large area; for the 'polymictic breccia' there is so far only one suitable analogue: explosion debris. (Kranz 1928: 296–7)[76]

There was also Wilhelm Ahrens, who had mapped the Laacher See area and who had actually written on the 'tuffs' of the Ries Basin (the suevite) in 1929. In this short paper, he distinguished a vent and an eruptive facies of suevite, which, however, could be distinguished for certain only in the outcrop and not in hand specimens.

But even in the outcrop, as the PhD-student Wilhelm Ackermann from Munich plainly stated: 'It won't be possible to distinguish between vent facies and tuff deposited in surface depressions, if there is no outcrop towards the depth. At present the [quarry of] '*Alte Bürg*' [see Chapter 11] is the only outcrop at which the geological context is clearly visible. Here it is a vent filling' (Ackermann 1958: 161).[77]

At the quarry of Aumühle (see Figure 1.3), the suevite showed a distinct stratification in the upper part and Ahrens gave a profile of the various strata. He assumed first a deposition of the massive suevite, followed by some further ashes in a small lake. He did not contemplate the possibility of reworking for the stratified suevite, but assumed that it had come from ash eruptions from other vents further away. Therefore, Ahrens assumed local, smaller eruptions and according to him 'so far nowhere is there any proof that tuffs may have come from the [central] Ries Basin itself' (Ahrens 1929).[78]

As the 'suevite-tuff' everywhere overlies the breccias, it was to be considered younger than the formation of the Ries crater or basin itself, possibly by just a few days; and, as Aumühle showed, there had been several eruptive phases. Thus the 'suevite-tuffs', according to Ahrens, had nothing to say on the central problem of the Ries, the formation of the crater basin. Tuffs pointed, however,

---

[76] 'Die Zerschmetterung der Griese in solcher Ausdehnung und die Verknetung verschiedenster Formationen in der Bunten Breccie kann nur durch vulkanische Tagwirkung bei vollster 'Befreiung' durch gewaltigste Explosion erfolgt sein, niemals durch kryptovulkanische 'Dachzerrüttung' mit oder ohne 'Riesberg'; auch in der Literatur über Bergrutsche ist mir ein derartig intensives Zerknallen und Durcheinanderkneten verschiedenster Gesteinsformationen bis jetzt nicht bekannt. Reibungsbreccien können auch tektonisch entstehen, solche Griesmassen, auf großen Flächen verteilt, aber wohl kaum; für die 'Bunte Breccie' gibt es bis dahin überhaupt nur einen Vergleich: Den Sprengschutt' (Kranz 1928: 296–7).

[77] 'Es wird im Gelände nicht möglich sein, zwischen Schlot- und Wannentrassen zu unterscheiden, wenn kein Aufschluß in die Tiefe vorliegt. Zur Zeit ist die 'Alte Bürg' der einzige Aufschluß, der den geologischen Verband klar aufgeschlossen zeigt. Hier handelt es sich um einen Schlottraß' (Ackermann 1958: 161).

[78] 'Es ist jedenfalls noch nirgends der Nachweis gelungen, daß Tuff aus dem Rieskessel selbst stammen kann'. (Ahrens 1929).

to the conditions underground at or about the time of the Ries event (Ahrens 1929).

Ahrens considered the petrography of suevite to have been well known for quite a long time: glass wisps, fragments of minerals, glass bombs and fragments of the country rock. The tuffs were welded and thus had been hot when deposited. Feldspar glass was common; quartz glass rarer. Everything was heavily altered and hydrated. The melting was unusual, unlike that recognised in other volcanic tuffs: '[o]n the contrary, in the Ries other forces must have acted and the recognition and more thorough investigation of these forces can be called the central problem of the Ries' (Ahrens 1929: 97).[79] Too true! But Ahrens as well as others unfortunately did not pick up this task.

Ahrens recognised that there was no juvenile material in the 'suevite tuff' and that it was only reworked and melted country rock: 'In order to achieve this, an enormous input of heat is necessary. The ascending of this heat was caused by the rising of some sort of mineralisers, fluid components, the nature of which we do not know' (Ahrens 1929).[80]

In the 1930s, Alfred Rittmann (1893–1980) was among the foremost German-speaking volcanologists, owing to his textbook, which he published in 1936 when he was still lecturer in mineralogy and petrography at the University of Basel (Rittmann 1936). From 1926 to 1934, Rittmann had worked at the Institute of Volcanology in Naples and in the 1940s he returned to Italy for geological investigations commissioned by the Italian Government. From 1949 onwards, Rittmann worked at the University of Alexandria in Egypt and in 1954 he became Professor of Mineralogy and Geology at the University of Cairo. In the 1960s he returned to Italy and was appointed Director of the Volcanological Institute at the University of Catania in Sicily. Until 1968 he also headed the International Institute of Volcanology, which was founded in Catania in 1960 through Rittmann's initiative.[81] Rittmann's friend and colleague Hans Pichler has recalled that Rittmann:

> too didn't have any new ideas [concerning the Ries]. He always said that lavas were missing, but he didn't question the [volcanic] interpretation in light of his volcanological experience. However, the fact was that Alfred never went there, he never investigated suevite personally; and one tends to write a lot in one's books, picking up the opinions of others without questioning, because you can't

---

[79] 'Im Ries müssen dagegen andere Kräfte gewirkt haben und die Erkennung und genauere Erforschung dieser Kräfte kann man geradezu als das Kernproblem des Rieses bezeichnen' (Ahrens 1929: 97).

[80] 'Hierzu ist natürlich eine gewaltige Wärmezufuhr erforderlich. Der Aufstieg dieser Wärme erfolgte durch das Aufdringen irgendwelcher Mineralisatoren, leichtflüchtiger Bestandteile, deren Natur wir vorderhand nicht kennen' (Ahrens 1929).

[81] <http://de.wikipedia.org/wiki/Rittmann> accessed 6 April 2011.

go there and check everything and this also happened with Alfred Rittmann. (Pichler interview)[82]

And indeed, the 1936 textbook by Rittmann, *Vulkane und ihre Tätigkeit* [Volcanoes and their Activity], mentioned the Ries Basin under the heading of 'gas volcanoes':

> Upon certain initiating eruptions, only gas is ejected. If the explosions are strong, they eject fragments of the penetrated rocks, so that around the penetration point a vent-clearing breccia is deposited. So-called *gas-maar volcanoes* are formed. Weaker explosions only brecciate and disrupt the beds, which are penetrated by the gases. Breccia-filled *penetration pipes* or *explosion funnels* [diatremes] are created. Probably, the Nördlinger Ries and the Basin of Steinheim have also been formed in this way. The Ries, which is nearly circular with a diameter of twenty-five kilometres, can be regarded as a *volcano-tectonic depression* without coverage by unconsolidated pyroclastics. (Rittmann 1936: 80)[83]

Rittmann's book was an attempt to classify all known types of volcanic edifices or structures and characterise them according to their eruption behaviour. The latter, especially in the more explosive cases, was highly speculative, as the contemporary lack of remote sensing techniques necessarily restricted the possibilities of observing such eruptions directly and surviving to tell the tale. Rittmann classified the Ries as a 'volcano-tectonic depression, without coverage by unconsolidated pyroclastics'. These highly hypothetical structures were regarded as extreme endmembers in his 'genetic systematics of simple central volcanoes' (cf. Rittmann 1936: 150). They were thought to be produced from

---

[82] '... der hat auch keine andere Meinung gehabt, der hat immer gesagt es fehlt an Laven und aber selber hat er das nicht in Frage gestellt, aufgrund seiner vulkanologischen Erfahrung. Das lag aber auch daran dass der Alfred nie da gewesen ist, er hat also den Suevit nie in Augenschein genommen und von daher man schreibt ja vieles, was man in eigenen Büchern schreibt, übernimmt man auch Meinungen von anderen unbesehen, weil man nicht hinfahren und das nachprüfen kann, das ist also auch bei Alfred Rittmann so gewesen' (Pichler interview).

[83] 'Bei gewissen Initialdurchbrüchen werden nur Gase gefördert. Wenn die Explosionen heftig sind, schleudern sie Bruchstücke der durchschlagenen Gesteine aus, so daß rund um die Durchbruchstelle eine Schloträumungsbreccie abgelagert wird. Es bilden sich die sogenannten *Gas-Maare*. Bei schwächeren Explosionen kommt es nur zu einer Zertrümmerung und Zerrüttung der von den Gasen durchbrochenen Schichten. Es entstehen brecciengefüllte *Durchschlagsröhren oder Sprengtrichter*. Vermutlich haben sich auch das Nördlinger Ries und das Becken von Steinheim auf diese Weise gebildet. Das Ries, das bei annähernd kreisrunder Form einen Durchmesser von 25 km aufweist, kann als *vulkano-tektonische Senke* ohne Lockerdecken angesprochen werden' (Rittmann 1936: 80).

enormous amounts of juvenile gases, not by melts (hence the lack of juvenile volcanic material), that had accumulated in the pegmatitic stage of magma development, leading to a highly explosive volcanism.

This explanatory endmember, a volcano-tectonic depression without an associated blanket of juvenile pyroclastics, probably was created for the sole purpose of classifying the Ries Basin as a special sort of 'volcano-tectonic depression'. The latter was Rittmann's term for large calderas, as can be seen from the examples he gave:

> Morphologically these volcanoes form a negative landscape relief. They are gigantic, often water-filled collapse-basins with steep margins, which are surrounded by country-wide, thick pumice and ash blankets, at the base of which in some places lie breccia tuffs. Lake Toba serves as an example ... (Rittmann 1936: 79–80)[84]

Even though Rittmann used the word 'collapse', he did not use it in the modern sense of the term, as becomes clear from his further explanations. By 'collapse', he did not mean subsidence of the roof of a magma chamber, when its lithostatic pressure surpasses the pressure within the chamber, often as a result of the discharge of large volumes of pyroclastics through ring fissures (Cas and Wright 1988: 233–5), but instead meant that during a caldera eruption the whole roof of the magma chamber would blow up:

> Finally, if it is the case that the covering rock strata resist the growing internal pressure, which is due to the development of the [crystallising] melt, until the remaining melt reaches its pegmatitic stage, it will reach a highly explosive initialising eruption, upon which nearly all the potential energy of the magma chamber will appear at once as kinetic eruption energy. Therefore the state of exhaustion after this enormous work of power will be final; all the more so as the gases, escaping with enormous force, tear off vast amounts of fragmented magma, which will be distributed in the wide surrounding area as a rain of pumice and ash. Such an initialising eruption will remain a singular event. It is not capable of building a volcanic mountain; only an extended blanket of ash and pumice will be formed. The missing volume in the magma chamber, caused by the expelling of these masses, will cause the sagging of the caldera, or the formation of a volcano-tectonic depression. ... To avoid misunderstanding it should be remarked here

---

[84]　'Morphologisch treten diese Vulkane als negative Geländeformen in Erscheinung. Es sind riesige, oft von Wasser gefüllte Einbruchsbecken mit steilen Rändern, die von ländergroßen mächtigen Bimsstein-und Aschendecken umgeben sind, an deren Basis streckenweise Breccientuffe liegen. Als Beispiel diene der Toba-See ...' (Rittmann 1936: 79–80).

that the amount of magma available to erupt will considerably influence the eruption history of a volcano. The three chosen examples [shield volcano, strato-volcano and caldera] occur in the described way only if the magma chamber exceeds a certain size. Should, however, the magma chamber be very small, then an early [with respect to the chemical evolution of the crystallising magma] initial eruption will form only a small lava-flow, while a later breakthrough will lead to a small tuff-cone ... and if the breakthrough happens very late, a shot channel [a diatreme] will be created and only a few ash or pumice materials will be erupted. In all cases, the energy within the magma chamber will hardly be sufficient to fuel on-going activity. ... [so that] instead of a caldera or volcano-tectonic depression, the late explosive breakthrough will only create a shooting channel, a maar of the type of the Laacher-See.[85] (Rittmann 1936: 117–20)[86]

Rittmann thus saw the Ries as a structure comparable to those of ash-flow calderas. The superficial similarity[87] of the Ries to such volcanic edifices is undeniable and would have been recognised as even greater if Rittmann and his contemporaries had already known about the fossil fumarole channels in the Ries

---

[85] A dormant volcano in the Eifel (Germany).

[86] 'Tritt endlich der Fall ein, daß die Deckschichten dem mit der Entwicklung der Restschmelze anwachsenden Innendruck so lange Widerstand leisten, bis die Restschmelze sich dem pegmatitischen Zustand nähert, so wird es schließlich zu einem hochexplosiven Initialdurchbruch kommen, bei dem fast die ganze potentielle Energie des Herdes auf einmal als kinetische Ausbruchsenergie in Erscheinung tritt. Der Erschöpfungszustand nach dieser gewaltigen Kraftleistung ist daher endgültig, dies um so mehr, als die mit ungeheurer Wucht hervorbrechenden Gase riesige Mengen zerspratztes Magma mitreißen, das als Bimsstein- oder Ascheregen in weitem Umkreis verstreut wird. Ein solcher Initialausbruch bleibt ein einmaliges Ereignis. Er ist nicht imstande einen Vulkanberg aufzubauen; es entsteht nur eine ausgedehnte Aschen- und Bimssteindecke. Der durch den Auswurf dieser Massen entstehende Volumendefekt des Herdes hat einen Calderaeinbruch, die Bildung einer vulkanotektonischen Senke zur Folge. ... Um Missverständnissen vorzubeugen, sei jedoch gleich hier bemerkt, daß die Menge des eruptionsfähigen Magmas die Ausbruchsgeschichte und die einzelnen Eruptionen eines Vulkans stark beeinflusst. Die drei gewählten Beispiele [Schildvulkan, Stratovulkan, Caldera] kommen in der beschriebenen Art nur dann zur Auswirkung, wenn der Herd eine gewisse Größe überschreitet. Ist der Herd dagegen sehr klein, so wird sich bei einem früh eintretenden Initialdurchbruch nur ein kleiner Lavastrom bilden, bei einem späteren Durchbruch entsteht ein kleiner Lockerkegel ... und wenn der Durchbruch sehr spät erfolgt, so wird ein Schußkanal ausgeräumt und nur wenig Aschen- oder Bimssteinmaterial gefördert. In allen Fällen wird die Herdenergie kaum ausreichen eine längere Tätigkeit zu unterhalten. ... statt einer Caldera oder vulkanotektonischen Senke hinterlässt der Explosionsdurchbruch des Spätstadiums nur einen Schußkanal, ein Maar von der Art des Laachersees' (Rittmann 1936: 117–20).

[87] As regards to those features that are accessible through traditional field work only, that is no seismic or geomagnetic studies, no geochemistry or mineralogy/petrology.

suevite and if people had realised that the present-day suevite outcrops are only relics of a formerly continuous fall-out blanket. Traditionally, it was assumed that erosion at the southern German *Schichtstufenland* ('Tableland' between the Rhine-Graben and the northern Molasse Basin of the Alps) happened basically only at the escarpments,[88] and so it was simply not realised that the suevite and the breccias below it had once formed a continuous ejecta blanket.

The volcanological ideas that Rittmann promoted had provided the theoretical backing for the Kranz–Löffler model of a central Ries explosion, which then represented state-of-the-art volcanology: 'Eruptive plutons' were held responsible for calderas or volcano-tectonic depressions:

> When magma rises through the sedimentary beds of the outer crust of the Earth, then it is sometimes the case that the resistance of the beds to be penetrated is greater than their weight. In that case, the melt will intrude into a bedding-plain. Once this intrusion has started, because of the increasing size of the plane of force, the hydrostatic pressure becomes more and more effective,[89] and the overlying beds will be raised and possibly domed up. If the intruding magma is very viscous, then a compact, lentil-shaped pluton is created, a laccolith. ... If the intrusions take place at a shallower depth, the overlying beds will be broken into blocks, and magmatic matter can come to light along the fissures, either explosively or effusively. In connection with sills, phreatic explosions (groundwater, rock moisture) are also apt to happen, which penetrate the overlying beds independently of fissures and form maars. In all these cases one can talk of eruptive plutons.

> Eruptive sill. ... [A] sill intrusion occurred in the upper valley of the Ukak close to the surface during the great 1912 Katmai eruption in Alaska. Along a plain 3 km wide and 52 km long the overlying beds were broken into blocks and in numerous places small pumice fragments and glassy sands were ejected through the fissures, covering the valley floor nearly 100 metres deep. Since then, extensive

---

[88]   Georg Wagner claimed (1931: 159) that erosion on the karst-plateau of the Swabian Alb was slowest compared to any other condition such as river action (as the erosion on the Alb plateau was basically only chemical, and mainly active underground within the clefts, fissures and cave systems). Thus, the landscape of the Swabian Alb was considered to be a 'dead' landscape, basically without geological processes. Even in 1986 on the occasion of a palaeontological field trip to the area, the author of this book heard the term 'volcanic embryos' used for the diatremes of the 'Swabian Volcano', with the substantiation that no tuff-ring was developed. People thought this to be a strange feature but did not attribute this lack to erosion; rather it was assumed that the tuff-ring was missing because it had never formed in the first place. That is the volcanic edifice was not 'completed', hence the term 'volcanic embryo'.

[89]   This is physical nonsense; to the contrary, the smaller the plane on which the force acts, the more effective it will be. This is how hydraulic machinery works.

fumarolic activity occurs at the numerous explosion funnels and fissures, which has provided the valley with its name of 'Valley of the 10,000 smokes'. At one single place, a continuous and highly viscous magma reached the light of day and piled itself into a dome with a diameter of 260 metres and a height of 60 metres, broken into big blocks, which was named Novarupta. (Rittmann 1936: 81)[90]

Ries geologists certainly knew about this or other, similar, descriptions of volcanic eruptions, but they did not have any opportunity to check the descriptions in the field and to compare the volcanic features with their field observations made at Ries Basin or Steinheim Basin. And so again, quite apart from the adequacy of contemporary volcanology or the lack thereof, the two German craters were termed 'volcanoes' without actual volcanological investigation of their structures.

---

[90] 'Eruptionsplutone // Beim Empordringen des Magmas in die Schichtgesteine der äußeren Erdkruste tritt manchmal der Fall ein, dass die Festigkeit der noch zu durchbrechenden Schichten größer ist als ihr Gewicht. Die Schmelze dringt dann in eine Schichtfuge ein. Hat einmal diese Intrusion begonnen, so wird durch die Vergrößerung der Angriffsfläche der hydrostatische Druck immer wirksamer, und das Hangende wird gehoben und evtl. aufgewölbt. Ist das eindringende Magma sehr zähflüssig, so entsteht ein gedrungener, linsenförmiger Pluton, ein Lakkolith. ... Wenn die Intrusionen in geringerer Tiefe stattfinden, wird das Hangende in Schollen zertrümmert, und auf den Spalten dringen magmatische Stoffe explosiv oder effusiv zutage. Bei Sills ereignen sich wohl auch gern phreatische Explosionen (Grundwasser, Bergfeuchtigkeit), die, unabhängig von Spalten, das Hangende durchschlagen und Maare bilden. In allen diesen Fällen kann man von Eruptionsplutonen sprechen. // Eruptionssill. ... während des großen Katmai-Ausbruchs in Alaska im Jahre 1912 [ereignete sich] eine oberflächennahe Sillintrusion im oberen Tal des Ukak. Auf einer 3 km breiten und 52 km langen Fläche zerbrach das Hangende des Sills in Schollen, und auf den Brüchen erfolgten an unzähligen Stellen Auswürfe von kleinkörnigen Bimssteinen und Glassanden, die den Talboden fast 100 m tief bedeckten. An den zahlreichen Sprengtrichtern und Spalten herrscht seitdem rege Fumarolentätigkeit, die dem Tal den Namen 'Valley of the 10 000 smokes' eintrug. An einer einzigen Stelle drang zusammenhängendes, sehr zähflüssiges Magma zutage und staute sich zu einer in große Blöcke zerborstenen Kuppe von 260 m Durchmesser und 60 m Höhe auf, die Novarupta genannt wurde' (Rittmann 1936: 81). The modern view interprets the deposits in the Valley of 10,000 Smokes as a welded ash-flow tuff generated from the vent of Novarupta (Cas and Wright 1988: 233). Large calderas, such as the Yellowstone Caldera (the size of which is about 55 times 72 km), which were even bigger than the Ries Basin, were merely seen as larger variants, which Rittmann called 'area volcanoes', thought to be formed by the extrusion of a huge pluton. But he also admitted that not much was known about these structures (Rittmann 1936: 83, 151).

Chapter 4
A Letter from Berlin

With the beginning of World War II, geologists in Germany had other things on their minds besides pressing the question of a possible impact origin of the Ries and Steinheim Basins, but this did not mean that there was no research relevant to the problem.

One prominent argument against an impact at the Steinheim Basin had always been the existence of a raised central mountain, which, as seemed evident, had been pushed up by some force from below.[1] What else could this force have been, besides volcanism? 'And in Steinheim Basin things had been quite clear; in the *Klosterberg* [the central mountain] Black and Brown Jurassic [Lower and Middle Jurassic] lies at the same level as White Jurassic [Upper Jurassic]. It is quite straightforward, it had been pushed up from below, therefore [it was] of volcanic nature' (Hölder interview).[2]

Meanwhile, however, World War II offered (again) the opportunity to closely and perilously observe that a central mountain in an explosion crater did not necessarily require magmatic forces:

> It was during the invasion in Poland. Whoever took part in it knows that there was no time for peaceful observation or to linger at one of the few outcrops. Only when the enemy by stronger bombardment or air raids forced us temporarily to seek cover on the ground, was one able to use the repose to look about somewhat. On such occasions I saw what shall be reported here. (Trusheim 1940: 317)[3]

Ferdinand Trusheim (1906–97) had obtained his doctorate from the University in Frankfurt am Main. From 1930 onwards, he worked as a geological assistant

---

[1]   Elastic rebound, as it had been observed in man-made explosion craters during World War I (Chapter 2), did not make it into the German literature and thus remained disregarded.

[2]   'Und im Steinheimer Becken war ja klar, der Klosterhügel, dort tritt der Schwarze und Braune Jura in der Höhe vom Weißen Jura auf. Das ist ganz naheliegend, das ist von unten hochgestoßen worden, also vulkanischer Natur' (Hölder interview).

[3]   'Es war beim Vormarsch in Polen. Wer ihn mitgemacht hat, weiß, daß keine Zeit war zu geruhsamer Beobachtung, zum Verweilen an einem der wenigen Aufschlüsse. Nur wenn uns der Feind durch stärkere Feuerwirkung oder Fliegerangriffe vorübergehend zwang, am Boden Deckung zu suchen, konnte man die Ruhepause benutzen, sich etwas umzusehen. Bei solchen Gelegenheiten sah ich das, worüber hier berichtet werden soll' (Trusheim 1940: 317).

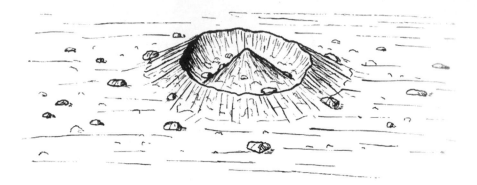

Figure 4.1   Drawing of a central mountain in a bomb crater of World War
             II. By Ferdinand Trusheim (1940: Figure 3).

and from 1936 as a lecturer at the University of Würzburg. In 1942, after his
military service during the first years of World War II, he became Professor
(without chair) of Geology and Palaeontology in Würzburg. However, he was
soon called up again. In 1943, he briefly commanded the military geologists of
the German army. Trusheim survived the war and spent some time as a prisoner
of war. From 1950 onwards, he worked in the oil industry (Freyberg 1974: 160;
Rose et al. 2000: 126).

    A keen and inquisitive observer, Trusheim wrote a little regarded but (as to
the impact question) highly significant paper about his geological observations
on the battlefields, which he published in the popularly oriented house journal
of the Senckenberg Museum in Frankfurt am Main:

> Another strange view was offered a few days later by impact funnels of German
> and Polish air bombs. The majority showed the usual form of a circular funnel
> with a ring dyke and blocks dispersed in the wider surroundings. Besides these
> ordinary funnels, however, every now and then there was one where additionally
> in the middle of the funnel a *central cone* arose [Figure 4.1] as Vesuvius rises within
> the Somma, but with the difference that the tip of the cone mostly did not reach
> the height of the surrounding ring dyke. (Trusheim 1940: 319)[4]

---

4   'Ein anderes, eigenartiges Bild boten einige Tage später Einschlags-Trichter deutscher
und polnischer Fliegerbomben. Die Mehrzahl von ihnen zeigte die übliche Form eines
kreisrunden Trichters mit Ringwall und den im weiten Umkreis versprengten Brocken. Aber
neben diesen normalen Trichtern sah man immer wieder solche, bei denen mitten im Trichter
noch einmal ein zentraler Kegel aufragte, ähnlich wie sich der Vesuv aus der Somma erhebt,
nur mit dem Unterschied, daß die Spitze des Kegels meist nicht die Höhe des umgebenden
Ringwalles erreichte' (Trusheim 1940: 319).

Trusheim's explanatory hypothesis for these small central mountains (Trusheim 1940: 320) is no longer convincing, but he clearly saw and pointed out the favourable implications for the impact theory at the Ries and Steinheim Basins. Unfortunately, he was completely ignored by the Ries experts:

> These small observations in air bombs ... called into our mind the remembrance of *mega-forms of the Earth surface*, which superficially are quite similar, but many thousand times larger. ... Some researchers consider them [the Ries and Steinheim Basins] to be volcanic explosion funnels; others, the result of mountain-building forces; still others as meteorite impacts. The *Steinheim Basin* hereby deserves a prominent role ..., especially because here, just as in our bombs, in the middle of the funnel is situated a central cone, a plug of rocks pushed up (or sucked up!) from the depths (the well-known Klosterberg). The promoters of a volcanic origin for the Steinheim Basin therefore must fall back on the assumption of a two-fold explosion; the first, which forms the funnel-proper and the ring-dyke, and a second, deeper one, which formed the Klosterberg-cone. The disciples of the meteorite theory can now argue with the metaphor of the air bombs and regard the funnel with dyke as an effect of the collision and the Klosterberg-cone as effect of the explosion of their meteorite. (Trusheim 1940: 320–21)[5]

Nearly a decade later, another effort to raise the issue fared scarcely better.

The Quiring–Kranz Correspondence

Heinrich Ludwig Quiring (1885–1964) was a widely travelled German geologist. Like most of his contemporaries he had served in the German military during World War I and in his Curriculum Vitae he proudly listed every battle in which he had taken part (Quiring 1955a). From 1933 to 1945, Quiring was

---

5   'Diese kleinen Beobachtungen an Fliegerbomben ... riefen uns damals die Erinnerung an *Großformen der Erd-Oberfläche* vor Augen, die äußerlich recht ähnlich, aber viel tausendmal größer sind. ... Manche Forscher halten sie [Ries und Steinheimer Becken] für vulkanische Sprengtrichter, andere für das Ergebnis gebirgsbildender Kräfte, wieder andere für Meteoriten-Einschläge. Eine besondere Rolle kommt hier dem *Steinheimer Becken* ... zu, vor allem deshalb, weil hier genau wie bei unseren Bomben, mitten im Trichter ein zentraler Kegel, ein aus der Tiefe hochgepresster (oder hochgesaugter!) Gesteins-Pfropfen steht (der bekannte Klosterberg). Die Verfechter einer vulkanischen Entstehung des Steinheimer Beckens müssen deshalb zu der Annahme einer zweimaligen Explosion greifen, einer ersten, die den eigentlichen Sprengtrichter und Ringwall, und einer zweiten tieferen, die den Klosterberg-Kegel entstehen ließ. Die Jünger der Meteoriten-Theorie könnten sich nunmehr auf das Bild der Fliegerbomben berufen und den Trichter mit Wall als Folge des Einschlages, den Klosterberg-Kegel als Folge des Zerknalls ihres Meteoriten ansehen'. (Trusheim 1940: 320–21).

a professor at the *Reichsamt für Bodenforschung* in Berlin (National German Geological Survey) and head of the geological research institute of the Saarland, for which he wrote reports on ore deposits and tunnel-building. His early geological interest was mainly in mining and later also in military applications. He was additionally interested in the early (antique and prehistoric) history of mining. His theoretical thinking was at times bold and fanciful. For example, he claimed that the northern Pacific Ocean was the scar of the formation of the Moon (Quiring 1948b, 1949a, b), and consequently rejected Wegener's continental drift theory.[6] In his list of publications (available for 1911 to 1955), there are papers on social Darwinism from 1930 onwards, which (judging from their titles) became increasingly racist in content. He also gave popular talks on the wireless on the same topic. He continued to publish in a nationalist, extreme right-wing journal[7] even after the war and up to his death (Quiring 1965). Nevertheless, at first there did not seem to have been any problems with 'denazification': shortly after World War II, in August 1945, Quiring managed to become chief geologist at the central office of the fossil fuel industry for the Eastern Sector in Berlin, possibly resulting from his good connections with the Russian authorities who had taken office in Quiring's villa in Falkenhain near Berlin in March 1945. From October 1945 onwards, Quiring also lectured at the Geological Institute of the Berlin Technical University. When he was dismissed for unknown reasons from his position at the central office of the fossil fuel industry in 1946, he took a professorship and became director of the institute for geology and palaeontology at the Technical University in Berlin-Charlottenburg. From about 1948 he became more and more interested in philosophical and increasingly esoteric issues (Quiring 1955a; Küppers 2003).

In November 1949, Quiring wrote the following letter to Walter Kranz, the main promoter of a central volcanic Ries explosion:

> Unfortunately, I had far too little to do with the Ries to have a personal opinion. This is the reason for my quote that "maybe" from the "meteor crater" of the

---

6   The idea of a violent rupture of the Moon from what is now the Pacific Ocean, and a subsequent breaking of the continents moving towards the large hole (causing the Atlantic to open), was developed by the British geologist and geophysicist Rev. Osmond Fisher (1817–1914) in the late nineteenth century. 'His idea retained much popular appeal until the middle of this century. By 1936, however, the geophysicist Beno Gutenberg (1889–1960), who had once favored the idea, calculated that it was dynamically impossible for the Moon to have escaped Earth so late in geological history' (Marvin 1986: 153).

7   The *Deutsche Hochschullehrer-Zeitung* was published between 1955 and 1972 by Herbert Grabert (1901–78), a protestant theologian by training, who increasingly included neo-Pagan elements in his theology. The journal was a forum for national-socialist scientists, who after 1945 no longer found employment at German universities (<http://de.wikipedia. org/wiki/Herbert_Grabert> accessed October 2002).

Ries blocks had been hurled over 60 km in distance. Since no true volcanic rocks – what is the suevite? – and only locally melted rocks have been found, the explosion, which you have assumed all the time and as it is obviously suggested by the round kettle (a true explosion funnel), can only have been the explosion of a meteorite. I do know the volcanic explosion craters of the Eifel ... so well – I have mapped there for 18 years – that my doubt concerning a volcanic origin of the Ries is well founded. Upon impact of a meteorite its kinetic energy is transformed into heat $M/2 \, v^2 = Ms \, (T-T_o) \, A$ and causes it to explode. Already for a velocity of 20 km/sec (meteorites have one of up to 197 km per second)[8] the temperature generated by the impact can reach 234,000°C. In comparison, your picric acid with an explosion temperature of 3200°C is a decidedly cool affair! Even if only 10 per cent of the calculated explosion temperature remains with the meteorite, it will result in <u>vaporisation</u> and disintegration of the molecules, maybe even the atoms, within less than a second. This is the reason, why in the surroundings of the Arizona Crater only a few fragments of the meteorite have been found, partly with inclusions of diamond bearing witness to the high explosion pressure. ... The melting and fragmentation of the rocks and perhaps also the suevite can be generated by volcanic as well as meteoritic explosion. (Letter from Quiring to Kranz, Berlin, 2 November 1949, GA 19404)[9]

---

[8]   In Quiring's time it was still assumed that meteorites were interstellar bodies, thus the excessively high velocity he attributed to them here. In 1935, the Estonian astronomer Öpik (Chapter 6) had declared that 'the interstellar origin of meteorites could be taken for granted' (Marvin 1996: 581), because they seemed too fast to belong to our solar system. 'In the 1930s and 1940s, painstaking photographic studies of meteors by Professors Fred L. Whipple and his colleagues at the Harvard College Observatory and C. C. Wylie at the University of Iowa showed elliptical solutions for the orbit of all the well-documented meteors and fireballs they measured. Öpik disputed their findings until 1959 when he abandoned the idea of hyperbolic orbits on his own evidence and wrote to Whipple apologizing for his previous criticis' (Marvin 1996: 581).

[9]   'Ich habe mich leider bisher zu wenig mit dem Ries beschäftigt, um ein eigenes Urteil zu haben. <u>So</u> ist auch mein Zitat zu verstehen, dass 'vielleicht' aus dem 'Meteorkrater' des Nördlinger Rieses Blöcke 60 km weit geschleudert worden seien. Da keine echten vulkanischen Gesteine – was ist eigentlich der Suevit? – und nur lokal aufgeschmolzene gefunden wurden, kann die Explosion, wie Sie sie von jeher angenommen haben und wie sie der runde Kessel (ein echter Sprengtrichter) offenbar verrät, doch nur eine Meteorexplosion gewesen sein. Ich kenne die vulkanischen Explosionskrater der Eifel ... so genau – ich habe dort 18 Jahre kartiert –, dass mein Zweifel an einer vulkanischen Entstehung des Rieses berechtigt ist. Beim Einschlag eines Meteors verwandelt sich seine Bewegungsenergie in Wärme $M/2 \, v^2 = Ms \, (T-T_o) \, A$ und bringt ihn zur Explosion. Die durch den Einschlag entstandene Temperatur kann schon bei v = 20 km/sec (Meteore haben bis 197 km Sekundengeschwindigkeit) 234 000°C erreichen. Da ist Ihre Pikrinsäure mit 3200° Explosionstemperatur eine wahrlich kühle Angelegenheit! Auch wenn von der berechneten Explosionstemperatur nur 10 % dem Meteoriten verbleiben, so ist <u>Vergasung</u> und Zerfall der

Contrary to all other Ries impactists reviewed thus far, Quiring in his letter derived the energy for the explosion at the Ries Basin from the kinetic energy of the impacting body. Nevertheless, it was probably not his original idea, as this mechanism had been discussed in the international literature by several authors (see discussion in Hoyt 1987: 185–7; Mark 1995: 38–9), mostly for lunar craters, whereas the transfer to earthly conditions seemed difficult because of the difference in atmosphere (see Chapter 6). Quiring's line of argument is rather similar to that of Herbert Eugene Ives (1919), who also concluded: 'Even if we assume that nine-tenths of this heat is given up to the surroundings, we still have … a temperature amply sufficient to gasify any known material, that is, *to produce an explosion*' (Ives quoted after Hoyt 1987: 198). Likewise, Quiring's comparison with chemical explosives is rather similar to Algernon Charles Gifford (1862–1948), who by way of illustration provided a table in which he compared calculated explosion temperatures of impacting bodies to those of known chemical explosives, among them picric acid, the explosive Kranz had used for his cratering experiment (Gifford 1924; see Chapter 6).

It is remarkable that Quiring recognised the suevite as non-volcanic rock and proposed its generation through impact. This is how his rhetorical question 'what is the suevite?' must be understood. This distinguished him even among many other impactists, who tended to regard suevite before 1961 (and occasionally even after) as indicative of secondary volcanism (Chapter 11), a result of magma rising in the Earth's crust fractured by the impact. In his letter, Quiring provided an intriguing 'back of the envelope' calculation using the principle of conservation of energy: Obviously, he realised that during the impact kinetic energy had to be transformed into heat and that the projectile does not just mechanically excavate a hole, but must also explode during the process. Here, for the first time, a German geologist demonstrated a true concept of the cratering mechanisms of impacts. Quiring thus understood that forces were involved that were much larger than what can be realised with conventional chemical explosives.

It is also noteworthy that Quiring hinted at a possible test for his hypothesis: He mentioned diamonds as witnesses to the high explosion pressure. Shock-produced diamonds had initially been observed in experimental explosion craters and were not published before 1961. However, it had been known for quite some time that some meteorite types did contain micro-crystalline diamonds, for example, the Canyon Diablo Irons associated with Meteor Crater

---

Moleküle, vielleicht sogar der Atome, im Bruchteil einer Sekunde die Folge. So kommt es, dass auch im Umkreise des Arizonakraters nur wenige Meteoreisenbruchstücke – z.T. mit Diamanteneinschlüssen, Zeugen des hohen Explosionsdruckes – gefunden worden sind. … Die Gesteinsschmelzungen, [Gries]bildungen, vielleicht auch die Suevite, können ebenso gut bei vulkanischer wie meteorischer Explosion entstanden sein' (letter from Quiring to Kranz, Berlin, 2 November 1949, GA 19404).

in Arizona (see Chapter 2). Diamonds in the Ries Basin had been sought in vain in the early 1960s. Then, in 1978, Russian scientists reported the existence of diamonds in the Ries Basin. This was confirmed in 1995. The maximum size of the diamonds is reported as 100 μm; thus, large ones are visible in thin-sections under a normal polarisation microscope.[10] Such diamonds, together with silica carbides, are now deemed to be a reliable hint to impact craters (Hough et al. 1995).

Major Walter Kranz (1873–1953; Figure 4.2), Quiring's correspondence partner, is seen as the founder of military geology in Germany. He began his military career in 1891 and became pioneer officer in 1893. He was stationed at various garrisons throughout Germany, including Swinemünde between 1909 and 1912. From 1895 to 1896, Kranz attended the Artillery and Engineering School in Berlin. Between 1901 and 1906, he studied at various universities, whenever time permitted. During World War I, Kranz was in active service as a pioneer officer and geological councillor for the German military. He was promoted to major in 1916 and obtained a PhD from the University of Munich in the same year. After the war, Kranz joined the geological department of the Württemberg Office of Statistics in Stuttgart, the predecessor of the Geological Survey of Württemberg. He also lectured on palaeontology, construction engineering and military geology. He retired in 1938. When his house was destroyed by bombing in 1944, Kranz moved to Bopfingen at the western rim of the Ries Basin (Carlé 1987: 79–81; Adam 1980; Rose et al. 2000: 117).

It seems that Quiring's letter had been an answer to a letter from Kranz, which itself had been provoked by an exchange of letters between Kranz and his editor [Heinrich?] Zeller. About a month earlier, Kranz had written to Zeller (whose initial letter is lost, but who must have asked Kranz why he did not address the issue of possible impacts in his latest manuscript). Zeller may have been familiar with Stutzer's talk in 1936. Kranz answered him:

> Apart from Stutzer only Werner and Quiring have published about the meteor [sic] theory. Right after Stutzer, three geologists have rejected this theory and I have assembled in *Petermanns Geographische Mitteilungen* [Kranz 1937b] all that speaks against it. According to the newest research as well, the Ries came into being by volcanic [forces] and not by meteor [sic] impact. In my new paper I can deal with this theory as well as with many other and older ones only quite briefly; otherwise its size would be too great, and for an extended public, too boring. (Letter from Walter Kranz to [Heinrich?] Zeller, 7 October 1949, GA 19390; reaction to a letter from Zeller on 6 October 1949)[11]

---

10 Currently (2014) on display at the Ries Crater Museum in Nördlingen.

11 'Ueber [sic] die meteor-Theorie [sic] haben ausser [sic] Stutzer nur noch Werner und Quiring publiziert, gleich nach Stutzer lehnten 3 Geologen diese Theorie ab, und ich habe

Figure 4.2   Walter Kranz developed, in the 1910s and 1920s, the central explosion theory to explain the genesis of the Ries 'volcano'. (Source: Carlé 1987: Figure 3).

Quiring had indeed mentioned the possible impact origin of the Ries Basin in a paper on the age, composition and genesis of the Moon (Quiring 1948a), where he showed himself to be one of the few German geologists (if not the only one) who cited Otto Stutzer approvingly. In contrast to many of his colleagues (Chapter 3), Quiring also demonstrated by his choice of references that he at least did not hesitate to draw on foreign literature (Quiring 1948a: 179).[12]

Although most of the paper seems rather fantastic from today's point of view, it also contains some more or less sensible reflections concerning the impact mechanism and the effect on the target, which would have been worthwhile to consider in connection to the Ries and Steinheim Basins:

> As a result of the collision stony meteorites are pulverized, iron meteorites are splintered and melted, and the targeted Earth-rocks are crushed to dust, as well as liquefied to pumice-like silica glass. Despite pulverization, splintering and melting, the living force (kinetic energy) of the impacting mass is sufficient to explosively blast out a rather deep, mostly almost circular, more rarely ellipsoid crater with a forced up rim. From meteor craters on Earth, rock masses and meteorite fragments have been hurled off to 10 km (Arizona), yes even to 60 km (Nördlinger Ries) distance. ...

> In 1928, Krejci uttered the opinion that a meteor hitting the Moon with cosmic velocity would be pulverized and melted, so that only a 'spot of lava' would be left over. His calculation of the heat equivalent does not consider that a body impacting on the Moon's surface would need time to be smashed and melted, just like a bomb, however sensitive its trigger may be. It can only explode after it has hit, but before and, yes, even during the explosion it will conform with its full mass to the law of inertia and excavate a hole, and only after the pulverization or melting will it again be expelled from the hole, widening it in the shape of a crater. Most of the time, however, splinters, dust and melt will lie partially in the hole. The faster the impacting velocity is, the deeper the hole will be and the wider the explosion crater, because the time between collision and explosion according to

---

in Petermanns Geographischen Mitteilungen zusammen gestellt, was dagegen spricht. Auch nach den neuesten Untersuchungen ist das Ries sicher vilkanisch [sic] entstanden, nicht durch Meteoreinschlag. Mit dieser Theorie wie mit vielen anderen älteren kann ich mich in meiner neuen Abhandlung nur ganz kurz befassen, sonst würde der Umfang viel zu gross [sic] und für weitere Kreise langweilig' (letter from Kranz to [Heinrich?] Zeller, 7 October 1949, GA 19390).

[12] Like Stutzer, Quiring was widely travelled and, due to his engagement with mining, applied and military geology, more interested in geological processes than in Earth history, which – despite his Nazi background – possibly made him less susceptible to the ideological influence of 'German Geology' (see Chapter 7). Quiring was no 'Earth historian'.

the data drawn from the Arizona crater is about 0.1 second. In all processes, time is an important factor that must not be neglected. (Quiring 1948a: 180, 183)[13]

Quiring's letter was obviously forwarded to Zeller to stress Kranz's argument:

> Enclosed you find my newest Ries *opusculum* and an exchange of letters with Dr Quiring, the latter of which I ask to be returned. From this can it be discerned, with what lack of true knowledge of the nature of the Ries-area and its literature the meteorite theory is being promoted. With Stutzer, it had obviously not been otherwise. (Letter from Kranz to Zeller, 19 January 1950, GA 19391)[14]

Although we do not know what Kranz wrote in answer to Quiring, a letter from Kranz to Georg Hermann Reich (1891–1976), Professor of Geophysics in Munich, may provide some insight. Walter Kranz wrote to Reich:

> Early in November 1949, Mr Quiring explained to me and calculated in a letter that the Ries has originated through the explosion of a stony meteorite, "The

---

[13] 'Durch den Aufprall werden Steinmeteore zerstäubt, Eisenmeteore zersplittert und geschmolzen, während die getroffenen Erdgesteine zu Staub zermalmt, auch zu bimssteinartigem Silicaglas verflüssigt werden. Trotz Zerstäubung, Zersplitterung und Schmelzung reicht die lebendige Kraft (kinetische Energie) der aufschlagenden Masse aus, einen mehr oder weniger tiefen, meist fast kreisrunden, seltener elliptischen Krater mit aufgepreßtem Rand explosionsartig herauszusprengen. Aus irdischen Meteorkratern wurden Gesteinsmassen und Meteoritenbruchstücke bis 10 (Arizona), ja vielleicht 60 km (Nördlinger Ries) Entfernung fortgeschleudert'. (Quiring 1948a: 180). 'Krejci hat 1928 die Ansicht geäußert, daß ein mit kosmischer Geschwindigkeit auf den Mond treffender Meteor zerstäubt und geschmolzen würde, so daß nur ein 'Lavafleck' übrigbleibe. Seine Wärmeäquivalentberechnung berücksichtigt nicht, daß ein auf die Mondoberfläche aufschlagender Körper zum Zerschellen und Schmelzen ebenso wie ein Geschoß mit noch so empfindlichem Zünder Zeit braucht. Er kann erst nach dem Aufschlag explodieren, wirkt aber bis dahin, ja auch noch während der Explosion, nach dem Gesetz der Trägheit mit seiner vollen Masse, schlägt ein Loch und fliegt erst nach der Zerstäubung oder Schmelzung wieder aus dem Loch heraus, es trichterförmig aufbrechend. Meist bleiben jedoch Splitter, Staub und Schmelze wenigstens z. T. im Loch liegen. Je größer die Auftreffgeschwindigkeit ist, um so tiefer wird das Loch, um so weiter der Sprengtrichter, da die Zeit zwischen Aufprall und Explosion nach den beim Arizonakrater gewonnenen Daten etwa 0,1 Sek. beträgt. Bei allen Vorgängen ist die Zeit ein wichtiger Faktor, der nicht vernachlässigt werden darf'. (Quiring 1948a: 183).

[14] 'Beifolgend mein neuestes Ries-opusculum und ein Schriftwechsel mit Herrn Dr. Quiring, welch letzteren ich zurück erbitte. Man kann daraus entnehmen, mit welchem Mangel an wirklicher Kenntnis der Natur des Riesgebiets und seiner Literatur die Meteoritentheorie verfochten wird. Bei Stutzer war es offenbar nicht anders' (letter from Kranz to Zeller, 19 January 1950, GA 19391).

melting and fragmentation of the rocks and perhaps also the suevite can be generated by volcanic as well as meteoritic explosion". He claimed that no true volcanic rocks and only locally melted ones have been found, and asked me in this connection "What is the suevite?" This was so characteristic of his expertise that I gave him a short lecture by letter about the volcanic basement of the Ries and its surroundings and told him that he should first inform himself about the known facts. – What do you think about this hypothesis? (Draft of a letter from Kranz to Reich, Bopfingen, 9 February 1950, GA 19412)[15]

Thus, Kranz had definitely opposed Quiring, and in his letters to Zeller as well as to Reich reproached Quiring because of his lack of knowledge concerning the Ries – especially because he completely misunderstood Quiring's rhetorical question concerning the origin of suevite. However, he obviously also felt a little uncertain and therefore asked the geophysical expert for counsel. This is how Reich responded:

Reading your letter about your communication with Prof. Quiring, I laughed heartily. Mr Quiring is well known to me as a colleague from Berlin. We all never took him seriously. It is a pity that a normally so intelligent man often lacks the most primitive exactness in dealing with an interesting problem. That I do not take Quiring's hypothesis seriously, I do not need to stress. (Letter from Reich to Kranz, Munich, 14 February 1950, GA 19413)[16]

And this was the preliminary end of the impact theory in Germany. It had enjoyed a kind of fashion in the 1930s, suspiciously eyed and quickly quenched

---

[15] 'Herr Quiring hat mir Anfang November 1949 brieflich auseinandergesetzt und berechnet, das Ries sei durch die Explosion eines Steinmeteoriten entstanden, "die Gesteinsschmelzungen, Griesbildungen, vielleicht auch die Suevite, können ebenso gut bei vulkanischer wie meteoritischer Explosion entstanden sein". Er behauptete, es seien keine echt vulkanischen Gesteine und nur lokal aufgeschmolzene gefunden worden und frug mich in diesem Zusammenhang: "was ist eigentlich der Suevit?". Das war für seine Sachkenntnis so bezeichnent [sic], dass ich ihm eine kurze briefliche Vorlesung über den vulkanischen Untergrund des Rieses und seiner nächsten Umgebung hielt und dazu meinte, man sollte sich doch erst einmal über das bekannte Tatsachenmaterial unterrichten. – Was halten Sie von dieser Hypothese?' (Draft of a letter from Kranz to Reich, Bopfingen, 9 February 1950, GA 19412).

[16] 'Bei Ihrer Mitteilung über den Schriftwechsel mit Herrn Prof. Quiring habe ich von Herzen gelacht. Herr Quiring ist mir als Kollege aus Berlin natürlich wohlbekannt. Wir haben ihn alle miteinander nie ernst genommen. Es ist schade, daß ein sonst so geistvoller Mann oft auch die primitivste Sorgfalt bei der Behandlung interessanter Probleme vermissen läßt. Daß ich die Quiring'sche Hypothese nicht ernst nehme, brauche ich daher nicht besonders zu betonen' (letter from Reich to Kranz, Munich, 14 February 1950, GA 19413).

by the automatic reaction of Ries experts and other conservative geologists. It had been briefly revived in 1949 by Zeller's questions and the ensuing exchange of letters between Kranz and Quiring, but by then it had already passed its prime and did not make it into publications again.

However, the automatic response, especially towards Quiring, becomes better understandable (even in hindsight) when reading more of Quiring's papers of the time, which bore some relationship to the impact question. It soon becomes obvious that Quiring was gifted with a rich imagination that not only bordered on the eccentric but ventured freely beyond: 'The genesis of the Moon is, like any volcanic eruption, caused by pressure release and spontaneous degassing – the magma follows the depressurized gases like water in a suddenly-opened bottle of mineral water – a sudden release of gravitational bonds by centrifugal-thermodynamic dispersal energy' (Quiring 1949b: 125).[17]

Quiring's ideas on the genesis of the Moon read like the mechanism of the heavens running amok. Compared to Quiring, Walter Kranz seemed much more 'down to earth', a typical, sensible and serious geologist. It does him credit that in his paper (Kranz 1937b; mentioned in his letter to Zeller), where he had 'assembled ... all that speaks against' a meteorite impact at the Ries and Steinheim Basins and which had been inspired by Stutzer's infamous talk and paper of the previous year, Kranz strictly reviewed factual arguments and refrained from straying into polemics:

> The larger a 'crater' is, the lower the probability may be that it is an earthly cosmic 'meteor'-crater: First of all, the analogous genesis of the gigantic lunar 'craters' is still debated and at the utmost 'possibly' meteoritic. And then, the morphological features on the Moon are so different from those on Earth, even when the lack of air and precipitation is considered, that they can be compared only with great doubt. Therefore, I refer to the *terrestrial* ones (and the 'true' meteor craters recognized on Earth are *comparatively small*). One of the largest (or the largest?), the one in Arizona, has a diameter of some 1200 m. The 'circumstantial evidence' points to its 'celestial' genesis, although even for it there remains still doubt, but – *in dubio pro reo* – I like to take it here as a true meteor crater in order to allow comparison. The Steinheim Basin is some 3 km in diameter, the Ries some 21 to 24 km, that is, it was in average 18 ½ times larger than the Arizona crater. (Kranz 1937b: 198)[18]

---

[17]   'Die Mondentstehung ist, wie jeder durch Druckentlastung und spontane Entgasung erzeugte Vulkanausbruch – das Magma folgt den entspannten Gasen wie Wasser in einer plötzlich geöffneten Selterswasserflasche – eine Sprengung gravitativer Fesseln durch zentrifugal-thermodynamische Streuenergie'. (Quiring 1949b: 125).

[18]   'Je größer ein 'Krater', desto geringer ist wohl die Wahrscheinlichkeit, daß es sich um einen irdisch-himmlischen 'Meteor'krater handelt: Einmal ist die entsprechende Entstehung der gewaltigen Mond'krater' immer noch umstritten, höchstens wahrscheinlich meteoritisch.

Kranz continued to comment on the conference talks and publications by Rohleder and Stutzer (Chapter 2) with the usual arguments: the occurrence of the mineral aragonite was seen as a sign of thermal springs, which must have been active for a much longer time than seemed reasonable for a meteorite impact. Aragonite was thus regarded as indicating volcanic activity. Kranz also pointed out that Stutzer and Rohleder themselves deviated in their interpretation of the Ries Basin: Rohleder considered the Steinheim Basin alone to be an impact crater, whereas Stutzer advocated an impact origin for the Ries Basin as well. Kranz then mentioned the 'Swabian Volcano', the contemporaneous 300 or so diatremes in the vicinity of the town of Urach, and the special, 'geologically predisposed' Ries area to show how unlikely it would be for a meteorite to hit just there (at this special location) and then (during a time when southern Germany was dotted with volcanoes).

The *Strahlenkalk* [ray-limestone; today recognised as well-developed shatter cones] of the Steinheim Basin, were interpreted by Kranz as dissolution structures caused by volcanic gases that were pressurised, hot and acidic.

Kranz claimed that the groundwater at the time of the Steinheim event was not in contact with the Earth's surface and 'thus no proper 'karst water' in the modern sense' (Kranz 1937b: 199).[19]

This somewhat cryptic remark allowed Kranz to infer that Stutzer's hypothetical meteorite did not have the chance to encounter groundwater (Stutzer had assumed that a water vapour explosion triggered by the interaction of an incandescent meteorite with groundwater was the cause of the crater basin). On the other hand, it also permitted Kranz to infer that the solid, 170-metre cover of Jurassic limestone might have allowed pressure to build up in deeper groundwater, volcanically heated from below as in a pressure cooker, until it gave way in a gigantic volcanically triggered explosion.

Kranz additionally pointed out that meteoritic material was missing from the two German structures, but he also fairly admitted that this material might have since been eroded from the surface. Kranz then took a sharp and (based on his previous argument) unexpected turn, as he did not want to exclude the possibility of a meteorite impact altogether. Like Barringer for Meteor Crater,

---

Und dann sind die morphologischen Erscheinungen auf dem Mond von den irdischen selbst unter Berücksichtigung des Fehlens von Luft und Niederschlägen so verschieden, daß sie nur mit starkem Zweifel verglichen werden können. Ich gehe also von den *irdischen* aus, und die auf Erden nachgewiesenen 'echten' Meteorkrater sind *verhältnismäßig klein*; einer der größten (oder der größte?), der von Arizona, hat rd 1200 m Durchmesser, die 'Indizien' weisen auf seine 'himmlische' Entstehung hin, selbst bei ihm gab es noch Zweifel, aber – in dubio pro reo – ich möchte ihn hier für einen echten Meteorkrater halten, um Vergleiche zu ermöglichen. Das Steinheimer Becken hat rd 3 km Durchmesser, das Ries rd 21–24 km, es war also im Mittel 18½mal größer als der Arizonakrater ...' (Kranz 1937b: 198).

[19] 'also kein eigentliches 'Karstwasser' im jetzigen Sinn'. (Kranz 1937b: 199).

he calculated from the crater size how much Ni-bearing iron might be buried in the Steinheim Basin, if Stutzer were correct:

> This would provide a new possibility to develop resources, including valuable minerals, from deep drilling in the Steinheim Basin. While such deep drilling projects have already been desirable from a geological point of view, they are now even more so, as they might be profitable economically as well. And only by such projects can it be decided which theory is correct. Meanwhile, because of the stated reasons, the solely volcanic blasting theory still seems to me to be the most probable. (Kranz 1937b: 200)[20]

Although for the smaller Steinheim Basin Kranz admitted the need for closer investigation based on the small chance that it could be meteoritic in origin (despite all arguments to the contrary), he saw much greater difficulties at the much larger Ries Basin. The sheer size of the Ries Basin did not allow a direct comparison to Meteor Crater in Arizona, the largest tentatively recognised meteorite crater known by Kranz. Apart from this problem of size, Kranz was on the other hand desperately aware of the lack of knowledge concerning caldera volcanoes, thus complicating the comparison of the Ries Basin with structures that were definitely volcanic:

> What are termed – after the Caldera at Las Palmas, which according to Gagel was formed by *erosion* (!) – as giant 'calderas', are of quite varied *volcanic* origin; and the 'caldera problem' is itself not yet solved). The Ries and Steinheim Basins, it seems, are in several respects built differently from a number of so-called 'calderas' …

> It remains to be seen whether giant 'calderas', like Aso and Kuttyaro in Japan, Oregon Crater, Lake of Bolsena and so on, have been formed mainly by subsidence. At the *Ries* (after my blasting experiment in 1911) it may be assumed, until proof of the opposite, that the enormous explosion that formed the central shallow, crater-shaped bowl has also blasted up to a certain extent the 5–6 km wide rim zone of the Ries Basin … . From a technical, blasting point of view, it

---

[20] 'Damit ergäbe sich aber eine neue Möglichkeit, Bodenschätze, auch wertvolle Mineralien, durch Tiefbohrungen im Steinheimer Becken zu erschließen. Wenn solche Tiefbohrungen schon bisher vom geologischen Standpunkt zu wünschen waren, so sind sie es jetzt noch mehr; sie könnten auch wirtschaftlich ergebnisreich werden. Und nur mit solchen kann entschieden werden, welche Theorie richtig ist. Einstweilen scheint mir aus den genannten Gründen die rein vulkanische Sprengtheorie noch immer die wahrscheinlichste'. (Kranz 1937b: 200).

seems not permissible to assume here exclusively or mainly an area of subsidence. (Kranz 1937b: 200–201)[21]

Kranz was aware that Ernst Werner had tentatively suggested a meteoritic origin for the Ries Basin (Chapter 2) and Kranz had read Kaljuvee's book. Kranz had rejected Werner's idea in the early 1920s, because no meteoritic material had been found at the Ries Basin and the enormous mass of debris surrounding the Ries seemed to Kranz impossible to explain by meteorite impact. For obvious reasons, Kranz did not take Kaljuvee seriously and he was only mentioned for the sake of comprehensiveness. Kranz continued to list the usual arguments and then pointed out the general similarity between the explosion mechanism suggested by Stutzer and that of his own theory. Why should there be the need for some device to trigger the explosion from outside? As Kranz saw it, he was using 'Ockham's razor'.

Kranz pointed out that suevite seemed to have been produced in several eruptions rather than a single event (Löffler 1926; Ahrens 1929) and that hot springs and their aragonite precipitate (travertine), just like in Steinheim Basin, required more time and thus a continuous latent heat for their formation. At the most, Kranz could imagine a meteorite impact that had caused subsequent volcanism. Although this scenario was even more unlikely than an impact at the Steinheim Basin, here too Kranz pleaded for deep-reaching drilling – 'for economic reasons' – because there remained a small chance that the Ries Basin might contain 'valuable mineral resources' (Kranz 1937b).

Although Kranz would not completely exclude the possibility of a meteorite impact, he clearly did so only to argue for the necessity of a scientific drilling campaign, which unfortunately, he failed to obtain. He certainly expected such a drilling campaign to support the volcanic theory and provide new data to understand the volcanic explosion mechanism. The impact idea was only used as the 'carrot in front of the donkey', because it might impress the administrators with a potential economic output (in which Kranz, however, did not truly believe).

---

[21] 'Was man – nach der Kaldera auf Las Palma, die nach Gagel durch *Erosion* entstand (!) – als Riesen'kalderen' bezeichnet, ist recht verschiedener *vulkanischer* Entstehung; und das 'Kalderaproblem' ist auch jetzt noch nicht gelöst). Ries (und Steinheimer Becken) sind offenbar in vieler Hinsicht anders gebaut als manche sogenannte 'Kalderen' ... Es bleibt abzuwarten, ob Riesen'kalderen', wie Aso und Kuttyaro in Japan, Oregon-Krater, Bolsena-See u.a., hauptsächlich durch Einsenkung entstanden. Beim *Ries* darf (nach meinem Sprengversuch 1911) bis zum Beweise des Gegenteils angenommen werden, daß die gewaltige Explosion, welche das mittlere flachtrichterförmige Becken schuf, auch die 5–6 km breite Randzone des Rieskessels bis zu einem gewissen Grad ausgesprengt hat ... . Sprengtechnisch erscheint es nicht zulässig, hier nur oder hauptsächlich ein Senkungsgebiet anzunehmen'. (Kranz 1937b: 200–201).

Despite all of the rebuttals that Kranz and others had conscientiously arranged against the impact hypothesis, there remained a constant trickle of people who kept the impact idea alive, making it hard to understand why Ries experts considered this idea so far off that it was never seriously discussed alongside the volcanic hypothesis.[22] Together, Kranz and Quiring had the key! Why did they not collaborate and try to figure out what really happened? The question becomes even more pressing due to the fact that Kranz himself was not completely averse to the impact idea as such, as he had published about impact craters in Estonia.

---

[22]   The situation had been quite different in other countries, such as Canada (see Plotkin and Tait 2011).

Chapter 5
Kaalijärv Crater and Köfels Landslide

Figure 5.1   Kaalijärv Crater on the Estonian island of Saaremaa in the Baltic
Sea. (Photo courtesy of Tõnu Pani, University of Tartu, Estonia).

Remarkably, Walter Kranz published a brief review paper in 1937 on the
Kaalijärv Crater (Figure 5.1) on the Estonian island of Saaremaa (or Ösel)
in the Baltic Sea. This is a recent impact site, possibly some 2,400 years old
(Rasmussen et al. 2000) and especially interesting because local songs and
stories collected in the Finnish epos Kalevala may contain fragmentary
eyewitness accounts of the event. These stories describe a young sun that
fell from the sky, causing uproar of the sea and a huge wildfire devastating
the region (Fromm 1985: 302, 309). The event described was the impact of
fragments of a large iron meteorite, which resulted in at least nine craters
on Saaremaa. The largest of these craters, Kaalijärv, is approximately 110
m in diameter and about 22 m deep. Its rim is raised 4 to 7 m above the
surrounding area. The energy released was about the same order of magnitude

as the Tunguska event. The other structures on Saaremaa range from 12 to 40 m in diameter.[1]

The first published interpretations of the Estonian craters as meteorite craters were by Ivan Reinwaldt[2] (1928) and by Rudolf Meyer and Alfred Wegener (Kraus et al. 1928):

> The plan for this endeavour followed from a correspondence between R. Meyer and A. Wegener, in which the former expressed the assumption that the crater near Sall might have been generated by the impact of a meteorite. The opportunity for investigation arose when, in autumn 1927, Wegener was called by the Herder Society to give holiday lectures in Riga. Upon his application, he and Meyer were granted a travel allowance of 300 *Reichsmark* by the *Notgemeinschaft der Deutschen Wissenschaften*[3], which is to be thanked for it. The Estonian government also supported us significantly, because following a petition by Meyer for permission to dig it advised its mining engineer Reinwaldt, who had already been at the crater occupied with drillings and diggings, to assist us in any way. Shortly before departure, E. Kraus and his assistant, the student N. Selle, joined our enterprise. (Kraus et al. 1928: 316–17)[4]

Both Rudolf Meyer and Ernst Carl Kraus had been professors at Riga at that time. Meyer had been an associate professor at the Riga Polytechnic in 1910. After World War I, he became a full professor at the Herder Institute, a private university in Riga, working in the field of fisheries and marine research. In 1935, he became the first director of the Institute for Geophysics and Meteorology

---

1   <http://muinas.structuur.ee/projektid/ecp/kaali/en/html/facts.html> accessed 4 October 2011.

2   Ivan Reinwaldt (1878–1941), an Estonian mining engineer, had been sent to Saaremaa to prospect for gypsum and salt as well as to investigate the origin of the craters. In 1937, he discovered the first meteorite fragments at two of the smaller craters (<http://muinas.structuur. ee/projektid/ecp/kaali/en/html/geology.html> accessed 4 October 2011).

3   The Notgemeinschaft der Deutschen Wissenschaften was the predecessor of the Deutsche Forschungsgemeinschaft (DFG), the German science foundation.

4   'Der Plan dieser Untersuchung entsprang einer Korrespondenz zwischen R. Meyer und A. Wegener, in welcher ersterer die Vermutung aussprach, daß der Krater bei Sall durch Aufsturz eines Meteoriten entstanden sein könnte. Die Gelegenheit zur Ausführung bot sich, als Wegener im Herbst 1927 von der Herder-Gesellschaft zu Ferienvorlesungen nach Riga aufgefordert wurde. Auf seinen Antrag bewilligte die Notgemeinschaft der Deutschen Wissenschaften in dankenswerter Weise für ihn und Meyer einen Reisezuschuß von 300 RM. Auch die estnische Regierung unterstützte uns sehr, indem sie auf einen Antrag Meyers auf Erlaubnis zu Grabungen ihren beim Krater mit Bohrungen und Grabungen bereits beschäftigten Bergingenieur Reinwaldt anwies, uns in jeder Hinsicht behilflich zu sein. Kurz vor der Abreise schloß sich E. Kraus mit seinem Assistenten stud. N. Selle der Unternehmung an' (Kraus et al. 1928: 316–17).

at the newly founded University of Latvia. In 1939, Meyer, like many others of German origin, 'repatriated' to the 'Third Reich' (Ervins Luksevics, Riga, e-mails to the author on 26 July and 2 August 2011).

Kraus (1889–1970) obtained his qualification for professorship from the Technical University in Munich. He served during World War I as a military geologist. In 1919, he moved to the University of Königsberg, where he became a professor in 1922. From 1924 to 1935, he was Professor of Geology in Riga, but then switched to the industry. In 1937, he became responsible for the Military Geology Office of the German military. In 1942, he again became Professor of Geology, this time at the Ludwig Maximilian University in Munich, a position he held until 1945 when he left the university during denazification. After denazification was officially ended in 1953, Kraus obtained full status as a professor emeritus and again lectured in Munich, also supervising PhD students.[5]

Kraus, Meyer and Wegener reported the results of their investigations and reviewed various previous hypotheses concerning the genesis of the Kaalijärv crater (a maar-type volcano, clay-deposits increasing in volume through the intake of water, a subterranean methane-explosion or a doline). Ernst Carl Kraus, the first author of the paper, opted for dolines as a cause for the smaller pits. However, he regarded the Kaalijärv crater as formed by a salt diapir mobilised by the weight of ice during the Ice Age. The rising salt would cause local elevation, followed by more intense erosion at that spot. The exposed salt was then dissolved, followed by sagging and crater formation (Kraus et al. 1928). Kraus's co-authors, Meyer and Wegener, admitted that, although this might well be a plausible scenario, they, however, rejected it. The morphological similarity to Meteor Crater in Arizona – including the raised beds at the crater rim, the missing volcanics and the fine rock-flour, which Kraus regarded as the residue of salt solution – in their opinion argued for an impact origin of the Kaalijärv crater. The main crater, together with the adjacent field of small craters, or rather percussion pits, also appeared similar to the crater-pocked area left by the 1908 Tunguska event in Siberia. In Tunguska too no meteoritic material had been found, although the fall of the meteorite had been observed[6] (Kraus et al. 1928: 364).

---

5   Freyberg 1974: 86; <http://www.geologie.geowissenschaften.uni-muenchen.de/geologie/geschichte/leiter/Kraus/index.html> accessed 5 April 2011.

6   Fragments of the Kaalijärv meteorite were unknown until some were found in 1937 by Reinwaldt, shortly after Kranz published his paper. It seems that most of the meteoritic iron is lost, because it was collected and excavated for many centuries by the locals and fashioned into weapons and other artefacts. Nevertheless, the Mineralogical Collection at the University of Tübingen did contain a 'meteorite from Ösel'. It had been part of a meteorite collection that the university received in 1869 as a present from the collector Karl Ludwig Reichenbach (1788–1869). It was mentioned specifically in letters that had been exchanged between Reichenbach and the university officials prior to the donation treaty (Engelhardt

Meyer and Wegener did not discuss the possible explosion mechanism of the proposed Kaalijärv impact. This was done by Reinwaldt. According to Reinwaldt, it was Wegener who had first brought up the idea of an impact, during his five-day visit with Meyer, Kraus and Selle in September 1927 (Reinwaldt 1928: 5).

In addition to the field evidence, Reinwaldt had been considering whether a falling meteorite would be capable of causing explosive phenomena at all. As a result of his theoretical reasoning, he unfortunately entangled himself in some of the physical pitfalls of the explosion mechanism:

> According to various observers, the velocity of meteorites when entering the Earth's atmosphere varies between two and sixty miles (3 to 96 kilometres) per second. 412 feet to four miles (6.5 km) per second have been observed as velocity upon impact. Thus, the velocity varies in a very wide range.

> Among other things, Merrill[7] also reflects on meteorite velocities in relation to the direction of movement of the Earth: If the meteorite overtakes the Earth, it may rest itself on the Earth's surface[8] nearly without damaging it; in the adverse case, it may impact with elementary force.

> ... Of course, these differences in velocities and fall conditions also lead to very different results if a meteorite collides with the Earth.

> Therefore, we want to imagine which phenomena accompany the fall of a meteorite that approached the Earth with maximum velocity; that is some 6.5 km/s.[9] After entering the atmosphere, it heats up when overcoming air resistance, slows its flight[10] and causes air waves in front, to the sides and behind it.

---

and Hölder 1977: 30, 75). This was obviously unknown to Kranz or anybody else involved in the discussion.

[7]   George Perkins Merrill (1854–1929), an American geologist and mineralogist who wrote many papers on meteorites (<http://en.wikipedia.org/wiki/George_P._Merrill> accessed 21 July 2011).

[8]   This is nonsense. If the meteorite follows the Earth from behind, with a relative velocity close to zero, it nevertheless possesses potential energy that upon falling onto the Earth is converted into kinetic energy (the meteorite accelerates). If we neglect the braking effect of air resistance, the meteorite will reach escape velocity (11.2 km/s) plus initial relative velocity upon impact. The larger the meteorite, the less its velocity will be affected by air resistance, because its mass (and thus kinetic energy) increases with the cube of the projectile radius while its cross-section, relevant to its air resistance, increases only with the square of the radius.

[9]   See the previous footnote. This is less than minimum velocity for a large enough projectile.

[10]   This contradicts even Reinwaldt's own reasoning, because he claimed this velocity to be the maximum observed upon impact (that is after it has been slowed down by air

... It is possible that the air, which is compressed in front of the meteorite, will penetrate into the rocks to be destroyed and will act explosively upon the sudden reduction of velocity; and this the more intense the harder the surrounding is. Upon impact thermal effects are unavoidable, as the glowing meteorite is bringing with it a considerable amount of heat, which is increased further by the heated air and the heat production of the impact. If the glowing meteorite comes into contact with the rock moisture or especially the groundwater, large amounts of water vapour must suddenly be generated, which additionally contribute to the explosiveness.

How might the meteorite itself behave in this case? The formation of a large temperature gradient between its surface (molten) and its interior (temperature of absolute zero) can develop tensions, which may blast it, as it is also often observed ...

The specific appearance of the phenomenon is dependent on many individual conditions: 1) on the type of meteorite, 2) on its size and form, 3) on the properties of the rocks that take in the meteorite (their hardness, bedding, water content and so on), 4) on the angle of impact and so on.

Also, if in the case of insignificant velocity the meteorite can rest itself on the Earth's surface without penetrating into it, so the impact in the opposite case can be accompanied by an explosion and form depressions of the type described above. (Reinwaldt 1928: 20–21)[11]

---

resistance, not before!).

[11] 'Nach den Angaben verschiedener Beobachter schwankt die Geschwindigkeit der Meteoriten beim Eintritt in die Erdatmosphäre von 2–60 Meilen (= 3–96 km) in einer Sekunde. Als Geschwindigkeit beim Einschlag sind 412' bis 4 Meilen (6,5 km) pro Sekunde beobachtet worden. Also schwankt die Geschwindigkeit in sehr großen Grenzen. // Unter anderem stellt Merrill Betrachtungen an über Meteoritengeschwindigkeiten im Verhältnis zur Bewegungsrichtung des Erdkörpers: wenn der Meteorit die Erde überholt, kann er sich auf die Erdoberfläche legen, fast ohne sie zu beschädigen; im entgegengesetzten Fall kann er mit elementarer Gewalt aufschlagen. // ... Diese Verschiedenheiten in den Geschwindigkeiten und Fallbedingungen bringen natürlich auch sehr verschiedene Resultate eines Zusammenstosses des Meteoriten mit der Erde mit. // Darum wollen wir uns vorstellen, welche Erscheinungen den Fall eines Meteoriten begleiten, der mit der maximalen Geschwindigkeit, also etwa 6,5 km/sec, sich der Erde nähert. Nachdem er in die Atmosphäre eingetreten ist, erhitzt er sich beim Überwinden des Luftwiderstandes, verlangsamt seinen Flug und erzeugt vor, neben und hinter sich Luftwellen. ... Es ist möglich, dass die vor dem Meteoriten zusammengepresste Luft in die zu zerschlagenden Gesteine eindringt und bei der plötzlichen Geschwindigkeitsabnahme des Meteoriten sprengend wirkt, und dabei desto stärker, je härter die Umgebung ist. Beim Aufschlag sind auch thermische Effekte unvermeidlich, da der glühende Meteorit bedeutende Wärmemengen mit sich bringt, die

Reinwaldt considered shockwaves to be the main cause of crater formation, greatly assisted by groundwater that flashed into steam upon impact. When Kranz took up the issue, only the latter effect remained.

Walter Kranz as Impactist

Kranz's paper is interesting in several respects. He freely admitted that he himself was not personally familiar with the craters. Instead, he voiced his opinion following descriptions of the craters in the geological literature. One of the sources he cited is the Estonian teacher Kaljuvee, who had described the Ries Basin as an impact crater (see Chapter 2). And indeed, Kranz too compared the Kaalijärv crater not only to Meteor Crater in Arizona but also to the Ries Basin and found that brecciation in Estonia and Germany was remarkably similar. At both locations Kranz considered it to be the result of a central explosion; impact-generated in Estonia and possibly Arizona, but volcanically triggered in Germany: 'Thereby, from an explosion-technology point of view, it would be optional, whether the explosions were 'triggered' by penetration of celestial explosives and detonators from above or of terrestrial ones from below' (Kranz 1937a).[12]

Kranz was reasoning from morphological analogy to his small, experimental explosion crater (see Chapter 1), but he did not understand the physics behind an impact. Rather, he was a practical man, falling back on his experience as a military pioneer in World War I (Kranz 1928):

---

durch die erhitzte Luft und die Wärmeproduktion des Aufschlagens noch vermehrt werden. In dem Falle, wenn der glühende Meteorit in Berührung mit der Bergfeuchtigkeit oder besonders mit Grundwasser kommt, müssen plötzlich grosse Mengen von Wasserdampf entstehen, die zur Sprengwirkung noch beitragen können. // Wie könnte sich dabei der Meteorit selbst verhalten? Die Entstehung eines großen Temperaturintervalles auf seiner Oberfläche (geschmolzen) und im Innern (T des absoluten Nullpunktes) kann Spannungen entwickeln, die ihn sprengen können, was oft auch beobachtet wird. ... // Der Habitus der Erscheinung hängt von vielen Einzelbedingungen ab: 1) von der Zusammensetzung des Meteoriten, 2) von seiner Größe und Form, 3) von den Eigenschaften der Gesteine, die den Meteoriten auffangen (ihre Härte, Lagerung, Wassergehalt usw.), 4) von [sic] Auffallwinkel des Meteoriten u. a. m. // Also, wenn bei unbedeutender Geschwindigkeit der Meteorit sich auf die Erdoberfläche auflegen kann, ohne in sie einzudringen, so kann der Aufstoss im entgegengesetzten Fall von einer Sprengung begleitet sein und Vertiefungen von obenbeschriebener Art bilden' (Reinwaldt 1928: 20–21).

[12] 'Dabei wäre es rein sprengtechnisch belanglos, ob die Explosionen durch Eindringen himmlischer Spreng- und Zündungsmittel von oben oder irdischer von unten 'gezündet' wurden' (Kranz 1937a).

Already years ago, I had wished that volcano researchers would obtain at least a minimum of explosion-technology experience on the pioneer test sites. And it would not be amiss if it were also introduced to the university education for geologists in general – during the World War many geologists were able to convince themselves of that. (Kranz 1928: 260)[13]

In his technical papers, Kranz distinguished two types of explosions: simple mechanical ones accompanied by pressure-release of compressed gases and chemical reactions accompanied by a rising of pressure; the former being much less effective than chemical explosions (Kranz 1928: 261). Whereas ordinary volcanic explosions were of the former, less effective type, this scenario seemed not reconcilable with the gigantic Ries-explosion, which required "triggering', chemical reaction and increase of pressure' (Kranz 1928: 262).[14]

The wide and shallow explosion funnels of the Steinheim and Ries Basins required a supercharged detonation of explosives – the chemical nature of which Kranz did not discuss – assembled under a dense cover of rock (Kranz 1928: 266, 282–3). However, the required charge could not be calculated, because such things were up to the experience of the explosives expert and for such large explosions there was simply no experience available. The larger the charge, however, the more effective the detonation became (due to a non-linear relationship) and so the charge would be smaller than intuitively expected (Kranz 1928: 286–7).

Like many of his contemporaries, Kranz was not aware that an impacting projectile must explode given enough mass and velocity. Just as Stutzer did for the Ries and Steinheim Basins, Kranz thought the Estonian crater was caused by mechanical fragmentation of the country rock and exploding water vapour or oxyhydrogen gas when a glowing meteorite encountered groundwater. He did not think about the balance of energy release and temperatures reached via his mechanical model when he speculated: 'Perhaps we could attribute an absence of iron meteorites near Sall ... to evaporation of celestial bodies on hitting the ground and exploding' (Kranz 1937a).[15]

This is precisely what happens at craters as big as the Ries or the Steinheim Basin, but only because kinetic energy of the meteorite is transformed into heat

---

[13]   'Ich hätte schon vor Jahren gewünscht, die Vulkanforscher möchten sich auf Pionier-Übungsplätzen wenigstens ein Mindestmaß an sprengtechnischer Anschauung erwerben, und es könnte nicht schaden, wenn das auch im Hochschulunterricht für Geologen allgemein eingeführt würde – im Weltkrieg haben sich ja wohl viele Geologen davon überzeugt' (Kranz 1928: 260).

[14]   "Zündung', chemische Umsetzung und Drucksteigerung' (Kranz 1928: 262).

[15]   '[Das] Fehlen von Eisenmeteoriten bei Sall könnte vielleicht auf ... [das] Verdampfen von Himmelskörpern bei Aufprall und Explosion hinweisen' (Kranz 1937a).

– and not, as Kranz thought, because of a water vapour explosion by simple heat transfer, which would never become hot enough.

Whereas Kranz could imagine a meteorite impact at the comparatively tiny Kaalijärv crater, even without evidence of meteoritic material in the surrounding area, he did not contemplate such an origin for the two much larger German explosion sites, possibly because there is no sensible way to scale up his Kaalijärv model to the much larger size of the German localities. Here, so he thought, the energy must have come from within the earth. What else could it have been?

Franz Eduard Sueß and the Köfels Landslide

Kranz had been inspired to write his Kaalijärv paper by the famous Austrian geologist Eduard Sueß (1831–1914). Sueß's son, Franz Eduard Sueß (1867–1941) was also interested in the impact phenomenon. He received his PhD in 1892 with a stratigraphic/palaeontological dissertation. Later in life, however, he became increasingly interested in crystalline rocks and mountain-building processes. In 1893, the younger Sueß became a geologist at the Austrian Geological Survey. From 1908 to 1910 he was Professor of Geology at the Technical University of Prague and later became Professor of Geology at the University of Vienna until his retirement in 1938 (Kölbl 1945; Waldmann 1953; Chapter 7).

At the same time Kranz was working on his Kaalijärv-paper, Sueß was also working on an impact paper. For Sueß it was the huge Köfels landslide in the Ötztal (Austrian Alps) (Erismann et al. 1977) that had attracted his attention.

The village of Köfels sits halfway up the mountain on top of the landslide mass in a conspicuous, terrace-like widening of the valley (Figure 5.2). In a fissure near Köfels, silica-rich melt solidified into vesicular, pumice-like masses (Figure 5.3); but elsewhere within the landslide debris denser glasses can also be found.[16]

Sueß began by reviewing the opinions of previous workers. Eberhard Fraas in the early 1920s had described the pumice-like masses as acidic lava. According to the 'volcanologist Reck'[17] the terrace of Köfels represented a subsided caldera, while Kranz had written of a gigantic central explosion (Sueß 1937).

Sueß was against comparing the Köfels terrace to the Ries Basin:

---

[16] The Köfels terrace is no longer interpreted as an impact crater, but as a gigantic landslide provoked by the extremely steep and high valley when the supporting glaciers had melted after the Ice Age. The Köfels melts are now interpreted as frictionites, melts generated by the heat due to the intense friction upon slumping. Such melts are also known from other large landslides, for example in the Himalayas (Erismann et al. 1977).

[17] For Hans Reck see Chapter 3.

Figure 5.2   The Köfels landslide in the Ötztal, Austria, was interpreted by
              Walter Kranz as caused by a volcanic explosion, whereas Franz
              Eduard Sueß opted for a meteorite impact. (Photo: Kölbl-Ebert).

Figure 5.3   Köfelsite, a vesicular, slack-like glass from the Köfels landslide. The
              glass was formerly thought to be the product of an obscure type of
              volcanism, while Franz Eduard Sueß interpreted it as impact melt.
              (Photo: M. Ebert).

On the origin of the Ries too, rather diverging opinions have been voiced. Following some 200 papers on the topic, some of the peculiarities that have remained controversial up to the present are now eventually clarified. For the purpose of this paper, however, it is only important that the Ries originated in a *completely different geological surrounding* than Köfels and thus under conditions that cannot be applied to Köfels and therefore cannot be used to explain Köfels. (Sueß 1937)[18]

Sueß subscribed to Kranz's genetic Ries model and saw in it a gigantic volcano that had experienced a central explosion. As to the interpretation of Kaalijärv and the other craters on Saaremaa as meteorite craters Sueß too remained sceptical, but not so for the Köfels terrace. Sueß considered this (then enigmatic) geological feature to be of impact origin and developed the following scenario:

> The harder the impact was, the more intense the sudden conversion of kinetic energy into heat. From those parts [of the meteorite] that were evaporated by highest pressure upon impact, the rest of the meteorite is reflected explosively. By the shock as well as the sudden heat spilling over [from the meteorite to the target] and possibly also from the tension built up by the evaporated moisture in the rocks, the mountain is also blasted from within. At the same time, it begins to glow and its surface is covered by a skin of gas-rich, bubbling melt. Strongly overheated and very mobile, it trickles over the slopes, tearing debris from the surface, and while it takes in more substances from the surface through melting, it partly gushes into the fissures torn up just a few moments ago.

> The shaken and fragmented rock mass begins to waver and, by breaking into large blocks and debris, it flows down and blocks the valley floor as a towering dam. (Sueß 1937: 151)[19]

---

[18] 'Auch über die Entstehung des Rieses sind recht weit auseinandergehende Meinungen geäußert worden. Im Anschluß an ein etwa 200 Arbeiten umfassendes Schrifttum vollzieht sich eine allmähliche Klärung gar mancher bis in die Gegenwart umstritten gebliebener Besonderheiten. Für die vorliegende Betrachtung ist es aber nur wichtig, daß das Ries in einer *ganz anderen geologischen Umgebung*, als Köfels, und deshalb unter Bedingungen entstanden ist, die für Köfels nicht gelten können, und daß es deshalb zur Erklärung von Köfels nicht verwendet werden kann' (Sueß 1937).

[19] 'Um so heftiger war der Anprall und um so ausgiebiger der plötzliche Umsatz der kinetischen Energie in Wärme. Von den im Aufschlag unter Höchstdruck vergasten Teilen wird der Rest des Meteoriten explosionsartig zurückgeschleudert. Durch den Schlag, durch die unvermittelt überströmende Hitze und wohl auch durch die Spannung der vergasten Gebirgsfeuchtigkeit wird der Berg auch von innen her zersprengt. Zu gleicher Zeit glüht er auf und seine Oberfläche überzieht sich mit einer Haut von gasreicher brodelnder Schmelze. Stark überhitzt und sehr beweglich rieselt sie über die Hänge, reißt den Schutt der Oberfläche

The paper by Sueß is of more than anecdotal interest, because the author also reviewed what he at that time knew and understood about impact mechanisms and the physics behind them. For example, Sueß was aware that '[t]he extraordinary effect [of an impact] is due to [the fact that] meteorites above a certain size reach the solid Earth with close to the full velocity they bring from outer space, as their kinetic energy is not consumed by air resistance' (Sueß 1937: 137–8).[20]

However, Sueß then astonishingly stepped back from this important insight, ignoring it as he continued to write at length about small bodies that are effectively braked within the Earth's atmosphere to some 70–100 m/s and penetrate soft ground for a couple of metres at the utmost (Sueß 1937).

In any case, it seems Sueß regarded rock as rigidly solid and unmovable by a meteorite, no matter its velocity:

> Tied to it is the question of the relation of mechanical work and heat in the transformation of the energy suddenly released upon impact. It depends on the condition of the target: depending on the specific gravity and rigidity of the resisting material, either work or heat will be more prominent (Sueß 1937: 138)[21]

As his impact scenario for Köfels shows, it did not occur to Sueß that the bolide would penetrate the ground prior to explosion. Instead, he envisioned transfer of heat and possible explosion only at the very point of contact, which would then immediately stop and reflect a shattered meteorite. In contrast, the underground target rock (as being a part of the solid Earth and thus constituting much more mass) would remain basically unaffected, apart from some melting of the surface at the point of contact and possibly some local blasting effect by water vapour expanding from within the heated rock – the heat having been transferred from the glowing meteorite and not generated within the target rock by compression.

---

mit sich und, indem sie aufschmelzend weitere Stoffe daraus aufnimmt, ergießt sie sich z.T. in die eben aufgerissenen Spalten. Die erschütterte und zerrüttete Bergmasse gerät ins Wanken und, indem sie in große Schollen und Trümmerwerk zerfällt, gleitet sie nieder und verbaut den Talgrund als hochaufragender Querriegel' (Sueß 1937: 151).

[20] 'Das Außerordentliche der Wirkung ist dadurch bedingt, daß Meteoriten über einer gewissen Größe mit nahezu der vollen aus dem Weltraume mitgebrachten Geschwindigkeit auf der festen Erde ankommen, da ihre Bewegungsenergie durch den Luftwiderstand nicht aufgezehrt wird' (Sueß 1937: 137–8).

[21] 'Daran knüpft sich die Frage nach dem Verhältnis des Umsatzes der mit dem Aufschlag plötzlich freiwerdenen Energie in mechanische Arbeit und in Wärme. Je nach der Beschaffenheit des Untergrundes, nach der Dichte und Festigkeit des widerstehenden Mittels wird Arbeit oder Erwärmung das Übergewicht gewinnen' (Sueß 1937: 138).

As his source, Sueß cited Öpik's English paper of 1936 and Fritz Heide's German essays on large meteorites (Heide 1933, 1934) (Chapter 6). From Ernst Julius Öpik (1893–1985), Sueß had learned that:

> The 'aerodynamic' pressure upon intrusion of the meteorite into a rock is 1000 times greater that the plastic limit of steel. By this process the bodies behave like liquids of various densities. ... During a deformation of such an order of magnitude, the meteorite cannot keep its connection; it will be torn into numerous fragments. (Sueß 1937: 138)[22]

From reading Sueß's paper, one gains the impression that he misunderstood Öpik. Sueß obviously thought that when the metallic mass of the meteorite behaved like a liquid upon hitting the target, this would preclude its penetrating the ground. However, despite what Sueß's quote suggested, Öpik did not write about impacting bodies of various densities, but instead clearly stated that 'the problem of a meteor impact is the case of the impact of a liquid drop of given density δ into a liquid medium of density ρ' (Öpik 1936: 5). Sueß did not see what to the trained physicist Öpik was self-evident: that according to the principle of '*action = reaction*', the forces working on the projectile were the same as those on the target, so that the target could not be expected to behave as rigidly and immovably solid no matter what was about to hit it.

Perhaps understandably, realisation also did not dawn on Sueß when he quoted Öpik literally: 'Further, the movement of deepest penetration hardly represents the final phase of the collision; subsequent movements, perhaps even an explosion due to heat action, may have diclocated [sic] the meteor fragments and mixed them with the other debris of the crater' (Öpik quoted in Sueß 1937: 139).

Although Öpik did speak of penetration – contrary to Sueß – he expressed himself rather cautiously concerning the explosive force of the process.

Sueß proceeded to estimate the temperatures reached at the hypothetical Köfels impact and the highest figure is 3,600°C (the boiling point of silicon dioxide as it was assumed at that time),[23] required for his 'bubbling' silica-rich melt. He continued by stating:

---

[22]   'Der 'ärodynamische' Druck beim Eindringen des Meteoriten in ein Gestein ist 1000mal größer als die plastische Grenze von Stahl. Bei dem Vorgange verhalten sich die Körper wie Flüssigkeiten mit verschiedener Dichte. ... Bei einer Deformation von solcher Größenordnung kann der Meteorit den Zusammenhang nicht bewahren; er wird in zahlreiche Stücke zerrissen' (Sueß 1937: 138).

[23]   Today, the boiling point of silica is given as 2,230°C (<http://www.inchem.org/documents/icsc/icsc/eics0808.htm> accessed 27 December 2011).

The heat equivalent of a mass of one million tons, corresponding to a diameter of 60–70 m, moving with cosmic velocity, some 60 km/s,[24] already provides (according to Spencer) temperatures of some million degrees centigrade. (Sueß 1937: 139)[25]

Thus, the melting would be no problem. In fact, the sudden heat at the point of contact must lead to sudden evaporation and local explosion, creating a circular crater:

> The angle of impact plays ... no role in the morphology of the crater, because the cavity is not formed by the impact, but by a central explosion: hence the circular form of the craters, the surrounding wall of debris, the up-pushing of beds at the crater rim ... and the expulsion of the rocks with meteorite fragments from the cauldron, which provokes the comparison with the mine craters of the war.

> Likewise the calculations of [Fritz] Heide lead to the result that already a meteoritic body of 10 m diameter has no chance to be left resting on the surface. It will be hurled back by the explosion and thereby totally or partially evaporated. (Sueß 1937: 139–40)[26]

Although Sueß spoke about craters, impact for him seemed to be basically a surface phenomenon. Whereas the projectile is evaporated, reflected and dispersed in the surrounding, the target rock in his model is shattered mechanically by the blow. It receives a coating of superheated meteoritic melt, which contains enough heat to incorporate a considerable amount of target material in a secondary process (hence the silica rich melt at Köfels and elsewhere). A transfer of heat from the incandescent cloud of expanding meteorite material helped

---

[24] That is within the realistic range, but three orders of magnitude larger than Sueß's examples above.

[25] 'Das Wärmeäquivalent für eine mit kosmischer Geschwindigkeit, d.i. etwa 60km/sec, bewegten Masse von einer Million Tonnen, entsprechend einem Durchmesser von 60–70 m, ergibt bereits (nach Spencer) Temperaturen von einigen Millionen Zentigraden' (Sueß 1937: 139).

[26] 'Der Einschlagswinkel spielt ... bei der Gestaltung des Kraters keine Rolle; denn die Höhlung wird nicht durch den Einschlag, sondern durch eine zentrale Explosion geschaffen. Daher die kreisrunde Form der Krater, der umgebende Schuttwall, der Aufstau der Schichten am Kraterrande ... und das Herausschleudern der Steine mit den Meteortrümmern aus dem Kessel, das den Vergleich mit den Minentrichtern des Krieges nahe legt. Im gleichen Sinne führen die Berechnungen von Heide zu dem Ergebnis, daß bereits ein meteorischer Körper von 10 m Durchmesser keine Aussicht hat, auf dem Boden liegenzubleiben. Er wird durch die Explosion zurückgeschleudert und dabei ganz oder teilweise vergast' (Sueß 1937: 139–40).

to fracture the target rock, and groundwater contained therein was likewise heated. For Sueß, heat was solely a property of the bolide, which, formerly in possession of kinetic energy, now contained an equal amount of thermal energy. The thermal energy could then be transferred to the target rock in contact with it, but Sueß did not realise that the energy release upon impact would be much more complicated, and that heat for example would also be generated within the target, which is utterly deformed and compressed while the bolide is still penetrating (Sueß 1937).

Despite his lack of physical understanding, Sueß did not finish before suggesting – like Quiring – a sensible diagnostic feature for impact craters. In melts of true impact craters he expected to find significant amounts of nickel from the metal vapours of the evaporated meteorite, even though the melts would be acidic due to additional melting of target rock; however, no such analysis had yet been undertaken at Köfels (Sueß 1937: 150) [or the Ries Basin].

Sueß had been pointed toward the significance of elevated nickel contents by the geochemist Victor Moritz Goldschmidt (see Chapter 7):

> I am especially interested in your interpretation of the pumice-occurrence at Köfels. Wouldn't it be possible to make quantitative analyses of nickel in the slags, compared with the original, unmolten rocks?
>
> In slags formed by the impact of meteorites, one finds in many cases a considerable elevation of the nickel contents, derived from the vaporized nickel-iron of the meteorite. The normal nickel contents of your gneisses will possibly be less than 0.001 per cent, whereas in such slags that have originated by the impact of meteorites, one could expect contents up to several tenths of a per cent of NiO. (Letter from Victor Moritz Goldschmidt to Franz Eduard Sueß, 3 July 1936, quoted after Franke 1983: 182)[27]

However, when Sueß gave a sample of the Köfels pumice to a colleague to be analysed, the result was negative (Franke 1983: 182).

Walter Kranz regarded the discussion about Köfels as still open. He himself opted for a weak volcanic explosion that triggered the landslide (Kranz 1939),

---

27 'Ganz besonders interessierte mich Ihre Deutung des Bimsstein-Vorkommens von Köfels. Wäre es nicht möglich, in den Schlacken quantitative Bestimmungen von Nickel vorzunehmen, verglichen mit den ursprünglichen ungeschmolzenen Gesteinen? // Bei den durch Aufprall von Meteoriten entstandenen Schlacken findet man in vielen Fällen eine wesentliche Zunahme des Nickel-Gehaltes, herstammend aus dem verdampften Nickeleisen des Meteoriten. Der normale Nickelgehalt Ihrer Gneise wird wahrscheinlich weniger als 0.001 % betragen, während man in solchen Schlacken, die durch Aufprall von Meteoriten entstanden sind, Gehalte bis zu Zehntel-Prozenten NiO erwarten könnte' (letter from Victor Moritz Goldschmidt to Franz Eduard Sueß, 3 July 1936, quoted after Franke 1983: 182).

while Otto Stutzer seconded Franz Eduard Sueß. Stutzer considered the Köfels landslide to be a surface feature, as the basement did not seem to be affected by the event and the Köfels pumice dyke was not a typical volcanic rock but had the same chemistry as the country rock (that is, it was molten and re-solidified gneiss). On the other hand, Stutzer thought the amount of debris was simply much too large to represent a mere ordinary landslide caused by gravitative failure of the rock at the steep mountain side (Stutzer 1936c).

Chapter 6
Impact Physics – Beyond Human Imagination

As we have seen, even for those who considered the impact idea for the Ries and Steinheim Basins as well as elsewhere, the physics behind an impact (see Melosh 1989) was alien to German-speaking geologists, despite the fact that since at least the late 1930s there had been trials to create feasible mathematical/physical models for smaller meteorite impacts and estimate their energy release. Until the 1960s, there were no initiatives in the German-speaking regions to develop a reasonable geological theory on the foundation of these physical models that would have been applicable to larger, enigmatic structures such as the Ries Basin. An important reason for this neglect was certainly the difficulty of imagining such a vast release of energy and its consequences.

In order to understand the problem, it is necessary to stray a bit into physics: small meteorites are effectively braked by friction within the atmosphere. After losing most of their cosmic velocity, they simply fall to the ground, just as a rock would do when dropped from an airplane. At the most, they mechanically excavate holes, so-called 'percussion pits'. However, very large cosmic bodies with a size of several tens of metres to several kilometres have so much mass compared to their cross-sections that they are hardly affected by air resistance. Their passage through the atmosphere takes only about one to four seconds, depending on their velocity and angle of impact. Upon impact, their energy is mostly transformed into heat, causing them to explode as soon as their momentum is spent and pressure is released. Thus, an impact crater is formed.

The mechanism of explosion has been described in the English literature since the turn of the twentieth century and was initially proposed and calculated in an approximate manner for lunar craters, where the physics is simpler because the Moon lacks an atmosphere. The model was later also applied to craters on the Earth.

Algernon Charles Gifford in particular, astronomer at the Hector Observatory in Wellington, New Zealand, made it clear (in the 1920s) that the sudden transformation of a projectile's velocity into heat would result in an enormous explosive force surpassing anything in human experience. This force, acting radially, also accounted for the general circularity of all lunar craters. Gifford explained that the impacting projectile 'is converted, in a very small fraction of a second, into an explosive compared with which dynamite and T.N.T. [tri-nitro-toluene] are mild and harmless' (quoted after Hoyt 1987: 185).[1]

Others too had argued for an explosive transformation of kinetic energy into heat, such as Gifford's friend (and fellow New Zealander) Alexander William Bickerton (1842–1929)[2] who had given lectures to that effect in 1915. The astronomer Ernst Julius Öpik and the physicist Herbert Eugene Ives had also published similar opinions in 1916 and 1919 respectively (Hoyt 1987: 185).

The physical principle used was simply the law of conservation of energy, and it was known since the mid-nineteenth century that 'heat' and 'work' are equivalent and interconvertible (Mark 1995: 50). Jules Verne's (1828–1905) novel *Journey around the Moon* nicely illustrates that by 1870 the idea had already entered popular culture, when the author has his protagonists discuss John James Waterston's (1811–83) idea that the heat of the Sun might be generated by impacting meteorites, which Waterston offered in 1853 (Verne 1966: 222; for Waterston see Mark 1995: 51). As Kathleen Mark wrote, 'although general acceptance of the fact that the impact of a high-speed meteorite upon the earth creates an explosion was slow, by the end of the 1930s it was widely accepted' (Mark 1995: 51).

The idea had by then already reached Germany, as becomes clear from a nice review paper on meteorites and meteorite craters in a popular astronomical journal (Heybrock 1934). The author was concerned with the as-yet-unsolved origin of tektites. He concluded, from the co-occurrence of tektites and meteorites around the Wabar (Arabia) and Henbury (Australia) craters, that tektites had most probably been 'a melt product of terrestrial desert sand and the great heat generated by the impact and reflexive explosion of iron meteorites' (Heybrock 1934: 52–3).[3] Heybrock also reported another interesting observation at the Henbury crater-field:

> A find from the smallest crater is remarkable. In more than two metres depth, there were four irons belonging together but separated by rust with a weight of altogether 200 kg. ... Around this crater 60 more, small iron pieces were found, but no glass. The situation allows the conclusion that the four iron pieces constitute the main mass of (and here the smallest) an iron meteorite, which was not affected by reflexive explosion, so that here the production of glass could not take place, which by the way as the investigation showed only happened around the largest

---

[1]   <http://www.teara.govt.nz/en/biographies/4g8/1> accessed 21 November 2011; compare Chapter 4.

[2]   <http://en.wikipedia.org/wiki/Alexander_William_Bickerton> accessed 21 November 2011.

[3]   ' ... die ein Schmelzprodukt aus terrestrischem Wüstensand und der durch den Einschlag und Rückexplosion von Eisenmeteoriten erzeugten großen Wärme darstellen' (Heybrock 1934: 52–3).

crater. This manner, that is the occurrence of tektites only in the surrounding of the largest crater, decidedly shows how much the production of glass is not only dependent on velocity but also on the mass of the impacting body. In connection with this, the absence of a reflexive explosion of the main mass in the smallest crater can be seen as a nice example for the necessity of cooperation of these two factors. It shows too, how varied the kinetic energy within a single meteorite swarm can be. (Heybrock 1934: 55)[4]

The Henbury craters had obviously been produced simultaneously by a meteorite swarm. Thus, the meteorites must all have had the same velocity upon entry into the Earth's atmosphere, but only the largest member of the swarm seemed to have had enough mass to bring the impact energy above a certain threshold to cause explosion.

As a simplified model, we can assume (just like Quiring: see Chapter 4), that the kinetic energy of the projectile ($\frac{1}{2} mv^2$) is converted into heat upon impact ($E_{kinetic} = E_{thermal}$). If we neglect air resistance, the minimum velocity corresponds to the escape velocity, which for the Earth is 11.2 km/s. That is, in any case, the velocity of the bolide is higher than the speed of sound (that is the velocity of seismic P-waves) in rocks. Thus, the energy cannot be dispersed as quickly as the bolide begins to penetrate the target rock, and it must be converted mostly into heat. The projectile and immediate target rocks are converted into incandescent plasma, which forcefully expands (it explodes).

This seems to be a quite straightforward model, accessible to anyone with a physical understanding on the level of an ordinary high school education. Nevertheless, the model did not appeal in Germany. As we have seen, even impactists such as Kaljuvee, Rohleder and Stutzer – or Sueß for Köfels and Kranz for Kaalijärv – promoted a totally different cause for the proposed explosion of impacting cosmic bodies. They all invoked superheated water vapour as the

---

[4]   'Bemerkenswert ist der im kleinsten Krater gemachte Fund. In über 2 m Tiefe befanden sich 4 zusammengehörende, durch Rost getrennte Eisen im Gewicht von insgesamt 200 kg. ... Rund um diesen Krater wurden noch 60 weitere kleine Eisenstücke gefunden, jedoch kein Glas. Die Sachlage berechtigt zu dem Schluß, daß die vier Eisenstücke die Hauptmasse eines (hier des kleinsten) Einzelmeteoriten darstellen, die von reaktiver Explosion nicht betroffen wurde, so daß hier auch die Produktion von Glas ausbleiben mußte, die im übrigen, wie die Befunde ergaben, nur um den größten Krater erfolgte. Dieses Verhalten, das Vorkommen von Tektiten ausschließlich in der Umgebung des größten Kraters, ist ungemein bezeichnend dafür, wie sehr die Produktion von Glas nicht nur von der Geschwindigkeit sondern auch von der Masse des auftreffenden Körpers abhängig ist. Das Ausbleiben der Rückexplosion des Hauptteils der Masse des kleinsten Kraters kann parallel damit als ein Schulbeispiel für die Notwendigkeit des Zusammenwirkens dieser beiden Faktoren angesehen werden. Es zeigt zugleich, wie verschieden die kinetische Energie innerhalb eines Meteoritenschwarms sein kann'. (Heybrock 1934: 55).

exploding agent, formed by contact of the hot, incandescent meteorite with groundwater.

As illustration, Kaljuvee (1933: 107) citing Reinwaldt (Chapter 5) reported a terrible accident in an iron foundry in which an iron nail covered with a few droplets of water fell into a vessel of molten iron, which then burst and exploded. However, for meteorite impacts, this argument from analogy is inadequate. Although expanding water vapour may certainly add to the explosion, the heat source is the kinetic energy of the projectile and not its temperature immediately before contact. In a meteorite descending through the Earth's atmosphere, only the thin outer surface is hot due to friction, whereas its interior still retains the cold temperatures of outer space. Consequently, in this explosion model we add a small amount of heat to a large amount of water, whereas Kaljuvee's foundry example describes a small amount of water encountering a large amount of heat.

The other way round, for example, dropping an incandescent iron nail into a bucket of water, will create a certain amount of fizz and steam but certainly no explosion to wrack the entire surroundings. Thus, for large impact structures, this model makes no sense. Why, however, was it so difficult to apply simple school physics?

Ernst Julius Öpik

In Europe, the physical modelling of impact structures as described above was done in the 1930s by two authors, the German mineralogist Fritz Heide (1891–1973) and the (then Estonian) astrophysicist Ernst Julius Öpik (1893–1985). Of these two Heide was the more widely read, as he published in German.

Öpik (Figure 6.1) had studied at Tallinn High School (Estonia) and Moscow Imperial University (Russia). After four years at Moscow Observatory, he became director of the astronomy department at Tashkent University. From 1921 to 1944 Öpik was an associate professor at Tartu University. During that time, from 1930 to 1934, he also travelled as a visiting scientist to Harvard University (USA). 'As former volunteer to the White Russian army, he vehemently opposed the Bolshevik Revolution and, when the Soviet occupation of Estonia was imminent, he moved, first to Hamburg, and lastly, in 1948, to Armagh Observatory [Northern Ireland]'. Öpik retired in 1981.[5]

Öpik's paper of 1916, written near the beginning of his career, showed that the conversion of energy upon impact caused an explosion creating a circular crater irrespective of the angle of impact. But as the paper was written in Russian during World War I, it had no great 'impact' (Mark 1995: 51–2; Hoyt 1987:

---

[5]   Quote and biographical information: <http://www.arm.ac.uk/history/opik/biog. html> accessed 7 June 2010.

Figure 6.1   The astrophysicist Ernst Julius Öpik published on impact physics. (Photo courtesy of Armagh Observatory).

196). However, Öpik returned to the topic in 1936. This time he published in English and was also read or reviewed to some extent in Germany and Austria (Chapter 5).

Öpik was well aware that a meteorite impact theory would influence the view of the lunar craters as well as of Meteor Crater in Arizona. He also mentioned Kaalijärv in Estonia as another known terrestrial impact crater (Öpik 1936: 3). As to impact physics, he did not repeat his considerations of 1916, instead proposing: 'Two different [alternative] methods for estimating the mass of the projectile, from penetration, and from the mass of crushed and ejected rock. ... In the case of the Arizona crater both methods point accordingly to a mass of about four million tons'[6] (Öpik 1936: 3).

While the first of these methods is rather hermetic to a non-physicist and involves numerous assumptions, the latter is rather simple and easy to understand:

> The mechanical work required to lift up the walls of the crater, throw out the fragments, and shatter and pulverize the rock, must represent a very small fraction of the total energy developed at impact. By setting this amount of work equal to the kinetic energy of the meteor, we get a minimum estimate of its mass. Judging from the distance at which fragments were found, it appears safe to assume that the mechanical work was equivalent to a lifting of all the mass involved to a height of 1200 metres. (Öpik 1936: 4)

Of course, this is just the usual and rather helpful manner of a physicist trying to simplify the problem so much that it becomes feasible. A non-physicist, however, might incorrectly take Öpik literally when he wrote about 'lifting' the mass. The image that comes to mind is rather that of a neat shovel than a gigantic explosion.

Öpik then calculated assuming realistic velocities of 20 to 60 km/s. However, the main problem with Öpik's estimates is his tendency to take crater shape and ejecta blanket as given. He was not aware of the potential severe alterations of the crater floor by elastic rebound after impact, by erosion or by sedimentation. For example, he took the flat floor of Kaalijärv – which was convenient for his physical argument – as a primary feature, and not due to the presence of lake sediments (Öpik 1936: 9).

Even though Öpik was well aware that the impacting projectile would move 'much faster than the velocity of sound of the medium' (Öpik 1936: 9), he seemed to think that the ensuing shock wave (which 'represents chiefly a sort of transfer of radial momentum over a continually increasing volume, with only a feeble transfer of matter': Öpik 1936: 9) would be able to disperse energy into

---

6   This corresponds roughly to an iron meteorite sphere of about 50 m radius.

the area surrounding the target rock fast enough to avoid highly focused damage and thus preclude more excessive explosions.

Because the two new methods were proposed by Öpik as alternative, independent approaches for calculating the size of an impacting projectile, throughout the paper he consequently avoided mentioning the more simple and straightforward approach of equating possible kinetic energies with impact heat (which he had treated already in his Russian paper of 1916). For a non-physicist such as Sueß, Kranz or other Ries geologists who only read this paper and not the earlier Russian one, this discretion precluded understanding of the basic physical problem; even more so because Öpik formulated cautiously when he mentioned fragmentation or explosion: 'because of the fluidity of the material there does not exist any definite surface of resistance; the impinging body changes in shape, the main feature of the change being evidently a flattening or broadening at right angles to the direction of motion, with a possible breaking into smaller pieces' (Öpik 1936: 5).

Consequently, there was 'little chance of expecting all that mass to be found somewhere in one large piece' (Öpik 1936: 11).

Öpik must have known that considering the energies involved both the projectile and its immediate target must have flashed into incandescent plasma during the process, and thus there could no longer be any brittle deformation of the projectile. Nevertheless, Öpik still expected there to be meteorite fragments that survived the process:

> at the final moment of the first phase of the impact (moment of deepest penetration), all this enormous mass of iron must have been flattened out over an area ... of about 330 metres diameter, with an average thickness of from five to seven metres; without doubt all this mass could not have kept together as a single piece, but must have been broken up into thousands of smaller fragments. Further, the moment of deepest penetration hardly represents the final phase of the collision; subsequent movements, perhaps even an explosion due to heat action, may have dislocated the meteor fragments and mixed them with the other debris of the crater. (Öpik 1936: 12)

Thus, according to Öpik, an explosion was only 'perhaps' to be expected, and only at the very extreme end of the scale. Professional understatement and cautious calculation of lower limits prevented the reader from gaining a sense of the event's true drama.

Fritz Heide

Hermann Wilhelm Friedrich 'Fritz' Heide (1891–1973; Figure 6.2) was more widely read than Öpik in Germany, as he published in German. Heide studied sciences from 1911 to 1913 in Munich and in 1913–14 in Jena. He spent World War I in the military, but after the war continued his studies in Jena. He was an assistant at the Mineralogical Institute in Göttingen from 1923 to 1929. In 1929, he '*habilitated*' and became a lecturer in Göttingen. From 1930 to 1969, he was Professor of Mineralogy, Crystallography and Petrography at the University of Jena (Franke 1983: 291).

In his first paper on the topic, Heide (1933) named four certain terrestrial impact craters (Meteor Crater in Arizona, Odessa Crater in Texas, Henbury crater field in Australia and the Wabar crater field in the Arabian desert; all four confirmed by the presence of meteoritic irons) and six possible ones where no meteorites had yet been found (Kaalijärv, Tunguska, Campo del Cielo in Argentina, a crater-field in North and South Carolina, Lake Bosumtwi[7] and Gwarkuh in Persia–Baluchistan). The Ries and Steinheim Basins were not mentioned:

> Secondly, so far it has not been possible in any of the ascertained and assumed meteorite craters to find the gigantic projectile itself. Only insignificant remains could be identified. Thus it seems that such gigantic meteorites do not remain intact upon impact. A short calculation shows indeed that in meteorites above a certain mass the kinetic energy upon impact is sufficient to vaporize them. If one assumes conservatively that only 10 per cent of the kinetic energy is used for the vaporization, we find that meteoritic iron of more than 10 m radius cannot persist through the impact. (Heide 1933: 381)[8]

One year later Heide (1934) gave a more extended outline of his ideas, and again he first reviewed the largest known meteorites as well as the topography and

---

[7]   This was the largest crater Heide at least tentatively acknowledged, with a diameter of 8 km. In later publications by Heide it was removed from the list. It is, however, now recognised as an impact crater (<http://www.passc.net/EarthImpactDatabase/bosumtwi. html> accessed 12 November 2012).

[8]   'Zweitens hat sich gezeigt, daß es bei keinem der sicheren und vermuteten Meteoritenkratern bisher möglich war, das Riesenprojektil selbst aufzufinden. Nur unbedeutende Reste konnten festgestellt werden. Es scheint also, als ob derartige Riesenmeteoriten bei ihrem Aufprall nicht bestandfähig bleiben. Eine kurze Rechnung zeigt in der Tat, daß bei Meteoriten von einer bestimmten Masse an die kinetische Energie beim Aufschlage ausreicht, um sie zu verdampfen. Nimmt man vorsichtigerweise an, daß nur 10% der kinetischen Energie zur Verdampfung Verwendung finden, so ergibt sich, daß Meteoreisen von über 10m Radius kaum den Aufprall überstehen werden' (Heide 1933: 381).

Figure 6.2   The German mineralogist Fritz Heide investigated why there are no large meteorites preserved on Earth. (Photo courtesy of Prof. Dr Klaus Heide, University of Jena).

geology of ascertained or potential terrestrial impact sites. Then Heide proceeded to the main question with which he was concerned: the largest known meteorite was the Hoba meteorite (Namibia) with a weight of 60 tons, whereas possible impact sites pointed to the existence of much larger masses that had never been found. Might this mean that they had been destroyed on impact?

To answer this question, Heide endeavoured to find out whether the impact energy was sufficient to vaporise the impacting iron masses. For this he needed to calculate two things: the energy of the impacting body, which for a given mass depended on its final velocity (which again was dependent on the initial velocity and the braking effect of the Earth's atmosphere); and the energy required to vaporise the same mass of meteoritic iron. In order to provide for other work done by the impact process (such as fragmenting and heating of the country rock, hurling the debris high into the air and so on), it was cautiously assumed that only 10 per cent of the projectile's energy was available to vaporise the iron, while the rest went elsewhere. Heide came to the conclusion 'that meteoritic irons of more than 10 m radius have all chance to vaporize upon hitting the Earth's surface' (Heide 1934: 250).[9]

So the answer to Heide's question (whether or not large meteorite masses are vaporised upon impact) was 'yes'; but what happened at the moment of impact in addition to this was beyond Heide's scope of interest:

> At the assumption of a vaporization of the impacting projectile, one question so far has been left out of consideration completely, which is of greatest importance; that is how the transformation of energy happens at the moment of the collision? Relevant experimental data are not yet available, and as far as I know, this question has also not yet been tackled theoretically in any detail. (Heide 1934: 250)[10]

Consequently, it was up to the reader to imagine the vaporisation of the iron mass. Even though there are craters, there is nothing in Heide's final conclusion that would enforce the image of a gigantic explosion if the potential reader was more inclined to envision a peacefully rising cloud of metallic steam. And contrary to Öpik, nowhere in Heide's text does it mention the hypersonic velocity of the projectile.

---

[9] ' ... daß Meteoreisen von über 10 m Radius alle Aussichten haben, beim Auftreffen auf die Erdoberfläche zu verdampfen' (Heide 1934: 250).

[10] 'Bei der Annahme einer Verdampfung des auftreffenden Projektils ist ja eine Frage bisher noch völlig außer acht gelassen worden, die von größter Bedeutung dabei ist, nämlich, wie findet die Umwandlung der Energie im Moment des Zusammenstoßes statt? Experimentelle Unterlagen darüber sind noch nicht vorhanden, und meines Wissens ist auch theoretisch diese Frage noch nicht eingehender bearbeitet worden' (Heide 1934: 250). This was true for the German literature, not however for publications in English.

Heide's third treatment of the issue (Heide 1934, 2nd enlarged edition: 1957) was a popular little volume on meteorites that contained a slim chapter on impact craters, in which he explicitly mentioned that the Ries and Steinheim Basins were definitely not impact craters, although he gave no reason for his assessment (Heide 1957: 50). Otherwise, the little chapter was nothing but a re-launch of his previous, more technical papers. He basically said the same things, but for the sake of simplicity, he said them differently.

Up to 1957, eight ascertained impact craters were known to have been formed by gigantic meteorites, which, however, had not been found: 'Observational facts and a short calculation lead to the conclusion that they *were destroyed nearly without trace upon hitting the ground*, that they exploded and evaporated into a cloud of gas' (Heide 1957: 50).[11]

While small meteorites were effectively braked by the Earth's atmosphere, the case was different for very large meteorites, whose energy was much greater:

> The energy of the impacting meteorite is used at once to chisel out the crater, to produce the earthquake and air waves, to warm up the rocks at the point of impact – we keep in mind that there are molten sandstone pieces to be found in meteorite craters – and additionally to warm up the meteorite itself. (Heide 1957: 52)[12]

This quote is typical for Heide's general style of writing. He constantly used diminishing terms, such as 'warming' instead of 'heating', so that the general drama of the impact process simply did not come across.

Heide continued to relate that it was possible to calculate the energy needed to vaporise the iron, that is, to heat it to 3,200°C: 'Meteoritic iron of 100 t and more ... is able to provide these 10 per cent [of energy] for the evaporation. This find, which does not change notably for other initial velocities, is in good accordance with our previous experience' (Heide 1957: 52).[13]

---

[11] 'Beobachtungstatsachen und eine kurze Rechnung führen zu dem Schluß, daß *sie beim Aufprall auf den Erdboden fast restlos zerstört worden sind*, daß sie explodierten, zu einer Gaswolke verdampften'. (Heide 1957: 50).

[12] 'Die Energie des auftreffenden Meteoriten wird nun einmal zum Herausschlagen des Kraters, zum Erzeugen der Erdbeben- und Luftwellen, zum Erwärmen der Gesteine an der Auftreffstelle – wir erinnern uns der eingeschmolzenen Sandsteinstücke aus den Meteoritenkratern – und weiterhin zum Erwärmen der Meteoriten selbst verbraucht' (Heide 1957: 52).

[13] 'Meteoreisen von 100t und darüber ... sind in der Lage, diese 10% für die Vergasung des Meteoreisens zu liefern. Dieser Befund, der auch für andere Eintrittsgeschwindigkeiten nicht größenordnungsmäßig verändert wird steht im guten Einklang mit der bisherigen Erfahrung' (Heide 1957: 52).

In his papers Heide, the mineralogist and expert on meteorites, was not interested in impact geology as such. He consequently dealt only with the question of why there are no large meteorites on Earth although geologists had proclaimed the existence of large meteorite craters. However, these same geologists tried to draw information from Heide's papers that Heide had never intended. This inadvertently led to misunderstandings. To the casual reader, Heide's mention of the evaporation point of iron might well convey the impression that the melting or evaporation temperatures of materials encountered in an impact crater gave a reasonable minimum temperature estimate of the whole process (something that Heide himself never intended to imply). And so Otto Stutzer, for example, estimated the impact temperature from the melting and vaporisation temperatures of material found in meteorite craters, completely neglecting the time factor (Stutzer 1936; see Chapter 2). Thus, he did not notice that he dealt with a process capable of melting several cubic kilometres of rock within mere seconds, and to achieve that you definitely need much more than just the melting temperature.

It is an education to read Heide's papers on the subject. Although with hindsight it is clear that Heide had some grasp of what actually happened at impact sites (and Quiring for example did understand him accordingly: see Chapter 4), for his contemporaries it was, however, possible to interpret him in a very different way.

It Cannot Be

While Öpik's English treatment missed the impact mechanism through conversion of kinetic energy (due to its previous publication in Russian during World War I) and Heide was preoccupied with a completely different question, other physicists, when confronted with initially huge energy estimates, tended to look for errors in their own calculation rather than realising what an impact might mean for its target site. They tried to downsize the impact's energy, because it simply would not fit their imagination. The British geologist and mineralogist Leonard James Spencer (1870–1959) 'calculated that ... 100 tons of iron traveling with a velocity of twenty-four miles per second would, on colliding with the earth, develop a temperature of 13,000,000°C. Even with a velocity of only ten miles per second, a temperature of about 200,000°C would be created'. But he 'simply could not believe such figures. "This result seems absurd", he wrote, "and no doubt many factors have been overlooked in this simple calculation"'. (Mark 1995: 55)

Considering the fact that even the astronomers and physicists doubted their models and felt unable to understand the implications of a large, hypervelocity meteorite impact (Mark 1995: 55), it becomes comprehensible that the average German geologist, isolated from the international discussion and with a marked deficit in knowledge of physics, was more than reluctant to harbour such seemingly fanciful scenarios. The deficit in physics was still to hamper the geologists understanding when the impact scenario was finally confirmed with geological and mineralogical means (see Chapters 9 to 12) in the early 1960s. Most of the traditional experts on Ries geology shared the opinion that the 'impactists' did not know the Ries Basin well enough to become valid partners in the discussion (Georg Wagner in Hölder 1962). However, on the other hand, understanding the physical processes related to hypervelocity impacts and eventually also the implications of the physical properties of high-pressure minerals such as coesite (Chapter 8), was clearly beyond the ability of most geologists involved in the debate at that early time: 'I have not yet seen proofs for explosion of 'bomb shells'. But can we carry on with our old-fashioned physical understanding?' (Letter from Georg Wagner to Bruno von Freyberg, 13 February 1965, UAT 605/343).[14]

The deficit in physics was not just a lingering educational problem, but had been furthered by Nazi ideology during the late 1930s and early 1940s, rejecting 'ahistorical' physical modelling for 'German Geology' that in turn became valued as the historical science *per se* (Beurlen 1935; Becksmann 1939).

---

[14] 'Ich habe noch keine Beweise für Bombeneinschläge gesehen. Aber kommen wir mit unseren altertümlichen physikalischen Vorstellungen weiter?' (Letter from Georg Wagner to Bruno von Freyberg, 13 February 1965, UAT 605/343).

Chapter 7
'German Geology'

Apart from the brief episode surrounding Quiring's letter, World War II seems to have been a major break in impact ideas in Germany. The fashion in 'impactist' papers concerning the Ries and Steinheim Basins, as well as the Kaalijärv Crater, the Köfels landslide or lunar craters (see Chapters 2 and 5) that characterised the early 1930s, was over by 1937. With increasing Nazi power and, finally, the beginning of the war, involvement in foreign impact discussions ceased completely, although it continued unabated in North America, leading, for example, to a research programme in the 1950s to search for impact craters on the Canadian shield (Plotkin and Tait 2011). The isolation of German geologists from this development was not only caused by external politics and the war. It was also self-inflicted, due to a general chauvinistic attitude favouring 'German Geology' as being something special that had nothing to learn from outside.

German Geology in the Romantic Period

The roots of this notion of 'German Geology' go back to the early nineteenth century, with German Romanticism (Fritscher 2002, 2012), and culminated in Nazism.

Geognosy, as it was termed in the early nineteenth century, was – apart from its roots in German mining – influenced by the theoretical–philosophical views of German idealism, and the main political question at that time was the possibility of a unified German national state replacing the patchwork of smaller and larger, more-or-less independent kingdoms and principalities (Fritscher 2012).

In the German parts of Europe, this meant that Geognosy:

> remained always implicitly restricted to the area of the German countries, or, better expressed: it aimed to ascertain the specific German geognostic conditions, or the typical German lithostratigraphy (which, of course, was explained by the fact that mapping, as a rule, was commissioned by the various German local authorities). This, however, did not mean that the stratigraphic conditions of other countries ... were not taken into consideration for comparative purposes (and thus also actively investigated). The central question, however – with the implicit focus on the territory of a (future) German national state – always remained that

of establishing a 'specifically German' lithostratigraphy. By comparison, English and French stratigraphy right from the start aimed at an internationally valid structuring of Earth history. (Fritscher 2012: footnote 48)[1]

The political goal of a unified 'German Nation' in the aftermath of the Napoleonic wars affected many intellectuals, including scientists. Lorenz Oken (1779–1851, philosopher and medical professor: Lang 1887) founded the 'Assembly of German scientists and physicians'[2] in 1822. In its journal, he planned to publish papers 'of a German kind'[3] and designed to further 'the scientific dialogue ... among us Germans'[4] (Fritscher 2002: 112).

For example, Oken offered *geological* arguments for the view that the territory on the western bank of the Rhine belonged to Germany, thus countering the view that the river formed a natural frontier (Fritscher 2002: 113).

This 'German Geology' was not, however, restricted to 'practical' applications in the political sphere, but also included more theoretical and philosophical aspects. It stood for a Romantic break with 'French' science, which was regarded as abstract and unemotional, turning instead to an illustrative, vivid 'German' science (Fritscher 2002: 116).

On the one hand, this movement was aiming for internationalisation in the sense of overcoming the particularism in the numerous German principalities of the time; on the other hand, it didn't go far enough as it struggled for internal unity only by demarcation from the world outside. The aim was not an international 'republic of letters', united by a common research interest that transcended national boundaries, but a unified national state (which was achieved in 1870). Thus, science was a vehicle for a non-scientific, political purpose (Fritscher 2002).

---

[1]   ' ... immer implizit weitgehend auf das Gebiet der deutschen Länder beschränkt blieb, oder besser gesagt: dass sie auf die spezifischen deutschen geognostischen Verhältnisse bzw. auf die typischen deutschen Gesteinsfolgen zielte (was sich natürlich auch daraus erklärt, dass die Aufnahmen in der Regel im Auftrag entsprechender deutscher Gebietskörperschaften erfolgten). Dies bedeutet wohlgemerkt nicht, dass nicht auch die stratigraphischen Verhältnisse anderer ... vergleichend in den Blick genommen (und eben auch aktiv erforscht) wurden. Die eigentliche Frage bleib dabei aber – mit implizitem Blick auf das Territorium eines (zukünftigen) deutschen Nationalstaates – immer die nach 'spezifisch deutschen' Lagerungsverhältnisse der Gesteine. Demgegenüber zielte die englische und französische Stratigraphie von Beginn an auf eine international gültige erdgeschichtliche Gliederung' (Fritscher 2012: footnote 48).

[2]   Versammlung der Deutschen Naturforscher und Ärzte.

[3]   'welche deutscher Art sind' (quoted after Fritscher 2002: 112).

[4]   'wissenschaftliche Unterhaltung ... unter uns Deutschen' (quoted after Fritscher 2002: 112).

Oken's Romantic notion of geology/mineralogy, which he termed 'geosophy', saw the genesis of minerals and rocks not as the aim of a research programme but as a foundation for mineral and rock classification (Fritscher 2002: 117). Consequently, he was unable to research genetic issues, because they were subject to pre-existing assumptions. Thus research became replaced by speculation.

Oken and others also wasted time replacing well-established scientific terms (which had foreign, often French, roots) with German neologisms, which hampered their international understanding (Fritscher 2002: 118). Obviously, this difficulty was not seen as a defect, since a German perspective was deemed sufficient and desirable. Outside Germany, there was supposedly nothing that was important, and consequently, it was not necessary to strive for a common scientific language.

Teleological notions, founded on Aristotelian roots, were also common. Like the ancient Greek philosopher, Oken assumed the existence of a special celestial matter (an elementary difference between the Earth and the Heavens). He did not carry this idea to its extreme, but the general tenor of his thoughts was undeniable (Fritscher 2002: 120).

While Oken's geosophy has hardly left any traces in geology and was rejected by contemporary geognosts and interested observers of the scene such as Johann Wolfgang von Goethe (1749–1832), it nevertheless manifested the general, partly unconscious, feelings of the time (Fritscher 2002: 125).

Geology in Nazi Germany

In 1933, the ultra-nationalist Nazi regime came to power; and it was in this political framework that 'German Geology' resurfaced – officially, so to speak. Immediately after the election that enabled the Nazis to take over, the scientific landscape in Germany was affected by a law[5] that allowed the 'removal' of undesired personnel at the German universities for racist or political reasons. At the University of Tübingen only eight members were dismissed, which was much less than at other German universities. However, this was not due to any resistance that Tübingen offered to the Nazi regime. Rather, at the anti-Semitic University of Tübingen, Jewish intellectuals already had hardly any chance of obtaining employment (Daniels and Michl 2010: 38).

From 1934 onwards, to gain *Habilitation* (the qualification for a professorship) a political assessment of the applicant and training in a camp for military sports (*Wehrsportlager*) became mandatory, in addition to the demonstration of academic expertise (Daniels and Michl 2010: 55).

---

5   Gesetz zur Wiederherstellung des Berufsbeamtentums; passed on 7 April 1933.

The German geologists were brought into line in 1936, when upon the desire of the working group of 'National Socialist university professors of geology' its chairman, Karl Beurlen in Kiel, Northern Germany, was made President of the German Geological Society (DGG) (ZDGG 81 (1936): 668).

Karl Beurlen (1901–85) studied geology and palaeontology at Tübingen University, where he obtained his doctorate in 1923 with a thesis on Upper Jurassic ammonites. Beurlen worked as an assistant in Tübingen from 1923 to 1925, later moving to the University of Königsberg from 1925 to 1934, where he *habilitated* in 1927. In 1934, he was appointed Professor and Director of the Institute of Geology and Palaeontology at the University of Kiel. In 1941, Beurlen was called to Munich as Professor of Palaeontology and Stratigraphy. In 1945, he was dismissed due to his close association with the Nazi regime and for a while earned his living as a construction worker. In May 1950, he finally took over a position with the Geological Survey of Brazil, which had invited European scientists to join its staff in order to improve its scientific standing. Beurlen was appointed a professor at the new University of Recife in northeast Brazil in 1957. After his retirement in 1969, he returned to Germany and lived in Tübingen (Tollmann 1986; Rieppel 2012; Grüttner 2004: 22; <http://fr.wikipedia.org/wiki/Karl_Beurlen> accessed 5 May 2011).

Beurlen's considerable influence in the German Geological Society (DGG), which increasingly represented all German geologists at universities, in industry and in the administration of the *Reich* [empire], is explained by the fact that he was deeply involved with national science politics and funding:

> The *Reich* Research Council, which has been established within the *Reich* Ministry of Science to integrate scientific research into the four year plan, has been ceremonially invested by *Reich* Minister Rust on 25 May, 1937, in the presence of the <u>*Führer*</u> [leader] <u>and *Reich* Chancellor</u> [Adolf Hitler] and the Minister–President, Colonel General Göring. After a programmatic speech by *Reich* Minister Rust, President of the *Reich* Research Council, General of the Artillery Professor Dr Karl Becker undertook the swearing-in of the heads of the individual sections. Geology, mineralogy and geophysics have been subsumed in the section for 'soil science'.[6] The Head of this Section is Professor Dr Karl Beurlen (Kiel). (ZDGG 89 (1937): 292)[7]

---

6   *Bodenkunde* means soil science, but in German *Boden* can also mean the underground. Also, in connection to the Nazi ideology at that time, it might also have had a strong nationalistic connotation as in 'German soil and ground' or in the Nazi terminology of 'blood and soil' *Bodenkunde* was probably preferred, as it is a German term (unlike 'geology', which is of Greek origin).

7   'Der beim Reichswissenschaftsministerium zum planmäßigen Einsatz der Forschung für den Vierjahresplan ins Leben gerufene Reichsforschungsrat wurde am 25. Mai 1937 in Gegenwart des <u>Führers und Reichskanzlers</u> und des Ministerpräsidenten Generaloberst

Meanwhile, more and more German scientists left the country because they were being persecuted for racist or political reasons. Lists of 'Displaced German Scholars' published in London in 1936 and 1937 contained more than 1,500 names, among them some 35 geoscientists (Martin Guntau, Rostock, pers. comm. 2002).

The celebrated geochemist Victor Moritz Goldschmidt (1888–1947), for example, lost his professorship in Göttingen due to his Jewish descent and left Germany for Norway, where he again had to flee the advancing German troops, travelling via Sweden to the United Kingdom.[8]

The vertebrate palaeontologist and 'founding mother' of palaeo-neurology Tilly Edinger (1897–1967), likewise of Jewish descent, held out in an unpaid position at the Senckenberg Museum in Frankfurt under the protection of its Director Rudolf Richter (1881–1957) until the pogrom night of 9 November 1938 made the danger of staying more than clear. Edinger managed to flee to England in May 1939 and emigrated to the USA a year later, where she became Curator of the Museum for Comparative Zoology at Harvard University (Kohring and Kreft 2003).

The stratigrapher and palaeontologist Curt Teichert (1905–96) left Germany in 1933 to accompany his Jewish wife. The ensuing odyssey led them via Denmark and Australia to the USA (Seibold and Seibold 2008) while Wilhelm Salomon-Calvi (1868–1941), director of the Institute of Geology and Palaeontology at the University of Heidelberg until 1934, was dismissed by the Nazi regime and emigrated to Turkey where he became professor at the Agricultural High School in Ankara. The palaeontologist Georg Walter Kühne (1911–91) was arrested in 1933 for 'communist activities'. He was demoted and so, when he was released from prison, he had to earn his living by helping out in a library and by dealing in fossils and minerals. He emigrated to England and was interned on the Isle of Man from 1940 to 1944. In 1944, Kühne became a lecturer in the Department of Zoology at University College London. He eventually received a PhD from the University of Bonn in 1949 and returned to Germany in 1951, where he lectured and *habilitated* in Berlin. He became Professor of Palaeontology at *Freie Universität* Berlin in 1966 (Guntau, Rostock, pers. comm. 2002).

When Salomon-Calvi died, Hans Cloos (1885–1951), Professor of Geology at Bonn, who considered himself an apolitical person, included Salomon-

---

Göring vom Reichsminister Rust feierlich eingesetzt. Nach richtungsweisenden Ausführungen des Reichsministers Rust ergriff der Präsident des Reichsforschungsrates, General der Artillerie Professor Dr. Karl Becker das Wort, um die Leiter der einzelnen Fachgliederungen auf ihr Amt zu verpflichten. Unter der Fachgliederung 'Bodenkunde' sind Geologie, Mineralogie und Geophysik zusammengefaßt worden. Leiter derselben ist Professor Dr. Karl Beurlen-Kiel' (ZDGG 89 (1937): 292).

[8]   <http://www.wikipedia.org/wiki/Victor_Moritz_Goldschmidt> accessed 7 April 2011.

Calvi's name in a 'list of deaths during the previous year' printed in *Geologische Rundschau*. But as Cloos recalled, in 1945, he received a complaint letter:

> "To put the emigrated Hebrew", wrote a reader [of the journal] from Thuringia, "on an honorary platter among German scholars and even before the fallen [soldiers] is an open provocation. In fact, it should be carefully considered whether other steps should still be undertaken! In the face of this 'derailment' it is no wonder if certain intellectuals are faced with the accusation of lacking a patriotic instinct". Such letters could not be counteracted, because behind them stood the power of the Gestapo,[9] but one could be pleased about them and keep them for a better occasion – like the present. (Cloos quoted from Seibold and Seibold 2000: 865)[10]

Denunciation could have led to serious trouble. Two of Cloos's PhD students, Hermann Korn (1907–46) and Henno Martin (1910–98), left Germany in 1935 for fieldwork in Southwest Africa (today Namibia) and did not return, in order to avoid unwanted political involvement. They 'absconded' into the Namibian Desert from 1940 to 1942 to avoid internment (Guntau, Rostock, pers. comm. 2002; Martin 1999).

This is but a small and somewhat random sample of people who successfully survived the Nazi period. The number of others who did not manage to avoid persecution, failed to emigrate and were eventually murdered, driven to suicide or simply dropped out of geology, is unknown. This chapter in the history of German geology still remains to be written. Only occasionally, thus far, do we get a few glimpses of the horror:

> Moos[11] died in January 1945 while still in Buchenwald Concentration Camp, after his mother starved to death in December 1944 in the Concentration Camp

---

9　　Short for *Geheime Staatspolizei*, the secret police of Nazi Germany.

10　　"'Den emigrierten Hebräer" so schrieb ein Leser aus Thüringen, "auf eine Ehrentafel zwischen deutschen Gelehrten und noch vor die Gefallenen zu setzen, bedeutet eine offene Provokation. Allerdings wäre zu erwägen, ob in dieser Angelegenheit nicht noch anderweitige Schritte zu unternehmen sind! Man braucht sich angesichts dieser Entgleisung nicht zu wundern, wenn gewissen Intellektuellen der Vorwurf der Instinktlosigkeit gemacht wird". Solche Zuschriften konnte man nicht bekämpfen, weil hinter ihnen die Macht der Gestapo stand, aber man konnte sich über sie freuen und sie für bessere Gelegenheiten – wie für die heutige – aufheben' (quoted after Seibold and Seibold 2000: 865).

11　　This is the petroleum geologist August Moos (1893–1945) and his wife, the palaeontologist and librarian Beate Moos (1902–84) (<http://kaliope.staatsbibliothek-berlin.de> accessed 4 October 2011). August Moos, who was Jewish, had obtained his doctorate degree in geology from the University of Tübingen in 1925. He was chief geologist at the Preußag Company when in 1937 his station there became increasingly precarious.

at Belsen. His wife and two children were also in Belsen and barely survived, but the son died afterwards from the after-effects. Mrs Moos is now a librarian in our survey-office in Hannover. (Letter from Alfred Bentz to Walter Kranz, 11 January 1947, GA 19340)[12]

On the other hand, there were also geologists who readily adapted themselves to the new regime. Beurlen, for example, used the forum of the DGG conferences, such as the annual meeting held in Aachen in August 1937, to convey his political messages and integrate German geologists into the general nationalistic efforts:

> In a short closing remark, Mr Beurlen stressed that the task faced by the [geoscience] section is of so wide a character that the full employment of all forces is necessary to its fulfilment. Such effort, however, will not only enable the sciences subsumed in the section to actively play their parts in the erection of the new *Reich*, but will place these sciences at the peak of research in the whole world. The rally closed with a *Sieg-Heil* [hail to victory] to the *Führer* [Adolf Hitler]. (ZDGG 89 (1937): 554)[13]

Beurlen explicitly tried to integrate the theoretical base of geology into the Nazi ideology, because he saw in geology's temporal frame an aspect that would be especially valuable to further the rising nationalistic thinking:

> In geology the sum and the wealth of facts are not so much and not singularly important. Rather the focus is on the orderly arrangement of the whole in space and time. The aspect of time is especially emphasized here, because this is what fundamentally distinguishes the geological view from all other sciences.

Alfred Bentz managed to get him a job at the Elwerath Company and had them send Moos to Yugoslavia for oil exploration. In 1944, however, Moos [together with his family?] was arrested and interned in the concentration camp, where he died (Kockel 2005: 294, 330).

[12] 'Moos ist noch im Januar 45 im KZ Buchenwald ums Leben gekommen, nachdem seine Mutter im Dezember 44 im KZ Belsen verhungerte. Seine Frau mit den 2 Kindern war auch in Belsen und kam mit knapper Not durch, aber der Sohn starb noch nachträglich an den Folgen. Frau Moos ist jetzt als Bibliothekarin bei unserem Amt in Hannover' (letter from Alfred Bentz to Walter Kranz, 11 January 1947, GA 19340).

[13] 'In einem kurzen Schlußwort betonte Herr Beurlen, daß die Aufgabe, die der Fachgliederung gestellt sei, so umfassenden Charakter habe, daß der volle Einsatz aller Kräfte zu ihrer Erfüllung notwendig sei. Solcher Einsatz aber werde nicht nur die in der Fachgliederung zusammengefaßten Wissenschaften befähigen, zu ihrem Teil am Aufbau des neuen Reiches aktiv mitzuwirken, sondern werde diese Wissenschaften an die Spitze der Forschung in der ganzen Welt setzen. Die Kundgebung klang in ein Sieg-Heil auf den Führer aus' (ZDGG 89 (1937): 554).

Thus, only in and through geology do the other sciences achieve their value in knowledge and education; since only on the basis of geology can a truly and deeply founded *Heimatkunde*[14] be built. The unity of blood and soil in a worldview can only be realized through geology, because only in that science, which is oriented towards a history of the Earth and life, can the inorganic and organic events be understood as a true unit. ... We as geologists and palaeontologists are all servants with the same task: the solution of the problems of historical development of the Earth and of life and the investigation of the German soil. (Beurlen 1937: 53–5)[15]

Although propagandistic messages such as 'Our works must be brief, informative and factually unobjectionable, if they are to serve the *Reich* and invoke respect for 'German ways and art' in foreign countries' (ZDGG 92 (1940): 64),[16] which litter the journal of the German Geological Society in that time, show at least a remnant of concern for the opinion of foreign geologists, other authors no longer cared whether anybody outside Germany would read their effusions. As in the Romantic period, some people strove to purge geological terminology of foreign terms. Karl Hummel (1889–1945) was reluctant to replace well-established terms (which even the German propaganda minister Josef Goebbels (1897–1945), according to Hummel, thought ridiculous) because it would be tedious to be forced to explain the new terms all the time to colleagues in

---

[14]   The word *Heimat* is almost untranslatable. It is a word to describe the place that you know best and feel all cosy about and at home. This may simply be the area around the house you grew up in, or it may be the village or town that you and your family live in and its surroundings with your favourite walks. *Heimatkunde* (see also Nyhart 2009), as a school subject, is thus everything there is to know about the vicinity of the school from local history and sociology to local botany, zoology, and geology. In Nazi Germany, this term was misused and extended to denominate the whole German national state.

[15]   'Denn in der Geologie kommt es nicht so sehr und nicht allein auf die Summe und die Menge der Tatsachen als solcher an, als vielmehr auf das Ordnungsgefüge des Ganzen im Raum und in der Zeit, wobei das Moment der Zeit hier besonders unterstrichen sei; denn dieses ist es, durch das die geologische Betrachtungsweise sich ganz grundsätzlich von allen anderen Naturwissenschaften unterscheidet. // In und durch die Geologie bekommen die übrigen Naturwissenschaften daher erst Bildungs- und Erziehungswert. Das zeigt sich ja schon daran, daß einzig von der Geologie her sich eine wirklich tief begründete Heimatkunde aufbauen läßt. Die Einheit von Blut und Boden in einem Weltbild ist ja nur von der Geologie her verwirklichbar, da nur in ihr, die auf eine Geschichte der Erde und des Lebens ausgerichtet ist, das anorganische und das organische Geschehen als eine wirkliche Einheit erfaßt werden kann'. (Beurlen 1937: 53). 'Wir sind als Geologen und Paläontologen alle Diener an der gleichen Aufgabe: die Klärung der Probleme des erd- und lebensgeschichtlichen Werdens und die Erforschung des deutschen Bodens'. (Beurlen 1937: 55).

[16]   'Kurz, inhaltreich und sachlich einwandfrei müssen unsere Arbeiten sein, wenn sie dem Reiche dienen und dem Ausland Achtung einflößen sollen vor 'deutscher Art und Kunst'' (ZDGG 92 (1940): 64).

order to be understood; but '[o]ne should reflect on whether it is useful to adopt newly-established foreign terms or instead a German term; in principle, a German researcher should always prefer a German term. ... because first of all, we write more for Germans than for foreigners ... ' (Hummel 1940a: 129).[17]

Hummel studied science in Freiburg, Heidelberg and Munich. He obtained his doctorate in 1913 in Freiburg, where he subsequently worked as an assistant, but he soon entered World War I as an artillery officer and military geologist. After his return in 1918, he continued his assistantship in Freiburg. In 1919, he switched to the University of Gießen, where he *habilitated* in 1920 and became professor in 1924. In 1934, he took over the directorship of the Gießen Institute, and from 1933 to 1937 was chancellor and pro-rector of the University of Gießen. Around 1935, he became head of the Gießen section of *Nationalsozialistischer Deutscher Dozentenbund*. In 1940, he again became an artillery officer and military geologist and died by suicide shortly before the end of the war (Heller 1974; Grüttner 2004: 80).

The increasing focus on 'inner-German' problems and an 'inner-German' audience 'meant in a nutshell the renunciation of the achievements and insights of earlier generations of scholars and the renunciation of the international character of science. ... which corresponded with the nationalistic hubris of the Nazi racist mania' (Guntau 2002: 135–6):[18]

> Every musing about the sciences being the path-maker for international understanding is no longer valid, when we have recognized that the final basis of scientific thinking rests in the way of the [German] People and Race. And likewise talk that we must pursue science to be respected as a cultured nation in foreign countries, as the people of poets and scholars, is also no longer convincing because we know that the German in his (for others, so eerie) dynamics of his life will always remain the 'barbarian' and the 'vandal', even though every German be a Hölderlin.[19] (Beurlen, published speech at Kiel University in 1935, quoted after Guntau 2002: 136)[20]

---

[17] 'Zu überlegen wäre, ob es zweckmäßig ist, neu einzuführende Begriffe mit einem Fremdwort oder mit einem deutschen Wort zu bezeichnen; grundsätzlich sollte ein deutscher Forscher dabei immer eine deutsche Bezeichnung vorziehen. ... denn erstens schreiben wir mehr für Deutsche als für Ausländer, ... ' (Hummel 1940a: 129).

[18] 'Das bedeutete im Kern den Verzicht auf die Leistungen und Erkenntnisse früherer Generationen von Gelehrten und auch den Verzicht der Internationalität von Wissenschaft. ... was mit der nationalen Selbstüberhebung des nazistischen Rassenwahns korrespondiert'. (Guntau 2002: 135–6).

[19] Friedrich Hölderlin (1770–1843), a well-known romantic poet.

[20] ' ... jedes Gerede, daß die Wissenschaft Wegbereiterin internationaler Verständigung sei, verfängt nicht mehr, nachdem wir erkannt haben, daß die letzte Wurzel wissenschaftlichen Denkens im Völkischen und Rassischen begründet liegt; und auch jenes Gerede, daß wir

From now on, Beurlen saw the purpose of geology and palaeontology as of any other science, 'to serve directly the [German] People's community and its existence'.[21] For Beurlen, that meant prospecting for mineral resources, soil science in the service of agriculture, protection of the coast from erosion and acquisition of new land from the sea (Beurlen 1935: 3).

Beurlen no longer saw any room for purely theoretical research alongside these practical issues. Basic research was only acceptable if it too was directed toward the overall purpose of serving the People's needs (Beurlen 1935: 4). Beurlen declared this 'practical geology' to be 'free, autonomous research, which ... demanded men who are national socialist within. ... [That] means that economic and material interests are of primary importance compared to [the purely] scientific endeavour for knowledge' (Beurlen 1935: 7).[22] Knowing quite well that this restriction to practical purposes, which hitherto had been of far less prestige than 'pure' science, would not sit well with his colleagues, Beurlen made a quick turn, claiming that 'man and for us that is German man does not pursue science for monetary gains but for knowledge' (Beurlen 1935: 9).[23] To escape the paradox, Beurlen then introduced the Nazi view of history, which for geology meant to distance itself from Lyell's uniformitarianism as well as from 'the today still leading text-book of Salomon' (Beurlen 1937: 9).[24] As we shall see later, this meant that in addition to its practical applications, geology was also supposed to strengthen the Nazi view of history and humankind. (Earth) history was to supersede what in Germany was termed general geology, that is the study of geological processes: 'Because in history there is basically neither cause nor consequence, there is no causality but only singularity and

---

Wissenschaft treiben müssten, um im Ausland als Kulturnation, als das Volk der Dichter und Denker anerkannt zu bleiben, auch das überzeugt nicht mehr, da wir wissen, daß der Deutsche in der für die anderen so unheimlichen Dynamik seines Lebens doch immer der ,Barbar' und der ,Vandale' bleibt, auch wenn jeder Deutschen ein Hölderlin wäre'. (Beurlen, 1935, quoted after Guntau 2002: 136).

[21] 'Die theoretische Wissenschaft der Geologie und Paläontologie hat ebenso wie jede andere Wissenschaft nur das eine Ziel, der Volksgemeinschaft und ihrer Existenz unmittelbar zu dienen' (Beurlen, 1935: 3).

[22] ' ... freien, autonomen Forschung, ... die aber auch Menschen verlangt, welche innerlich Nationalsozialisten sind'. ... '[Das] bedeutet, daß wirtschaftliche und materielle Interessen den Primat vor wissenschaftlichem Erkenntnisstreben haben, ... ' (Beurlen 1935: 7).

[23] 'Der Mensch, und das heißt für uns der deutsche Mensch treibt Wissenschaft nicht um des Erwerbs, sondern um der Erkenntnis willen'. (Beurlen 1935: 9).

[24] 'dem auch heute noch führenden Salomon'schen Lehrbuch' (Beurlen 1937: 9). Beurlen was referring to the four-volume textbook *Grundzüge der Geologie*, edited by Wilhelm Salomon-Calvi (1922–26).

time-relatedness' (Beurlen 1935: 9).[25] For Beurlen this Nazi-style geology was to become 'autonomous, independent science', which was self-sufficient. In contrast, he disqualified research on geological causes and consequences by posing the rhetorical question 'For what do we need a geology that has no other purpose but to dissolve itself in physics and chemistry?' (Beurlen 1935: 9).[26]

As is well known, German nationalism and racism rapidly led to the catastrophe of the Holocaust and to World War II; and geology too was involved in the war effort. On 17 January 1941, in an effort to accelerate the geological education of young men needed for military geology, as well as for resource prospecting, a national diploma certificate was introduced at all German universities that offered geology. The idea was to ensure a uniform level of education needed by the German military and the national geological survey. Young geologists, it was hoped, could then go directly into the field without further training on the job (ZDGG 93: 49). At the same time, the national diploma explicitly denied access to the examinations to students who were unable to prove 'that the applicant or his wife (if married) were of German or type-related blood' (ZDGG 93: 52).[27]

Defining 'German Geology'

On 15 July 1938, the General Assembly of the German Geological Society was opened in Munich. The first scientific session was dedicated to the 'history of science' with talks by Karl Beurlen, Robert Gangolf Schwinner and Ernst Becksmann (1906–86) (ZDGG 90 (1938): 535), who set out to rewrite history in Nazi mode. Schwinner took the opportunity to air his racist attitude and obviously found colleagues to support him:

> Not much of importance came out of [the conference]. Everyone – as always – was totally convinced of the importance of his own little special area and feared in the following discussion to get his Highness hurt. Schwinner abused the Jew Eduard Süss [sic] and, in a spiteful and also dishonourable way, prepared the way for the general assumption that Alpine geology was a purely German issue

---

[25] 'Denn in der Historie gibt es im Grunde nicht Ursache und Wirkung, gibt es keine Kausalität, sondern nur Einmaligkeit und Zeitgebundenheit' (Beurlen 1935: 9).

[26] 'autonomen, selbstständigen Wissenschaft'; 'Was soll eine Geologie, die kein anderes Ziel kennt, als sich in Physik und Chemie aufzulösen?' (Beurlen 1935: 9).

[27] '... daß der Bewerber und gegebenenfalls seine Ehefrau deutschen oder artverwandten Blutes sind' (ZDGG 93: 52).

with which the accursed exotics have nothing to do at all. (Letter from Herbert Rohleder to Carl Troll, Munich, 16 August 1938, BGIB)[28]

Eduard Sueß (1831–1914), the most celebrated of all Austrian geologists, was in fact half-Jewish by his mother Eleonore née Zdekauer. Thus, Eduard Sueß's son, the Vienna geology professor Franz Eduard Sueß, who had written about the Köfels landslide, was classified as 'second degree half-breed' when the Nazis took power in Vienna on 12 March 1938. This for him meant the loss of Austrian citizenship, and consequently, in December 1939, he lost his membership in the Austrian Academy of Sciences. As to his professorship, he had reached the usual age of retirement; otherwise he would have been dismissed. Franz Eduard Sueß died a natural death in Vienna in 1941 (Johannes Seidl, University Archive Vienna, via Marianne Klemun, e-mail to the author 29 March 2011; Kölbl 1945).

Schwinner's talk was ostensibly about the history of tectonics in the eastern Alps, but as Rohleder has correctly noted it was basically a means to display his chauvinism not only towards Eduard Sueß but also to everything foreign in geology; especially the idea of tectonic nappes (Schwinner 1940).

To Schwinner's delight, Sueß had been unable to successfully propagate his ideas about the eastern Alps, and so everything was well until two Frenchmen revived nappe tectonics:

> Again, it was simply the appeal of the foreign and of that which must have been the most successful under the given conditions. Could 'nappism' have been launched, for example, from Berlin? But from Paris! By this word [Paris], the average Swiss 'intellectual' gets almost devout eyes. And Termier[29] is *so* French: smooth and logical, logical to the point of absurdity.

---

[28]   ' ... viel wesentliches kam dabei nicht heraus. Jeder war – wie stets – von der Wichtigkeit seines kleinen Spezialgebietes bis aufs äusserste überzeugt, & fürchtete in der folgenden Diskussion einen Zacken aus der Krone gebrochen zu kriegen. Schwinner schimpfte auf den Juden Eduard Süss, & bahnte in einer ebenso gehässigen wie unwürdigen [W]eise[?] den Weg zur allgemeinen Annahme, dass Alpengeologie eine rein deutsche Angelegenheit sei, mit der di [sic] verfluchten Exoten überhaupt nichts zu tun hätten' (letter from Herbert Rohleder to Troll, Munich, 16 August 1938, BGIB). About a month later, Rohleder left Germany for London, from whence he sent one last, printed note to Troll, dated October 1938, before contact broke off for nearly seven years. The note reads: 'This is to inform you that Dr Herbert P. T. Rohleder has changes his name by Deed Poll to Dr Herbert P. T. Hyde' (BGIB).

[29]   Pierre-Marie Termier (1859–1930), French geologist, professor in Paris and from 1911 onwards director of the French Geological Survey. He was the first to interpret the Tauern-Alps as a tectonic window (<http://de.wikipedia.org/wiki/Pierre_Marie_Termier> accessed 6 April 2011).

If alien protein is injected into the body, this causes fever. And it can only be called a fever, how the ingression of these alien thoughts has made themselves felt. Nappism was soon established as a sect, accompanied by all the bad taste and vice of a modern organized sect such as the Salvation Army ...

And today? After fifteen more years have passed –? After the given <u>account of one human age of Nappism</u>, one must conclude that this alien injection has been no good for German geology, and now must be exterminated finally! (Schwinner 1940: 266–9)[30]

Beurlen's and Becksmann's talks, in the same session as Schwinner's, were mainly to define what was to be regarded as specifically 'German Geology'. Beurlen (1939), like Schwinner, harped on Eduard Sueß's Jewish descent when he wrote about Diluvialism in the seventeenth and eighteenth centuries, which had appealed to people because a changing sea level seemed more rational than movements of the solid land: '[M]oreover, large flood catastrophes and sea storm catastrophes in historical experience happen again and again, and the biblical deluge story seemed to confirm the evidence. Let us remember that Suess still reckoned strongly with eustatic sea level changes!' (Beurlen 1939: 241).[31]

In Beurlen's view, what distinguished 'German Geology'[32] from non-German geology or other sciences was its dynamic outlook, its development by German or at least 'Germanic' researchers and a quasi-religious or esoteric attitude towards Nature:

---

[30] 'Es war einfach wieder der Reiz der fremden Art, und zwar jener, der im gegebenen Fall am wirkungsvollsten sein mußte. Hätte der Nappismus z.B. von Berlin aus lanziert [sic] werden können? Aber von Paris! Bei diesem Wort bekommt der durchschnittliche Schweizer 'Intellektuelle' förmlich andächtige Augen. Und Termier ist <u>so</u> französisch: glatt und logisch, logisch bis zum absurden. // Wird dem Körper artfremdes Eiweiß injiziert, so gibt das Fieber. Und nur als Fieber kann bezeichnet werden, wie das Eindringen dieser artfremden Denkweise sich auswirkte. Der Nappismus stand bald als Sekte da, mit all den Geschmacklosigkeiten und Untugenden einer modern aufgezogenen Sekte: Heilsarmee etwa'. (Schwinner 1940: 266–7). 'Und heute? Nach weiteren 15 Jahren —? Nach der gegebenen Bilanz über ein Menschenalter Nappismus muß man zu dem Schlusse kommen, daß diese artfremde Injektion der deutschen Geologie nicht gut getan hat, und nun endlich ausgemerzt werden sollte!' (Schwinner 1940: 269).

[31] ' ... zudem ja große Überschwemmungskatastrophen und Sturmflutkatastrophen in der historischen Erfahrung immer wieder auftreten und die biblische Sintfluterzählung den Befund scheinbar bestätigte. Erinnern wir uns, daß noch Suess stark mit eustatischen Meerespiegelschwankungen rechnete!' (Beurlen 1939: 241).

[32] 'German Geology' was no isolated phenomenon. There were likewise 'German Physics', 'German Chemistry', 'German Biology', 'German Mathematics' and so on. The humanities were also affected, most notably archaeology (Beyerchen 1982; Focke-Museum 2013; Mehrtens and Richter 1980; Rieppel 2012).

And if we look at the bearers of the development of geology, so will we notice that Germans appear most predominantly. Steno[33] was a Dane and thus admittedly not a German, but he still belonged to the Germanic world. ... Obviously, the inherent desire to understand natural reality from a historical development was strongest and most prominent in the German realm, while in Romance countries the static physical worldview seemed to suffice.

Here a difference in expression of view and thinking is expressed, which lie deep in human nature and that of the [German] *Volk* [People]. In contrast to the biblical cosmogony of the Creation narrative, which accepts the created world shaped in a singular creation event as a given, there is the 'cosmogonic' view of the Germanic world based on the idea of a gradual unfolding of the cosmos, such that the formative forces reside within the phenomena themselves. Thus within all the phenomena of reality there lives a divine being, and Schiller, in a last echo of the Greek Gods, mourns about the 'desouling' and secularisation of Nature by the Creation doctrine, and the German Natural Philosophy, prominent in Goethe, repeatedly leads to a sort of Pantheism. The worldview that stands behind this Germanic cosmogony leads necessarily to a science built on experience. A reality developing in diverse phenomena and forms, as an expression of the force present within the phenomena themselves, can only be understood in their causality by trying to experience the multitude of phenomena and their modes of expression'. (Beurlen 1939: 246–7)[34]

---

[33]   Nicolas Steno (1638–86), Danish anatomist and later Catholic priest, published important, foundational works on fossils and geology. He is often hailed as one of the founding fathers of geology.

[34]   'Und wenn wir uns die Träger der Entwicklung der Geologie ansehen, so stellen wir fest, daß hier ganz vorwiegend Deutsche uns entgegentreten. Steno war Däne, damit also zwar nicht Deutscher, aber doch dem germanischen Raum angehörig. ... Offensichtlich war das innere Bedürfnis, die natürliche Wirklichkeit aus einem historischen Entwicklungsgang zu verstehen, am stärksten und ausgeprägtesten in deutschen Raum, während im romanischen Raum das statische physikalische Weltbild zu genügen schien. // Hier prägt sich eine sehr tief im Menschlichen und Völkischen liegende Verschiedenheit der Anschauungs- und Denkformen aus. Im Gegensatz zu der biblischen Kosmogonie des Schöpfungsberichtes, die nach dem einmaligen Schöpfungsakt die geschaffene Welt als gegebenen Zustand hinnimmt, ist die kosmogonische Vorstellung der germanischen Welt dadurch bestimmt, daß eine allmähliche Entfaltung und Entwicklung des Kosmos stattgefunden habe derart, daß die wirkenden Kräfte in den Erscheinungen selber liegen – daher denn in allen Erscheinungen der Wirklichkeit ein göttliches Wesen lebt und Schiller in einer letzten Nachwirkung in den Göttern Griechenlands über die Entseelung und Entgöttlichung der Natur durch die Schöpfungslehre klagt und die deutsche Naturphilosophie, deutlich bei Goethe, immer wieder in eine Art Pantheismus ausmündet –. Die Weltanschauung, welche hinter dieser germanischen Kosmogonie steht, führt zwangsläufig in eine auf Erfahrung aufbauende Naturforschung hinein: Denn eine in vielfältige Erscheinungen und Formen sich

Beurlen saw the roots of geology not in mining but as a continuation of the revolution initiated by Galileo Galilei – a revolution which, according to Beurlen, was carried essentially by German scientists against scholastic theology (Beurlen 1939: 248). The foundations of the fulfilment of this revolutionary action were laid by Abraham Gottlob Werner (1749–1817), Georg Christian Füchsel (1722–73), Karl Ernst Adolf von Hoff (1771–1837) and other German geognosts of the Romantic Period – 'that is the time, which of all its relevant outputs is characterized by the fact that a specifically German way of thinking was most prominent and became conscious of itself' (Beurlen 1939: 250).[35]

The further development of geology, however, then encountered Lyell's doctrine of 'uniformity', which was quickly corrupted as it became elevated to a form of dogma. According to Beurlen, Lyell always had regard for uniformly acting processes, but never asked about the origin and development of things (Beurlen 1939: 250). Thus 'research, for Lyell, is in itself already a fulfilment, because the Earth's history and origin do not interest [him], as all events at all times are expression of essentially the same conditions' (Beurlen 1939: 251).[36]

According to Beurlen, Lyell's view excluded historical thinking, and:

> in consequence of this tendency finally the view could originate that all rock types can form at any time ...

> It is remarkable that the progressive pursuit of this trend went hand in hand with the endeavour to reduce Earth-historical time measurement, if possible, to physical processes, eliminating the phylogenetic and organic fundaments of the geological timescale. (Beurlen 1939: 251)[37]

---

entwickelnde Wirklichkeit, als Ausdruck der in den Erscheinungen selber sich auswirkenden Kraft kann in ihrer Eigenart und vor allem in ihrer Ursächlichkeit nur erkannt werden, indem ich die Vielfältigkeit der Erscheinungen und ihre Äußerungsformen durch Erfahrung zu fassen suche'. (Beurlen 1939: 246–7).

[35] '... der Zeit also, die in all ihren entscheidenden Äußerungen dadurch gekennzeichnet ist, daß das spezifisch deutsche Denken besonders stark hervortrat und zum Bewußtsein seiner selbst kam' (Beurlen 1939: 250).

[36] '... ist diese Forschung für Lyell schon Erfüllung in sich selber, da Erdgeschichte und Entstehung gar nicht interessieren, sondern sämtliches Geschehen zu allen Zeiten Ausdruck des grundsätzlich gleichen Zustandes sind' (Beurlen 1939: 251).

[37] 'Im Verfolg dieser Tendenz konnte schließlich die Ansicht entstehen, daß alle Gesteine zu jeder Zeit sich bilden konnten ... // Es ist bemerkenswert, daß mit dem fortschreitenden Vordringen dieser Bestrebungen Hand in Hand ging das Bestreben, auch die erdgeschichtliche Zeitmessung nach Möglichkeit auf physikalische Vorgänge zu reduzieren und die stammesgeschichtliche organische Grundlegung der geologischen Zeitmessung auszuschalten' (Beurlen 1939: 251). Beurlen here complains about the new methods of

Beurlen concluded his paper with his usual nationalistic 'pep talk': 'Let us uphold and keep the venerable and proud German tradition of German Geology ... alive. Geology will regain its status and leading position in the world and in Germany, which it formerly had and which is its due as a truly specific German science!' (Beurlen 1939: 252).[38]

Ernst Becksmann obtained his doctoral degree in 1930 from the University of Kiel. Two years later he qualified for a professorship, and in 1939 he became professor without chair in Heidelberg. In the 1950s and 1960s, he worked for the Geological Survey in Baden-Württemberg, first in Heidelberg and later in Freiburg im Breisgau, where he also held several lectureships in geology and mineralogy in the 1960s (Küppers [2007]; UA Freiburg 1996: 121, 203–5, 207, 252; GA3/23, GA3/24).

Becksmann, even more clearly than Beurlen, argued basically the same line, distinguishing 'German Geology' from the supposedly flawed non-German ways of pursuing geology: namely the biblical-neptunistic, supposedly 'Jewish' tradition and British uniformitarianism (Beurlen 1939; Becksmann 1939).

Becksmann (1939), like Beurlen, distinguished three ways of pursuing geology:

- 'Physicalising': concerned with ahistorical processes, the main promoter being Charles Lyell
- 'Rationalistic' or 'chronological': systematising knowledge and thus creating an array of unconnected facts, which, according to Becksmann's definition, were likewise ahistorical. This was exemplified by Neptunism, which Becksmann saw as being closely tied to the biblical/Jewish/Christian worldview
- Romantic or holistic: the only true way to comprehend history, understanding the 'coming-into-being' as the cumulative sum of history.

The latter, according to Becksmann, was 'German Geology', whose origin he saw in the National Romantics of Heidelberg, which he sharply distinguished from the Romantics in Jena, who, through Goethe and Abraham Gottlob Werner, were allied to Neptunism.

For this 'German Geology', the entity on which it focused was not the historical event but the so-called *erdgeschichtliche Gestalt* [Earth-historical figure], an area of land (*Erdboden*: literally 'Earth-soil' or 'Earth-ground';

---

radiometric dating of rocks, which seemed to replace the relative dating of strata by means of guide fossils.

[38] 'Sehen wir zu, daß wir die alte stolze deutsche Tradition der deutschen Geologie ... lebendig erhalten und weiterführen; dann wird die Geologie die Geltung und führende Stellung in der Welt und in Deutschland wieder gewinnen, die sie einmal hatte und die ihr als einer ganz spezifisch deutschen Naturwissenschaft gebührt!' (Beurlen 1939: 252).

again we see links with Nazi blood-and-soil mysticism), shaped and re-shaped in the historical coming-into-being, and which, *per se*, brings together in its present appearance the totality of its history. It can only be understood if the completeness of a region's history is taken into consideration. An event, on the other hand, can have various effects, depending on the 'historical formation' of the place where it occurs. Thus, according to Becksmann, it makes no sense to concern oneself with ahistorical processes because their effects would always be tied to specific locations.

Becksmann saw the 'historically formative causation' of Earth history not as contingent but somehow teleologically directed by 'fate', a view that corresponded directly to Nazi ideology in sociology. Thus, Earth history received some sort of non-religious, naturalistic purpose. The whole argument of Becksmann was thus an effort to translate Nazi ideology from the 'German People' [*Volk*] to geological entities under the heading of 'holistic' considerations:

> To the circle of the Heidelberg Romantics ..., however, history is not a succession of 'ideas'[39] and 'deeds'[40] but a lively connection held together by 'blood;' it is a 'service to the forefathers'. On this soil the feeling must have awoken that a People constitutes a community that has come into being by history and is connected by blood, in which the individual is rooted and resting, grown into by fate. On this soil, from which the truly historical question has risen to full blossom ... originated thus in contrast to idealism the concept of an historical entity. (Becksmann 1939: 747–8)[41]

The nationalistic attitudes of sciences and humanities at the German universities:

> are likewise effects of this foundational experience, as are the question of race and especially the political contests of our time, which ... are concerned with the political, voluntary uniformity of a nature-given body of the People [*Volk*], with a unity of state and People and with the creation of the 'life-historical entity' of the People as bearer of history. From this political foundational experience a view of history must grow that is now no longer contemplative and retrospective, as it was in the Romantic era, but is active as well as oriented towards the future,

---

[39] Ahistoric: Plato, Lyell.

[40] Unconnected events: Neptunism, biblical/Jewish view.

[41] 'Dem Kreis der Heidelberger Romantik ... dagegen ist Geschichte nicht eine Folge von 'Ideen' und 'Taten', sondern ein lebendiger, blutmäßig verbundener Zusammenhang, ein 'Ahnendienst'. Auf diesem Boden mußte das Gefühl dafür wach werden, daß Volk eine geschichtlich gewordene, blutverbundene Gemeinschaft ist, in der der Einzelne wurzelt und schicksalhaft verwachsen ruht. Auf diesem Boden, aus dem die wahrhaft historische Fragestellung zu voller Blüte sich erhob ..., erstand damit, im Gegensatz zur Gestaltidee, der geschichtliche Gestaltbegriff' (Becksmann 1939: 747–8).

which, from the historical fate of past and present, sees its duty in the creation of the future[42]. (Becksmann 1939: 752)[43]

This foundation of the specific 'German' historical thinking in Nazi ideology allowed Becksmann, citing Beurlen, to elevate geology from among the multitude of sciences and style it as *The* Science (Becksmann 1939).

Hummel too, whom we have already encountered as the man who endeavoured to purge 'German Geology' of foreign technical terms, expressed his animosity towards the foreign, Lyellian, geology of his time:

> I stress explicitly that the 'principle of actualism' (the hypothesis that the processes of rock formation have at all times been the same as today) must be clearly distinguished from the actualistic or ontological way of research. Actualistic research is the fundamental and irreplaceable foundation of all Earth-historical research, the application of which is beyond any doubt and which in the recent debates has not been doubted by anybody (including Beurlen). The 'principle of actualism', however, which was introduced into geology by Lyell and his successors is, according to our new evidence, a false hypothesis, which has hampered the progress of Earth-historical research for decades. The common failure to appreciate the difference between these two terms has been a leading cause for the passion of the debate concerning this question. (Hummel 1940b: 459; footnote)[44]

---

[42]   Just a month after Becksmann gave this talk, World War II began with the German attack on Poland.

[43]   ' ... sind genau so Auswirkungen dieses Grunderlebnisses wie die Rassenfrage und vor allem der politische Kampf unserer Tage, der ... um die politische, willensmäßige Geschlossenheit eines naturgegebenen Volkskörpers, um die Einheit von Staat und Volk, um die Schaffung der lebensgeschichtlichen Gestalt des Volkes als Trägerin der Geschichte geht. Aus diesem politischen Grunderlebnis heraus muß eine Geschichtsauffassung erwachsen, die nun nicht mehr nur kontemplativ und rückwärtsgewandt wie die der Romantik war, sondern zugleich aktiv, zukunftsgerichtet ist, die aus dem historischen Schicksal von Vergangenheit und Gegenwart die Verpflichtung zur schöpferischen Gestaltung der Zukunft sieht' (Becksmann 1939: 752).

[44]   'Ich betone ausdrücklich, daß der 'Grundsatz des Aktualismus' (nämlich die Hypothese, daß die Vorgänge der Gesteinsbildung zu allen Zeiten dieselben gewesen seien wie heute) scharf getrennt werden muß von der aktualistischen oder ontologischen Forschungsweise. Die aktualistische Forschungsweise ist die wesentlichste und durch nichts zu ersetzende Grundlage aller erdgeschichtlichen Forschung, deren Anwendbarkeit über alle Zweifel erhaben ist und die von niemand (auch nicht von Beurlen) bei den Auseinandersetzungen der letzten Zeit in Zweifel gezogen worden ist; der von Lyell und seinen Nachfolgern in die Geologie eingeführte Grundsatz des Aktualismus dagegen ist eine nach unseren neuen Erkenntnissen unrichtige Hypothese, die jahrzehntelang den Fortschritt der erdgeschichtlichen Forschung gehemmt hat. Die häufige Verkennung des Unterschiedes

Despite Hummel's affirmation, however, the debate about actualism and what it meant for 'German Geology' demonstrated that not all German geologists at that time were prepared to subordinate scientific thinking to ideology.

German Geology 'Down the Drain'

In 1940, Karl Beurlen became Professor and Director of the Institute as well as the Bavarian State Collection for Palaeontology and Historical Geology in Munich, where his infatuation with Nazi ideology was apparent even at a glance, as he used to come to work 'in the brown uniform. He appeared in the 'gold pheasant' uniform, even at the Institute, according to what I was told by Spiegler' (Pichler interview).[45]

Because of the imminent bombing of Munich from about 1943 onwards, curators of the palaeontological and geological collections began to evacuate parts of the collections in an undercover action and against Beurlen's will:

> Georg Spiegler[46] ... was very sceptical of the National Socialism, and during the War, when the bombardments began in 1941, the cities closer to England were involved [were bombarded], and he probably realized that eventually it [the bombing] must reach Munich; and it was in '43 that the bombardments came closer. Then he thought about evacuating the [geological] collection and put the idea to Karl Beurlen, but was aggressively admonished by Beurlen. Beurlen even threatened him with a disciplinary action, because of defeatism. Yes, that is how it

---

zwischen den beiden Begriffen hat an der Schärfe der Auseinandersetzung über diese Fragen wesentliche Schuld gehabt'. (Hummel 1940b: 459; footnote).

[45]   ' ... in der braunen Uniform. In der Goldfasanenuniform ist er aufgetreten, sogar im Institut, so hat mir das der Spiegler erzählt' (Pichler interview). *Goldfasan* [gold pheasant] was a mocking term for high-ranking civilian Nazi party members who wore a brown uniform with gold trimming.

[46]   Georg Spiegler (born 1903) learned electrical metrology from 1917 to 1921 and worked as an electro-technician from 1921 to 1929 at the Ing. Dröge company in Munich. In 1929, he became employed at the Anthropological Institute of the University of Munich. From 1930 to 1947 he was a technical employee, first at the Bavarian State Collection for General and Applied Geology and from 1936 at the Institute for General and Applied Geology of the University of Munich. In 1941, he was called for a short military service as a soldier in a signal battalion. In 1947, Spiegler was promoted to *Beamter* [civil servant]. In 1954, he took exams at an Administrative and Economic Academy and retired in 1968 (Universitätsarchiv München: Personalakte Georg Spiegler PA-allg.-398).

could be termed. "Munich won't be bombarded!" ... Then, Spiegler, to whom we owe a great deal, took matters into his own hands. (Pichler interview)[47]

Spiegler approached one of the palaeontological curators, Therese zu Oettingen-Spielberg (1909–91),[48] who provided access to storerooms (essentially unused beer cellars on the estate of her family in the Ries Basin), and the minerals and rock specimens were successively evacuated to that locality (Pichler interview):

> The material was packed into boxes and driven there in wood-gas cars. (By then wood-gas was already being used, as there was no petrol or diesel for such purposes.) And ... every time when Beurlen was gone – he often was gone, as he also had political tasks – Spiegler evacuated the odd box, and every time such a cabinet was empty, he 'lost' the key; and Beurlen did not have the time to venture into the collections. But those specimens that were above in the showcases, Spiegler had to leave there. However, he went so far ... as to exchange individual fine-looking pieces with somewhat less valuable material. And in this way [most of] the specimens of the [geological] collection left Munich, and survived the war. (Pichler interview)[49]

---

[47] 'Der Georg Spiegler ... war also sehr skeptisch gegenüber dem Nationalsozialismus und er hat wohl auch im Krieg schon gesehen, als die Bombardierungen anfingen, 1941 waren es die näher an England gelegenen Städte, dass irgendwann auch München drangewesen ist, und '43 war es, als die Bombardierungen näher kamen. Da dachte er daran, die Sammlung auszulagern und er hat es dem Karl Beurlen mal näher gebracht, und wurde vom Beurlen richtig zusammengeschissen. Der Beurlen hat ihm sogar ein Disziplinarverfahren angedroht, wegen Defätismus, ja, so kann man es sagen. "München wird nicht bombardiert!" ... Da hat also Spiegler, dem dafür sehr zu danken ist, der hat es in eigene Hände genommen' (Pichler interview). Pichler's anecdote is corroborated by a letter from Spiegler's superior, the geology professor Albert Maucher (1907–81) to the Dean, Professor Clusius, 18 September 1946 (Universitätsarchiv München, Personalakte Georg Spiegler PA-allg.-398).

[48] Princess Therese zu Oettingen-Spielberg held a doctorate in zoology, and up to the 1970s worked as curator at the *Bayerische Staatssammlung für Paläontologie und Historische Geologie* in Munich (letter from Richard Dehm to Albrecht Fürst zu Oettingen-Spielberg, 4 November 1991, *Fürst zu Oettingen-Spielberg'sches Archiv*).

[49] 'Das wurde in Kisten verpackt und dorthin geschafft mit Holzgasautos natürlich; damals gab es schon Holzgas, es gab ja kein Benzin oder Diesel für solche Fahrten. Und ... immer wenn Beurlen auf –, er war sehr oft weg, weil er auch politische Aufgaben wahrzunehmen hatte, hat also Spiegler immer kistenweise ausgelagert und hat, wenn so ein Schrank leer war, den Schlüssel verlegt, und Beurlen hatte keine Zeit in die Sammlung zu gehen. [Spiegler] musste aber die oben in den Vitrinen liegenden Stücke musste er lassen, er hat es aber auch da so weit getrieben, ... dass er einzelne schöne Stücke ausgetauscht hat mit etwas minderwertigem Material. Auf diese Weise ist das Material der Sammlung rausgekommen aus München und hat den Krieg überdauert' (Pichler interview).

Practically everything that was left, that is part of the geological and about 80 per cent of the much larger palaeontological collection, which was under closer scrutiny by Beurlen, was destroyed in April 1944 when the building was completely devastated by fire after a heavy bombing attack on Munich.[50]

By the end of 1944, the situation in Germany had deteriorated so far that the German Geological Society was unable to hold its general assembly, and thus the people who had guided its fate through the last few years (among them Karl Beurlen as President) remained in office without election until matters might improve (ZDGG 96 (1944): 237).

In the spring of 1945, the German Geological Society was disbanded by the allied forces until further notice and no journal could be issued. Not until 1947 did the business of the German Geological Society resume 'in modest circumstances' (ZDGG 97; 98; 99: prefaces). Thanks to several sponsors, however, the 1945 volume of *Zeitschrift der Deutschen Geologischen Gesellschaft* was issued in October 1947. The journal then slowly caught up with the missing issues until 1949, with a double-volume that contained the centenary volume for 1948 and the regular one for 1949.

The end of the War, of course, affected not only scientific societies but the universities as well. The University of Tübingen, for example, was shut down for the entire summer semester of 1945. However, examinations were still allowed, and it was also possible to take private lessons in the homes of the lecturers (Zauner 2010: 940).

Denazification, or *épuration* as it was called by the French forces who occupied Tübingen, was initially conducted with less efficiency than in the American sector. Only eight professors were classified in May–June 1945, among them the geologist Edwin Hennig, Beurlen's former PhD supervisor (see Chapter 3). A second wave of suspensions followed in July 1945 and affected all those who had become members of the NSDAP before 1933 or who had risen within the Nazi regime to higher positions. This time no geologists were affected and Hennig was allowed to return (Zauner 2010: 943–8).

The situation became critical for Tübingen University when in August 1945 it was decreed that 'state employees who had joined the NSDAP before May 1933 or had taken on political functions in it or had been members of the SS,[51] were no longer to receive payment' (Zauner 2010: 948–9).[52]

---

[50] <http://www.palaeontologie.geowissenschaften.uni-muenchen.de/ueber_uns/geschichte/index.html> accessed 27 March 2012.

[51] Short for *Schutzstaffel*. The SS was a major paramilitary organisation of the Nazi Party and responsible for many of the crimes against humanity during Nazi time (<http://en.wikipedia.org/wiki/SS> accessed 7 April 2011).

[52] 'Staatsbediensteten, die vor dem 1. Mai 1933 der NSDAP beigetreten waren oder politische Funktionen in ihr übernommen oder der SS angehört hatten, [durften] keine Gehälter mehr ausbezahlt werden' (Zauner 2010: 948–9).

Consequently, at the beginning of September in 1945, the French administrators dismissed all members of the NSDAP as well as those that had applied for membership. For a short time, this affected 53 per cent of all teaching personnel (63 per cent of the professors and other '*habilitated*' personnel). The decision was partly withdrawn in October, and at the end of that month only five professors and one *Dozent* [*habilitated* lecturer] were dismissed, again including Edwin Hennig (Zauner 2010: 950–53).

In May 1946, three former professors from the 'Reichsuniversity Strasbourg' who late in 1944 had fled to Tübingen received permission to continue with their lecturing, among them Richard Dehm (see Chapters 3, 9 and 10) (Zauner 2010: 963).

In July 1948, an amnesty was declared for all 'followers' of Nazism. Hennig, who by then had reached retirement age, was to retire and to lose his right to vote [he was not allowed to stand for any elective office] for some time. Georg Wagner (see Chapters 9 and 10), however, who had also been suspended for some time, was allowed to return to his university position in 1948. In 1951, state employees who had been dismissed due to the denazification were permitted to return and apply for a position similar to the one they had held previously, or at least receive pensions. Consequently, Edwin Hennig now officially became *professor emeritus*, but was no longer interested in resuming his lecturing activities. In July 1953, denazification was officially ended by a 'concluding law' (Zauner 2010: 973–92).

It is difficult, without delving deeply into university archives, to assess or evaluate the conduct of individual German geologists during the Nazi period. In obituaries and other biographical printed sources, the time between 1933 and about 1948 is often simply omitted. If one is lucky, some people who had problems with denazification become recognisable through the conspicuous gaps in their CVs after 1945, but we must also be prepared to encounter erroneous statements such as the claim that Hennig had been a professor in Tübingen between 1917 and 1951 (Engelhardt and Hölder 1977: 261).

Of course, a chauvinistic or even plainly racist attitude in certain people did not suddenly disappear with the end of the war. For example, an overtly anti-American paper by Karl Beurlen and Ulrich Lehmann (1916–2003)[53] was published in the journal of the German Geological Society in 1947. The paper

---

[53] In 1935 Lehmann began to study geography and history, first at the University of Jena, and later at Humboldt University in Berlin, where he added philosophy and English Studies to his curriculum. Inspired by Hans Stille (1876–1966), then a geology professor in Berlin, Lehmann switched to geology. In 1944, he '*habilitated*' in palaeontology, with Karl Beurlen as his supervisor. In 1951, Lehmann became a lecturer at the University of Tübingen. In 1957–58 he was a guest scientist at Southern Illinois University, and in 1958 he became a professor (without chair) at the University of Hamburg (<http://de.wikipedia.org/wiki/Ulrich_Lehmann> accessed 14 November 2011).

had been submitted in February 1944, but as the volume for that year (ZDDG 96) could not be published until three years later, it sat there waiting. There may have been some reworking of the manuscript, as palaeontology was no longer represented as the sole and glorious product of Germany but as part of the European tradition. However, the paper attacked the general achievements of the American palaeontologist Henry Fairfield Osborn (1857–1935).[54] Can we, as the question was rhetorically posed, learn from this man, deemed to be the greatest palaeontologist of the USA? The authors' answer was 'no', as they judged Osborn's work to be neither methodical, thorough, nor logical:

> Here simply by authority a new opinion is brought forward, and not in a short preliminary notice but in a monograph that surpasses in size every normal and usual scope! It displays either thoughtless superficiality and cursoriness or the same excessive arrogance with which today's Roosevelt–America wants to command and decree over Europe, without having the least idea about Europe. (Beurlen and Lehmann 1944; printed in 1947: 234)[55]

According to 'European criteria of palaeontological work', Osborn's work 'could only be termed quite unscientific' (Beurlen and Lehmann 1944; printed in 1947: 234).[56]

Nevertheless, to the authors' annoyance, Osborn's publications seem to have been appreciated by European palaeontologists. Why this was so was not quite clear to Beurlen and Lehmann. Certainly, it could not be attributed to simple unquestioning belief in authority:

> although experience shows that in the USA it is comparatively easy by corresponding advertisement to create a public opinion against which there is then no more criticism, in our old Europe there is on average a greater rational and critical independence of the individual, which does not accept propagated authority so naively or blindly. (Beurlen and Lehmann 1944; printed in 1947: 235)[57]

---

[54]   <http://en.wikipedia.org/wiki/Henry_Fairfield_Osborn> accessed 6 April 2011.

[55]   'Hier wird einfach kraft Autorität eine neue Ansicht vorgetragen, und das nicht in einer kurzen vorläufigen Mitteilung, sondern in einer Monographie von einem jedes normale und übliche Maß überschreitendem Umfang! Das ist entweder gedankenlose Oberflächlichkeit und Flüchtigkeit oder aber die gleiche maßlose Überheblichkeit, mit welcher das heutige Roosevelt-Amerika über Europa bestimmen und verfügen möchte, ohne eine Spur von Ahnung von Europa zu haben'. (Beurlen and Lehmann 1944; printed in 1947: 234).

[56]   'europäischen Maßstäben paläontologischer Arbeiten'; 'nur als ganz unwissenschaftlich bezeichnet' (Beurlen and Lehmann 1944; printed in 1947: 234).

[57]   'Denn wenn es auch erfahrungsgemäß in USA. verhältnismäßig einfach ist, durch eine entsprechende Reklame eine öffentliche Meinung über etwas zu machen, an der es dann keine Kritik mehr gibt, so ist doch in unserem alten Europa im Mittel eine größere geistige

In the light of the situation in Nazi Germany, this last sentence claiming independence from (and a critical attitude towards) authority as their share seems to represent the ultimate peak of self-delusion in the authors.

Such flawed science as Osborn's, so the authors postulated, was the fruit of a 'liberalist' attitude. And by criticising Osborn, the authors wanted to be understood as making a criticism of empiricism as a whole (Beurlen and Lehmann 1944; printed in 1947: 236):

> American palaeontology with its most prominent experts has originated from the school of German palaeontology. The pupil, however, has not surpassed the teacher, who, living in modest quarters and blinded by the seeming wealth of the matured pupil, has pretended this to himself. Osborn displays how much superficiality and vacuity lurks behind the brilliant façade. Therefore, let us distance ourselves from the admiration for American palaeontology, which has no authorization, and let us remember our own capabilities and our distinguished tradition. (Beurlen and Lehmann 1944; printed in 1947: 236)[58]

As can be seen, a previously chauvinistic attitude tended to linger on, albeit sometimes in a new disguise. It lay behind Georg Wagner's criticism of Edward Chao as well as Richard Dehm's apologetic vindication of the German Ries research (see Chapter 10), and it was by no means restricted to German nationalism, but affected all levels of encounter with 'strangers':

> In Georg Wagner's person a contrast was embodied that is more or less forgotten [without consequence] today. Even though he felt at home in Swabia, he always remained a Franconian inside.[59] When, however, during ... an infamous newspaper campaign around 1950 against the alleged 'foreign' infiltration of the University

---

und kritische Selbständigkeit des einzelnen vorhanden, die eine propagierte Autorität nicht so naiv unbesehen hinnimmt' (Beurlen and Lehmann 1944; printed in 1947: 235).

[58] 'Die amerikanische Paläontologie ist mit ihren hervorragendsten Vertretern aus der Schule der deutschen Paläontologie hervorgegangen. Der Schüler hat aber seinen Lehrer nicht überflügelt, wie dieser, der in kleinen Verhältnissen lebt, geblendet von dem scheinbaren Reichtum seines großgewordenen Schülers, selbst sich eingeredet hat. Wie viel Oberflächlichkeit und Leere hinter dieser blendenden Fassade steht, zeigt uns Osborn. Lösen wir uns daher aus der Bewunderung für die amerikanische Paläontologie, die keinerlei Berechtigung hat, und besinnen wir uns auf unser eigenes Können und die große Tradition' (Beurlen and Lehmann 1944; printed in 1947: 236).

[59] Franconia is a northern province of the state of Bavaria; meaning that Wagner was no native of Swabia, where Tübingen is situated. People would be able to tell that from his dialect. Also, there used to be a historically grown rivalry between Württemberg (Swabia) and Bavaria, of which Franconia nowadays is a part (much to the dismay of some especially patriotic Franconians).

of Tübingen by non-Swabian professors, ... he found himself pushed to the side of the strangers, his alarm was great, although it actually happened as a joke. By no means did he want to be grouped in the same category as the Prussians,[60] even though he had close friendships with like-minded researchers from East Prussia to Ostfriesland![61] (Engelhardt and Hölder 1977: 154)[62]

This mental attitude explains why at that time basically all professors in Tübingen and Munich had been trained by those two universities. They never looked further to broaden their horizons.

Scientific and linguistic isolation of Nazi Germany, beginning around 1937 and continuing through the early post-war years, prevented any competent knowledge of English. For this reason, the international literature was rarely held by the institutes' libraries (see Chapter 3). Thus, the German Ries experts had been isolated from international impact crater research since about 1937. At first this was due to the political conditions in Germany, World War II and the general problems of the post-war years. However, after the universities began to recover in the 1950s, the continuing isolation or lack of interest in the wider world was self-inflicted. With self-consciousness about the merits of a 'German Geology' – an attitude that still owed much to the propaganda of the 1930s and the war years – people saw no need to concern themselves with foreign literature or to reflect on their own methodology. Because 'German Geology', as it had been defined in the late 1930s and early 1940s, was about the individual history of a given piece of land, where processes had led to specific, unique phenomena influenced by the local geohistory, there was considered to be no need to or even sense in comparing the Ries Crater or the Steinheim Basin to foreign places of a similar appearance. The local history was all that was thought necessary to understand the two craters – a fallacy, as it turned out.

---

[60]   In this context meaning anything in the northern half of Germany. For a Bavarian, to be taken as a Prussian is even worse than being confused with a Swabian. It might have the same effect as showing a red rag to a bull.

[61]   That is all along the coasts of the Baltic and North Seas.

[62]   'In Georg Wagner verkörperte sich noch ein heute weithin vergessener Gegensatz. So heimisch er sich im Schwabenland auch fühlte – er blieb stets ein Franke. Als er sich freilich in einer ... wenig rühmlichen, um 1950 geführten Zeitungskampagne gegen die angebliche Überfremdung der Universität Tübingen durch nichtschwäbische Professoren ... auf die andere, 'verfremdete' Seite gedrängt sah, war sein Schrecken groß, obgleich es eigentlich im Scherz geschehen war. Denn mit den Preußen wollte er sich erst recht nicht in einen Topf werfen lassen, ungeachtet seiner herzlichen Freundschaft mit ihm gleichgesinnten, forschenden Männern von Ostpreußen bis Ostfriesland!' (Engelhardt und Hölder 1977: 154).

Chapter 8

Setting the Stage

While the period of Nazi rule, as well as World War II and its aftermath, caused a significant disruption in Germany of the impact concept of crater formation, impact research continued unabated in North America, even though the detailed scientific investigation at Meteor Crater was still hampered by a lack of interest among officials of the United States Geological Survey (USGS) as well as the reservations of the local mining company under Barringer. Thus, detailed geophysical investigation of this classic among impact craters was not undertaken until the early 1970s (Hoyt 1987: 328–30). Rather, impact research was fuelled by an increasing interest in the origin of lunar craters, which by most astronomers and geologists were still considered to be of volcanic origin: 'It cannot be said that [the photographs of the Moon] give a final answer to the problem of the origin of the chief craters, though they certainly provide immensely strong support for a volcanic hypothesis (and ... I personally doubt whether the impact theory can survive for much longer)' (Patrick Moore, British astronomer born 1923, in 1966 after the first lunar surface photos made by the Russian Luna 9 lander; quoted after Torrens 1998: 183).

In Germany the same scepticism prevailed, despite, for example, Alfred Wegener's impact experiments concerning an impact origin of lunar craters:

> When I studied [geology in the 1930s], the craters of the Moon were regarded as volcanic features ... just as at the Ries ... . Perhaps, at that time, meteorite impacts had been discussed a little, but they were not taken seriously; and if anybody compared the Ries and Steinheim Basins with lunar craters, it was of course primarily in a volcanic light. (Hölder interview)[1]

But a constant stream of people, such as Dietz (1946), Baldwin (1949) and others, opted for an impact origin of the craters of the Moon and kept the American discussion going.

---

[1] 'Die Mondkrater waren in meiner Studienzeit vulkanische Erscheinungen; ... genau wie beim Ries ... . Meteoriteneinschläge wurden damals vielleicht schon ein bisschen diskutiert aber nicht ernst genommen. Und wenn man das Ries und das Steinheimer Becken mit Mondkratern verglichen hat, dann natürlich zunächst unter vulkanischen Gesichtspunkten' (Hölder interview).

However, the possible impact nature of the Ries and Steinheim Basins was likewise not forgotten, as the notion had entered the international literature thanks to Rohleder, who had published one paper about it in English (Rohleder 1933b) that had in turn been quoted in the seminal work of Baldwin. Baldwin gave brief descriptions of Steinheim as well as of the Ries Basin (1949: 107–10), although Rohleder had denied an impact origin of the latter due to the presence of what seemed to be volcanic rocks. But according to Williams (1941) in a book on calderas and their origin, no magma was associated with the actual Ries explosion:

> To produce a broadly flaring funnel of so great proportions, it is obvious that the explosion focus must be very shallow. But at shallow depths magma can hold little gas in solution and can only produce explosions of weak to moderate intensity such as would puncture the roof by a series of diatremes. No new magma was associated with the first explosions of the Ries; they seem to have been low temperature, phreatic eruptions. All the more is there reason for doubting the theory of Kranz.

> Bentz, Branca, Reck, and others agree that after the main explosive phase the summit of the dome collapsed along ring fractures, and it was primarily this engulfment which formed the caldera. Rittmann also seems to adopt this view when he refers to the Rieskessel as a 'volcano-tectonic sink'. After the collapse, dikes of suevite were injected close to the margins of the basin and mild explosions of suevite pumice ensued. Apparently the magma was produced by gas fluxing of the granitic basement. (Williams 1941 as quoted in Baldwin 1949: 109)

Baldwin pointed out that Williams denied the Kranz-type central explosion only because no such big volcanic explosions were known. In meteorite craters, however, no such problem existed, as any amount of energy could be produced (Baldwin 1949: 108–10).

Baldwin's book would prove rather influential when the USGS, in about 1960, began to develop interest in lunar 'geology' following the *Sputnik*-shock in October 1957. When in 1961 President John F. Kennedy (1917–63) opened the 'Race to the Moon', the USGS founded an 'astrogeology' working group, mainly funded by the National Aeronautics and Space Administration (NASA), which later was also responsible for the geological support of the Apollo missions (Barsig and Kavasch 1983: 29–30).

Visit by a 'Crazy' American

Robert Sinclair Dietz (1914–95) (Figure 8.1), geologist and oceanographer, worked for his PhD at the Scripps Institution of Oceanography (SIO) but received his degrees from the University of Illinois. He spent World War II in the US Army Air Corps, partly as a pilot. In 1946 Dietz founded the Sea Floor Studies Section of the Naval Electronics Laboratory in San Diego and became its director, a position he held until 1963. From 1950 to 1963 he was also an adjunct professor at the SIO. He took two larger breaks from these responsibilities; first in 1953 as a Fulbright Scholar at the University of Tokyo and from 1954 to 1958 working at the Office of Naval Research in London. During the latter time period, he also visited the Ries and Steinheim Basins.

In 1963 Dietz became a member of the oceanographic and geological studies group of the US Coast and Geodetic Survey, later integrated into the National Oceanic and Atmospheric Administration (NOAA). In 1975 Dietz retired from the NOAA but continued his research work and lectured at various American universities.

Dietz acquired an international reputation for promoting seafloor spreading and plate tectonics, but also for his interest in impact craters. He proposed more than 130 impact locations based on the presence of shatter cones, which Dietz thought to be a diagnostic feature of impact processes.[2]

Dietz's interest in impact craters stemmed from his student days, when he frequently visited the Kentland structure where he had seen well-developed shatter cones that he thought had been caused by forces from above, not below; the Kentland structure thus was not 'cryptovolcanic' but represented the underground of a heavily eroded meteorite crater (Mark 1995: 123):

> Dietz defined shatter cones as "striated percussion fracture cones" or, more simply, as "conical fragments of rock characterized by striations that radiate from the apex". In his opinion, cryptovolcanic or crypto-explosion structures, which he called "astroblemes" – literally, "star wounds" – were created by the hypervelocity impact of large meteorites. ... Hypervelocity impact must create an intense, high-velocity shock (or pressure) wave which spreads out from the impact point through the surrounding rocks. Therefore, Dietz pointed out, indications that a large volume of rock has been intensely and naturally shocked would constitute evidence of meteoritic impact. Perhaps the presence of shatter cones could provide such evidence. (Mark 1995: 123)

---

[2]   <http://scilib.ucsd.edu/sio/archives/siohstry/dietz-biog.html>,   <http://geology. asu.edu/resources/museum/dietz.html>, both accessed 19 September 2002; Bourgeois and Koppes 1998: 142–4.

Figure 8.1   Robert Sinclair Dietz promoted shatter cones as macroscopic clues to impact processes. (Photo courtesy of Scripps Institution of Oceanography Archives).

Figure 8.2   *Strahlenkalk*, a shatter cone from the Steinheim Basin. (Photo: M. Ebert).

Dietz documented occurrences of shatter cones and found that they were not related to volcanic phenomena or any other 'ordinary' geological processes; instead, he reported similar features from quarries and military test sites (for chemical explosives as well as nuclear bombs), while the similarity between natural shatter cones and the structures caused by artificial explosions increased with increasing detonation velocity (Dietz 1960: 1782). This confirmed the genetic relationship between shock waves and shatter cones, even though until 1959 they were only known at five potential impact craters, among them the Steinheim Basin and Lake Bosumtwi. The latter occurrence had been reported by Herbert Rohleder (Rohleder 1936; see Chapter 2).

On 24 June 1955,[3] Robert Dietz visited the Steinheim Basin, the type-locality of cryptovolcanic sites as Baldwin (1949) had stated, and proclaimed the *Strahlenkalk* – 'ray limestone', a conspicuous but strange surface feature of limestone fragments commonly found there – to be shatter cones and thus evidence for a meteorite impact (Figure 8.2). He took a guide from the

---

[3]   According to Dietz (1959), he visited the Steinheim Basin in 1956 and 1957, but Seibold's field notebook confirms that the first visit was in 1955.

University of Tübingen: Eugen Seibold (1918–2013), a sedimentologist and micro-palaeontologist, who had received his PhD degree in 1948 from the University of Tübingen with a stratigraphical thesis based on locations close to the north-western margin of the Ries. Seibold remained in Tübingen as an assistant at the Institute for Geology and Palaeontology from 1950 to 1951, as a professor without chair from 1954 to 1957, in succession of Georg Wagner (see Chapters 9 to 11) whose student he had been. In 1958 Seibold became a professor at the University of Kiel (northern Germany) (Engelhardt and Hölder 1977).

Because of language problems, the tour proved to be less fruitful for Ries geology in Germany than it might have been if the guide had been able to properly discuss Dietz's argument. Seibold well remembered the excitement that seized Dietz when he saw the *Strahlenkalk*, but after Dietz left again he did not leave any trace in Tübingen. Nobody seemed to have valued his ideas:

> If some strangers came who wanted to see the field, the Swabian Alb, the Black Forest, the Ries and so on, then Schindewolf,[4] who then was the boss, obviously liked to draw on a native and so I was allowed to guide very interesting people. Mr Schindewolf had connections all over the world, and then came Bob Dietz, quite surprisingly, and said that he would love to see the Ries and the Steinheim Basin too. "There" I thought, "it's really impressive that an American knows about the Steinheim Basin!" Well, we fixed a date and drove away in the morning to Steinheim, and all the way there he enthused about the possibility that it was due to a 'me-toe-write' [Seibold imitates Dietz's heavy American accent]. Well, at that time my English was not yet trained at all, and it took me until we got to Geislingen[5] until I realized 'me-toe-write' – that's meteorite! And then he began [to talk] and came to [mention] *Strahlenkalk*, that of course I knew, and then we said, we'll drive to Steinheim – and I thought: "such a crazy American" (I did not know him) ... – and then we arrived in Steinheim and he wanted to see the *Strahlenkalk* and we had enormous luck because, I think, the school [foundation] had just been dug out, and there was *Strahlenkalk* lying around, and then he packed it into the car at once and was completely happy and started to tell me of America. There were some impact craters, I think the Arizona crater, which is clearly confirmed, and others as well, where he had reasoned with the help of *Strahlenkalk* whether they might be a meteorite [crater], and then ... I showed him everything south of Nördlingen [the southern part of the Ries Basin] and brought forward all the arguments that I knew from Georg Wagner and Mr Löffler, and said: "Well, the suevite to some extent looks like the volcanic vents

---

4   Otto Schindewolf (1896–1971), successor to Edwin Hennig (after 1948) as Chair of palaeontology (Zinnstein 2005).
5   Some two hours by car from Tübingen.

on the Swabian Alb, with the filling and so on" ..., and then we returned happily home and he tried to convince me. (Seibold interview)[6]

Dietz had possibly learned of the *Strahlenkalk* from the English paper by Rohleder (1933b), where this feature is briefly described:

> A strange phenomenon within the shattered Upper Jurassic limestone is afforded by certain striated surfaces, which were first described by Branca and Fraas, as resembling some problematic algae like Cancellofycus, Taonurus, etc. Such marks have never been found elsewhere, and are generally regarded as effects of volcanic explosions. W. Kranz calls such fragments "volcanic bombs", in spite of the fact that they consist of pure white limestone. (Rohleder 1933b: 490)

Dietz not only saw the shatter cones in the field, but also reviewed collection material at the Tübingen institute, which was shown to him by Seibold (Dietz 1959: 501).

In 1957 Dietz revisited the Steinheim Basin and two years later published his important papers on shatter cones in crypto-explosive structures, in which he argued for his opinion that these structures were in fact impact craters (Dietz 1959, 1960).

---

6 'Wenn irgendwelche Leute von auswärts kamen, die das Gelände sehen wollten, von der Alb, vom Schwarzwald, vom Ries u.s.w. dann hat der Schindewolf der damals Chef war natürlich gerne auf einen Einheimischen zurückgegriffen und so dass ich eben sehr interessante Leute führen durfte. Herr Schindewolf hat ja Verbindungen in der ganzen Welt gehabt, und da kam Bob Dietz daher ganz überraschend und sagte, er würde gerne das Ries sehen und auch das Steinheimer Becken. Da dachte ich, das ist ja schon toll, wenn ein Amerikaner das Steinheimer Becken kennt! Gut also, da haben wir was ausgemacht, sind morgens los gefahren nach Steinheim und auf dem ganzen Weg hat er mir vorgeschwärmt, dass das sehr wahrscheinlich auf ein 'me-toe-write' zurückgeht. Nun war mein Englisch damals überhaupt noch nicht geübt, und ich brauchte also bis Geislingen bis ich dahinterkam: 'me-toe-write' – das ist Meteorit! Und dann hat er also losgelegt und dann kam der Strahlenkalk, das wusste ich natürlich, und dann sagten wir gut, dann fahren wir nach Steinheim – und ich dachte, so ein verrückter Ami (ich kannte ihn nicht) ... – und dann kamen wir nach Steinheim und er wollte also Strahlenkalk [sehen] und wir hatten ein irres Glück, denn ich glaube, die Schule wurde gerade ausgehoben, und da lagen tatsächlich Strahlenkalke da und dann hat er das natürlich gleich in's Auto gepackt und war also hell glücklich und hat mir dann also erzählt von Amerika. Da gab es ja einige Einschlagkrater, ich glaube der Arizonakrater, der ja ganz klar ist und auch noch andere, wo er mit Hilfe von Strahlenkalk darüber nachgedacht hat, dass das vielleicht ein *meteorite* ist und dann ... habe [ich] ihm da südlich von Nördlingen alles gezeigt und hab dann die Argumente, die ich vom Georg Wagner und vom Herrn Löffler kenne, vorgebracht und gesagt, ja der Suevit der sieht ja in manchen Dingen aus wie die Vulkanschlote auf der Schwäbischen Alb mit der Füllung usw. ... und dann sind wir also fröhlich heim und er hat dann versucht mich zu überzeugen' (Seibold interview).

As to the Ries Basin, Dietz admitted that they had not found any shatter cones in this structure,[7] but he recognised the great structural similarities between the two craters and for this reason he regarded both structures as impact craters 'formed simultaneously in a double holocaust' (Dietz 1959: 502).

> Probably the wildest paper Dietz wrote was in 1958 ... suggesting that Earth's ocean basins may have had their origin by giant asteroid impacts early in Earth's history, and using the nature of lunar maria to support his hypothesis. This paper received little attention, and within three years, Dietz had proposed a different origin for Earth's ocean basins – sea-floor spreading – and was working to distance himself from what he considered the 'fringe' element, who as he briefly had, proposed asteroid impacts to explain mega-scale features and events on Earth. (Bourgeois and Koppes 1998: 139)

Dietz was not the only one who systematically searched for impact craters. Should the craters of the Moon really be of impact origin, then the Earth too must have been subject to this bombardment. The greatest chance of detecting its traces would be on old surfaces, where ample time would have allowed craters to accumulate. The closest accessible space that answered this description was the Canadian Shield, with its vast expanse of Precambrian rocks. In the mid-1950s, thousands of air-photographs provided by the Canadian air force were examined by Canadian scientists for possible impact craters that afterwards could be investigated from the ground, such as the Chubb Crater, which was visited by expeditions in 1950 and 1951 (Mark 1995: 133–8; Plotkin and Tait 2011).

Coesite

Meanwhile, in addition to Dietz's macroscopic evidence for impact craters, microscopic/mineralogical evidence had reached a level of knowledge that allowed it to be transferred from basic research to practical application in impact research.

In 1953, under laboratory conditions, a new mineral had been discovered by the American industrial chemist Loring Coes (born 1915), later named 'coesite' in his honour. Coesite was a high-pressure modification of quartz. Its different crystal structure led to a somewhat higher density. Although the mineral was formed under extremely high pressure, once formed it proved to be rather stable under normal conditions. Thus, it seemed possible that it might also exist in nature and so far had been overlooked. However, a first check in

---

7   The first clear evidence of shatter cones in the Ries Basin was published by Rudolf Hüttner (1969: 165–7).

eclogite-xenoliths, brought to the Earth's surface from deep within the mantle by kimberlith magmas, failed to produce coesite, and it was also not found at nuclear test sites (Pecora 1960; Mark 1995: 114–15):

> With coesite, it was like this: ... When Eugene M. Shoemaker investigated this structure [Meteor Crater] within the scope of his PhD thesis, the meteorite researcher Harvey H. Nininger[8] suggested that he look for the high-pressure modification coesite, which hitherto had never before been found in nature, in the Coconino sandstone, which consists of quartz grains and is situated at the bottom of the crater. With the cooperation of the mineralogist Edward C.T. Chao, who worked for the USGS in Washington, Shoemaker succeeded in detecting coesite by X-ray methods within samples of the Coconino sandstone from Meteor Crater. Thus, the effect of pressures over 30 kbar on the genesis of the crater was confirmed, and the origin by impact ascertained. (Letter from Wolf von Engelhardt to the author, 16 March 2000)[9]

The details of the story, however, are quite complicated. Eugene Merle Shoemaker (1928–97) obtained his M.Sc. degree at the California Institute of Technology. An M.A. degree from Princeton University followed in 1954. In 1948 he had started to work for the USGS, doing fieldwork in Colorado and Utah (Carlé 1987; Barsig and Kavasch 1983: 30).

During that time, the USGS prospected in these states for uranium supporting the US Atomic Energy Commission (AEC), which was the successor of the Manhattan Project. Uranium at that early time was mainly needed for the continued production of nuclear bombs, as the Manhattan Project had basically used up all of the available metal. Shoemaker had for ten years been part of the prospecting staff when, in 1958, the massive uranium exploration of the USGS came to a preliminary stop. Shoemaker asked for and was granted an extended break to pursue his PhD at Princeton University, as he planned to develop some of his mapping work for the USGS into a PhD thesis. Some time later,

---

[8]   Harvey Harlow Nininger (1887–1986) was an American meteorite collector and educator (<http://en.wikipedia.org/wiki/Harvey_H._Nininger> accessed 27 March 2012).

[9]   'Mit dem Coesit verhält es sich folgendermaßen: ... Als Eugene M. Shoemaker im Rahmen seiner Dr.-Arbeit die Geologie dieser Struktur untersuchte, schlug der Meteoritenforscher Harvey H. Nininger ihm vor, in dem am Grund des Kraters anstehenden, aus Quarzkörners [sic] bestehenden Coconino-Sandstein nach der bis dahin noch nie in der Natur gefundenen Hochdruckmodifikation Coesit zu suchen. Unter Mitarbeit des am U.S. Geological Survey in Washington tätigen Mineralogen Edward C.T. Chao gelang Shoemaker der röntgenographische Nachweis von Coesit in Proben des Coconino-Sandsteins vom Meteor Crater. Damit war die Wirkung von Drucken über 30 kbar bei der Bildung des Kraters nachgewiesen und die Entstehung durch einen Impakt sichergestellt' (letter from Wolf von Engelhardt to the author, 16 March 2000).

however, he received an attractive offer from the USGS to map the distribution of elements, which might point to ore deposits, within the sediments in the Colorado Plateau. So again he took a break, this time from his PhD coursework at Princeton.[10]

Meanwhile, since about 1956, the United States had difficulties with plutonium production; plutonium at that time was again needed for the production of nuclear weapons and the extensive nuclear testing of that era.[11] In order to mitigate that problem, the AEC thought about producing plutonium by means of a gigantic underground nuclear explosion and subsequent retrieval of the plutonium through conventional mining technology (known as the MICE project). First, however, a feasibility study was deemed necessary, as it was feared that the explosion might get out of hand and erupt like an artificial volcano. Shoemaker, who had worked in the area and was especially an expert on the diatremes of the Colorado Plateau, was called in to lend his expertise. He started to familiarise himself with the processes related to large explosions. Shoemaker also investigated two comparatively small nuclear test craters in Nevada that were available for scientific studies, mainly in order to learn more about the geochemistry of plutonium and to find out where the newly produced plutonium would go. Shoemaker mapped and measured the test craters and investigated their mineralogy and geochemistry. It quickly dawned on him that in contrast to the wishful thinking of the AEC people, the plutonium would not nicely [but supercritically] pool in a cavity created by the explosion or line its walls, but that it would become finely dispersed in a large volume of shocked debris. In order to be able to scale up his findings from the small nuclear craters made by bombs equivalent to a few kilotons of TNT, to a bomb releasing a thousand times more energy, Shoemaker went to nearby Meteor Crater (still within the scope of the MICE project), which he also mapped in order to obtain some guidelines for his scaling problem. The experiment to produce large amounts of plutonium never became a reality, but Shoemaker's investigations were to become valuable in quite different respects.[12]

The time after World War II was also a time when another new technology was about to mature: rockets. In 1956, Russian engineers succeeded in getting the first artificial satellite into Earth orbit, and in the ensuing enthusiasm for American space exploration, Shoemaker, who later claimed that he had always

---

[10]   Interview of Dr Eugene Shoemaker by Ronald Doel on 16 June 1987, Niels Bohr Library and Archives, American Institute of Physics, College Park, MD USA, <www.aip.org/history/ohilist/5082_2.html> accessed 25 February 2011.

[11]   Between 1945 and 1963 (when the international ban on nuclear testing was ratified), the USA carried out 259 tests with nuclear explosions (Franke 1983: 31).

[12]   Interview of Dr Eugene Shoemaker by Ronald Doel on 16 June 1987, Niels Bohr Library and Archives, American Institute of Physics, College Park, MD USA, <www.aip.org/history/ohilist/5082_2.html> accessed 25 February 2011.

been interested in the Moon and the possibility of travelling into space, got the USGS to develop an interest in lunar geology by telescope; and so several lines of expertise came together in the person of Eugene Shoemaker: lunar geology and the craters of the Moon, knowledge of crater mechanics, interest in shocked material, mineralogy/geochemistry and knowledge about Meteor Crater.

Shoemaker planned to look for coesite in his samples from Meteor Crater. Much later he no longer remembered whether he got the idea from Harvey Nininger or whether he reached it independently. His colleague, however, Beth M. Madsen, who was the person responsible for the XRD machine, was off for a few days of vacation and meanwhile, by some circuitous means, one of Shoemaker's samples had been sent to Edward Chao at the USGS laboratory in Washington DC, who hit on the same idea; Chao looked for and indeed found coesite. Shoemaker was shocked when he heard about it, but as soon as his colleague Madsen returned, they went through all other samples and thus were able to assess the distribution of coesite in Meteor Crater. Chao, Shoemaker and Madsen published jointly in 1960.[13]

In 1960 a phase diagram for coesite was published by Boyd and England. For the formation of coesite pressures of at least 20 kilobars and temperatures between 700°C to 1,700°C were necessary, which at the Earth's surface could only be provided by meteorite impacts. Consequently, the presence of coesite seemed a good indicator for an impact (Mark 1995: 118–19; Figure 8.3).

Meanwhile, because of all these distractions, Shoemaker had completely neglected the writing of his PhD thesis and nearly missed the deadline, and so, by agreement with his supervisor, he simply used what was uppermost on his desk. He had planned to present a talk at the International Geological Congress in Copenhagen on *Penetration mechanics of high velocity meteorites, illustrated by Meteor Crater, Arizona,*[14] and as the MICE project was secret, he was forced to submit his paper one year in advance and apply for travel permission. The first draft was much too long and had to be shortened for the talk, but the long version became Shoemaker's PhD thesis (Shoemaker 1960).[15]

---

[13] Interview of Dr Eugene Shoemaker by Ronald Doel on 17 June 1987, Niels Bohr Library and Archives, American Institute of Physics, College Park, MD USA, <www.aip.org/history/ohilist/5082_3.html> accessed 25 February 2011.

[14] Guest and Greely (1979: 245).

[15] Interview of Dr Eugene Shoemaker by Ronald Doel on 17 June 1987, Niels Bohr Library and Archives, American Institute of Physics, College Park, MD USA, <www.aip.org/history/ohilist/5082_3.html> accessed 25 February 2011.

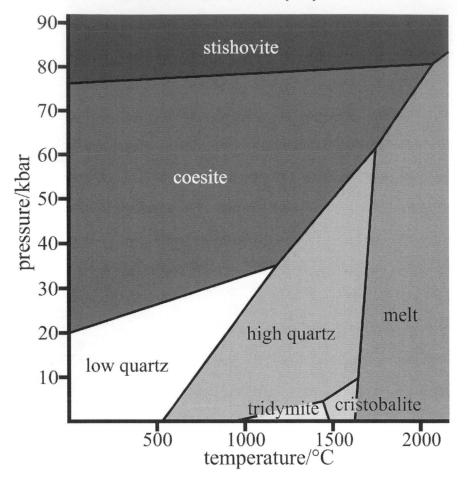

Figure 8.3   Phase diagram of SiO$_2$ showing the temperature and pressure
conditions under which the high-pressure modifications coesite
and stishovite form. (Redrawn using data from <https://www.
uwgb.edu/dutchs/Petrology/Silica%20Poly.HTM> accessed 12
November 2013).

Meanwhile in Germany

While impact research continued unabated on the North American continent,
in Germany after World War II the popularity of impactist thinking of the 1930s
was stunted. Nevertheless, the geology of the Ries and Steinheim Basins was not
just left to stasis. In Germany too, things began to move again, inspired by the
experience of a most horrible war.

After World War II, imagining larger explosions than had previously been conceivable suddenly seemed possible. This was due to the shock caused by the nuclear bombs dropped on Hiroshima and Nagasaki. This experience of a force larger than conventional chemical explosives by several orders of magnitude led to tentative speculations about new, albeit again terrestrial mechanisms, which can be found in private letters as well as in printed sources:

> By the way, how is it with the possibility of falling back on processes of nuclear physics for triggering the Ries explosion? In this context, perhaps one should look again into the nature and origin of the somewhat dubious radium bath in Bissingen in the Ries foreland. At least, one should not a priori decline a natural formation of plutonium, even when this might be possible only under exceptional conditions and very rarely. However, the Ries always remains an unusual case, no matter which hypothesis one is inclined to follow. If, however, one tries to split up the whole complex of questions into countless local and local-most causes, doing so in my opinion only proves the defect of the observer, who obviously does not see the forest due to all the trees, and that is unsuitable for any synthesis. (Letter from Alfred Bentz to Walter Kranz, 11 January 1947, GA 19340)[16]

Kranz, perhaps inspired by this letter, tried to include the possibility of a naturally occurring nuclear explosion in a paper for the popular journal *Naturwissenschaftliche Rundschau*. The editor, Hans Walter Frickhinger (1889–1955), however, rejected the manuscript: 'I felt especially alienated by the mentioning of atomic energy as a possible agent of the explosion', wrote Frickhinger. 'This mentioning to me seems in no way advisable' (letter from Hans Walter Frickhinger to Walter Kranz, 13 January 1949, GA 19367).[17]

---

[16] 'Wie steht es übrigens mit der Möglichkeit, Vorgänge der Kernphysik zur Auslösung der Riessprengung heranzuziehen? In dem Zusammenhang sollte man vielleicht doch einmal die Natur und Herkunft des etwas dubiösen Radiumbades Bissingen im Vorries einer Nachprüfung unterziehen. A priori sollte man wenigstens die Möglichkeit einer natürlichen Bildung von Plutonium nicht ablehnen, wenn dies auch nur unter ekzeptionellen Bedingungen und sehr selten der Fall sein dürfte. Ein solcher außergewöhnlicher Fall ist und bleibt das Ries aber immer, welcher Hypothese man auch zuneigt. Versucht man aber, den ganzen Fragenkomplex in eine Unzahl lokaler und lokalster Ursachen aufzuspalten, so beweist dies m. E. nur den Mangel des Beobachters, der offensichtlich den Wald vor Bäumen nicht sieht und zu jeglicher Synthese ungeeignet ist' (letter from A. Bentz to Kranz, 11 January 1947, GA 19340).

[17] 'Besonderen Anstand nahm ich dann noch an der Erwähnung der Atomenergie als möglichen Agens der Explosion. Diese Erwähnung scheint mir in keiner Weise ratsam zu sein' (letter from Hans Walter Frickhinger to Walter Kranz, 13 January 1949, GA 19367).

But Bentz and Kranz were by no means the only people to harbour such speculations. The geographer Carl Troll (1899–1975),[18] a friend of Herbert Hyde (né Rohleder), offered a printed comment to a joint Ries paper by Wagner and Löffler (1949) in the respected German geological journal *Geologische Rundschau*:

> The description by the speaker of the disturbed bedding, the structural metamorphosis of the rocks, fossils and pebbles, and not least the geomorphological situation, was so remarkably vivid that one could hardly follow it without viewing the process in one's head. I take it that normal tectonics must definitely be excluded. At shallow depth, an explosion of obviously exceptional force must have occurred and pushed huge slices in a flat trajectory radially outside. The crystalline basement was melted, but effusive material itself was not erupted by the explosion. One is tempted to ask whether it was not a natural example of atomic fission from a crystalline basement containing uranium. At any rate, atomic fission could explain the enormous destruction by pressure and the creation of heat without invoking a plutonic origin. (Printed remark by C. Troll following Wagner and Löffler 1949; offprint in UAT 605/343)[19]

Clearly, these ideas were nonsense from a physical point of view, but at that early time, hardly anything was known by the public about the mechanisms of nuclear physics. In any case, such speculations allowed an opening of the horizon of what it was possible to think. They broadened the mind about the possibility of much

---

[18]   Troll studied a broad range of science subjects in Munich. In 1921 he obtained a PhD in botany but *habilitated* in 1925 in geography. From 1922 to 1927 he worked as an assistant at the Geographical Institute in Munich, followed by a research journey to South America. In 1930 he became Professor of Colonial and Overseas Geography in Berlin. From 1936 to 1938 he was Professor of Economic Geography in Berlin, and from 1938 until his retirement in 1966 he was Professor and Director of the Geographical Institute at the University of Bonn. During the early years of his professorial career, he undertook extensive expeditions to Africa as well as to the Himalayas (<http://de.wikipedia.org/wiki/Carl_Troll> accessed 27 March 2012).

[19]   'Die Schilderung der gestörten Lagerungen, der Strukturmetamorphose der Gesteine, Fossilien und Gerölle und nicht zuletzt der geomorphologischen Situation durch den Referenten war so greifbar anschaulich, daß man ihr kaum folgen konnte, ohne den Vorgang im Geiste zu schauen. Normale Tektonik muß also endgültig ausscheiden. Eine Explosion von offenbar beispielloser Brisanz in geringer Tiefe muß vor sich gegangen sein und riesige Schollen im 'Flachschuß' radial nach außen geschoben haben. Das Grundgebirge wurde aufgeschmolzen, aber Effusivmaterial wurde bei der Explosion selbst nicht gefördert. Man ist versucht zu fragen, ob es sich nicht um ein natürliches Beispiel von Atomzerfall aus einem etwa uranhaltigen Grundgebirge gehandelt hat. Atomzerfall könnte jedenfalls die enorme Druckzertrümmerung und Hitzeentfaltung ohne eine plutonische Ursache erklären' (printed remark by Carl Troll following Wagner and Löffler 1949; offprint in UAT 605/343).

larger explosions than had been conceivable from experience with chemical explosives alone. On the other hand, they show how deeply entrenched was the conviction of an earthbound cause. Any wild speculation was admissible as long as it involved a terrestrial mechanism.

In the USA, in contrast, nuclear test sites and their craters fuelled the imagination of impact geologists in a more concrete way. The nuclear tests, their adverse influence on the environment notwithstanding, provided a unique possibility of investigating the influence of shock waves on various rocks and correlating the observed metamorphic features with the known amount of energy used for their generation (see Hoyt 1987: 321).

On the other hand, new field evidence from the Ries crater remained equivocal and was consequently interpreted within the traditional volcanic paradigm. From 1948 onwards, Alfred Bentz as well as Hermann Reich initiated several campaigns of seismic and geomagnetic research at the Ries Basin (Vidal 1969: 9). During the winter of 1953–54, the petroleum company *Deutsche Erdöl AG*, of Hamburg, drilled in the Ries centre near the village of Deiningen. The borehole reached a depth of some 350 m below ground-level, with nearly 100 m of core recovery. Below the younger lake sediments 'rock suspected to be suevite' was found, with glassy components as well as breccias (Preuss 1964). This evidence was well reconcilable with the Kranz–Löffler model of a central volcanic explosion, but less so with the Munich idea of 'local outbursts'. In the 1960s the data from this borehole was reviewed in the light of the impact theory, but in the 1950s it had brought nothing to ruminate about, as it had not been deep enough to show that there was no 'laccolith' present underneath the Ries or that the disturbance petered out towards increasing depth.

And thus, when in 1959–60 the Belgian Air Force officer and impact enthusiast René Gallant (1906–85) enquired in Tübingen about a possible impact origin of the Ries Basin, Helmut Hölder 'completely rejected that it might have had a possible meteoritic origin' (Torrens 1998: 179).

All in all, the two schools of Munich and Tübingen remained pitted against each other, seemingly unable to fruitfully discuss matters at the Ries Basin, as became public every now and then when proponents of both schools happened to meet at conferences or field trips. Such a clash of opinions was described by Reinhold Seemann, who in 1948 attended the annual meeting of the *Oberrheinischer Geologischer Verein*, a society of geology enthusiasts, professionals and amateurs, in Baden-Württemberg. The scientific sessions took place in Heidenheim, and on the following day the society went on a Ries excursion by bus:

Figure 8.4   Richard Löffler explaining an outcrop at the Ries Basin. After Walter Kranz's death, Löffler was the main promoter of a central volcanic explosion hypothesis for the Ries Basin. (Source: Carlé 1987: Figure 2).

Volcanologically infected by the talk of Löffler[20] on Wednesday evening, otherwise however mostly ignorant [or disoriented] participants left on Thursday (6 May) morning via Nattheim to Zöschingen at first, where Löffler [Figure 8.4] showed the debris masses resting on marine molasse sediments and pointed out their chaotic setting. When he thereby slightly touched [my work], I pointed out the order, which Moos[21] had found there in the succession of the allochthonous beds and sought to argue for the tectonic concept, which always is to be advanced together with the other one (the volcanic and morphological). When Löffler rejected this theoretical discussion, Dr Weber[22], from Munich (who had mapped the Wemding map sheet) rebuked him for obviously intending beforehand to treat the problem in a very biased way. Löffler energetically fought against this rebuke. As he became completely personal in this, I left it to Dr Weber to oppose Mr L. This he now did all day in all the outcrops, unfortunately somewhat too self-conscious and no less self-opinionated, so that it often created dramatic scenes. That was rather unpleasant, especially for those participants who were not informed about the wider connections. Additionally, L. kept clinging to comparatively small details, and in discussing these, much precious time was lost and the participants were greatly fatigued. I only joined in once more, when Georg Wagner claimed that in the finely brecciated outcrop north of Demmingen the brecciation of the Upper Jurassic limestone could not possibly be of tectonic origin, because a force of more than 1000 kg/cm^2 (?) is needed – The same argument that he also used at the colloquium in Tübingen, and which you too thought was noteworthy. My retort that here we have brecciation of massive limestone clamped in a bedded facies under moving pressure (by repeated flexing of the beds) he simply overruled by crying loudly "impossible", and since in general Löffler too used the tactics of loud yelling, I no longer joined in. Dr Weber was more than adept in this battle, even surpassing the two. From the yelling I only discerned the insecurity of the opponents. As far as I understood Mr Weber, he supported the old laccolith theory by Branca/Fraas with the ascent and descent of a deep magmatic body that caused the tectonic brecciation. – Benz [sic][23], at the end in Harburg (in the quarries of the concrete factory), joined in the debate and stressed the unfruitfulness of this type of debate. I only took pity on the participants, most of whom did not understand what this was all about. The Buchberg, where something about the tectonic concept could still have

---

[20]   Richard Löffler (see Chapters 1, 3 and 9).

[21]   August Moos (1893–1945) (see Chapter 7).

[22]   Emil Weber (born 1908) obtained his doctorate from the University of Munich in 1934 and was an assistant at the Palaeontological Institute (of which Richard Dehm had been the director since 1950). From 1956 onwards, Weber was a curator at the Bavarian State Collection for Palaeontology and Stratigraphy, still under the directorship of Dehm (Freyberg 1974: 167).

[23]   Alfred Bentz (see Chapter 3).

been said, was regrettably passed by in pouring rain. In the evening, despite all the unpleasantness, I did not have the feeling of inferiority, but the contrary. The fact that I was no longer alone[24] – everybody in Munich rejects Kranz's explosion theory – and that Wagner and Löffler really had no factual arguments to bring up, gave me the confidence that my work nevertheless had not been in vain. (Letter from Reinhold Seemann to Helmut Hölder, 12 May 1948, GA 3/325)[25]

---

[24]   This suggests that Reinhold Seemann, one of the promoters of a tectonic Ries origin (see Chapter 3), did not know about the Munich School of Ries genesis before this field trip, pointing again to the extremely local confinement of geological discussion in Germany at that time.

[25]   'Die durch den Vortrag von Löffler am Mittwoch Abend schon vulkanologisch infizierten, sonst aber wohl meist ganz unorientierten Teilnehmer fuhren also am Donnerstag (6. V.) in der Früh los, über Nattheim zunächst bis Zöschingen, wo Löffler die auf mariner Molasse liegenden Trümmermassen zeigte u. deren Durcheinander hervorhob. Als er mich dabei antippte, wies ich auf die Ordnung hin, die Moos dort in der Aufeinanderfolge der ortsfremden Schichten gefunden hatte, u. suchte den tektonischen Gedanken zu begründen, der mit dem anderen (dem vulkanischen u. morphologischen) zusammen eingesetzt werden müsse. Als Löffler diese theoretischen Erörterungen zurück wies, machte ihm Dr. Weber aus München (der Bl. Wemding kartiert hat) den Vorwurf, dass er von vorneherein offenbar beabsichtigte, das Problem ganz einseitig zu behandeln. L. wehrte sich energisch gegen diesen Vorwurf. Da er dabei ganz unsachlich war, überliess ich es Dr. Weber Herrn L. zu opponieren. Dieser tat es nun den ganzen Tag in allen Aufschlüssen, leider ein wenig zu selbstbewusst u. nicht weniger rechthaberisch, so dass es oft zu dramatischen Szenen kam. Das war wenig erfreulich bes. für diejenigen Teilnehmer, die über die grösseren Zus. hänge gar nicht im Bilde waren. Dazu kam, dass L. an verhältnismäßig kleinen Dingen hängen blieb, durch die Diskussion darüber viel kostbare Zeit verlor u. die Teilnehmer sehr ermüdete. Ich griff nur noch einmal ein, als Gg. Wagner bei dem Griesaufschluss nördl. Demmingen behauptete, die Zertrümmerung des Weissjurakalkes könne unmöglich tektonisch sein, weil man dafür eine Belastung von mehr als 1000 kg/cm² (?) braucht – Dasselbe Argument, das er im Tüb. Kolloquium vorbrachte u. das ja auch Dir bedeutsam erschien. Mein Einwand, dass es sich ja hier um Zertrümmerung von in geschichteter Fazies eingespannten Massenkalken unter bewegtem Druck (durch widerholte Schichtverbiegungen) handle, überschrie er einfach mit 'unmöglich' u. da überhaupt, auch von Löffler die Taktik der Lautstärke angewandt wurde, tat ich nicht mehr mit. Dr. Weber war diesem Kampf gewachsen, ja den beiden sogar noch über. Ich hörte aus dem lauten Geschrei nur die Unsicherheit der Gegner heraus. Soweit ich Herrn Weber verstanden habe, vertritt er die alte Branca-Fraas'sche Lakkolith-Theorie mit der Auf- u. Abwärtsbewegung eines tiefenmagmatischen Körpers, die die tektonische Zertrümmerungen im Gefolge hatten. – Benz [sic] griff auch noch am Schluss bei Harburg (in den Steinbrüchen des Zementwerkes) in die Debatte ein u. betonte die Unfruchtbarkeit dieser Art von Diskussionen. Mir taten nur die Teilnehmer leid, die meist gar nicht verstanden, worum es ging. Am Buchberg, wo man noch einmal etwas über die tekton. Gedanken hätte sagen können, ging es bedauerlicherweise im strömendem Regen vorbei. Trotz allem Unerfreulichem hatte ich am Abend nicht das Gefühl des Unterlegenseins, im Gegenteil, die Tatsache, dass ich nicht mehr allein stand – München lehnt geschlossen

While thus fruitless disputes continued to dominate the local scene, American impact research, with its newly discovered mineralogical evidence for impact craters, was about to enter the German stage.

die Kranz'sche Sprengtheorie ab – u. Wagner + Löffler wirklich keine sachlichen Gründe vorzubringen hatten, gab mir die Sicherheit, dass meine Arbeit doch nicht umsonst gewesen war' (letter from Reinhold Seemann to Helmut Hölder, 12 May 1948, GA 3/325).

Chapter 9
The Tide is Turning

Figure 9.1   Eugene Merle Shoemaker (right) and Edward Ching-Te Chao
(left), geologists at the USGS, discovered the high-pressure
mineral coesite at the Ries Basin in 1960, thus providing the vital
clue to the impact nature of this structure. (Photo courtesy of Dr
Wulf-Dietrich Kavasch).

In the spring of 1960, the high-pressure mineral coesite had already been
discovered at Meteor Crater by Chao, Shoemaker and Madsen, and Shoemaker
was about to give an oral presentation on his research at the 21st International

Geological Congress in Copenhagen. However, before he went to Denmark, Shoemaker (Figure 9.1), together with his wife and his mother, undertook a holiday tour through Germany and part of the Alps to visit various sites of geological interest. These sites included the Eifel maars (the type-locality of maar-volcanoes, the variety of volcanic edifice that Shoemaker had worked with on the Colorado Plateau), the diatremes of the Swabian Volcano, the Köfels landslide, the Steinheim Basin (type-locality of 'cryptovolcanic explosions') and the Nördlinger Ries. All of these localities were known to Shoemaker from the geological literature, and he now wished to see them for himself. During his work for the MICE project (Chapter 8) Shoemaker had visited all of the crypto-volcanic sites in the eastern United States, which had been described by Walter H. Bucher (1888–1965; see Chapter 10). Some of them had already been reinterpreted by Dietz and others as remnants of impact craters.[1]

The suevite quarry in Otting outside the eastern edge of the Ries crater, which Shoemaker had identified from the geological map, was the first site he visited at the Ries structure. The party arrived in a VW camper late in the evening:

> When I saw the shocked, crystalline rock components of suevite my heart beat faster. There were all the typical phenomena well-known to me from the Arizona-Crater, which are based on shock metamorphism due to impact. Quickly I took three suevite samples as the darkness fell, and then we camped in the wood nearby. On the next day we drove to Nördlingen, and I sent the samples to Chao in Washington. Within a few days they reached him and quickly he had found coesite by X-ray tests. (Shoemaker quoted in Carlé 1987: 88)

Edward Ching-Te Chao (Figure 9.1), Shoemaker's colleague from Washington D.C., was born in 1919 in Soochow (China). He studied geology at the National Southwestern Associated University of Kunming, where he obtained a BSc-degree in 1941. From 1941 to 1945, he worked for the National Geological Survey in the Chinese province of Szechuan. During the Chinese civil war he immigrated to the USA, where he resumed his studies with a focus on petrology and geochemistry at the University of Chicago. He finished his PhD in 1948, and in 1949 began to work for the USGS. In the mid-1950s, he obtained American citizenship (Carlé 1987; Barsig and Kavasch 1983).

Shoemaker stayed for two weeks in the Ries area, where he saw numerous outcrops and took additional samples. He also visited the sites that had been described as volcanic vents in the literature, but he became convinced that these localities represented not vents, but rather suevite-filled fractures in the breccia

---

[1]   Interview of Dr Eugene Shoemaker by Ronald Doel on 17 June 1987, Niels Bohr Library and Archives, American Institute of Physics, College Park, MD USA, <www.aip.org/history/ohilist/5082¬_3.html> accessed 25 February 2011.

below the suevite, or that the suevite and other rocks had come to lie adjacent to each other as a result of faulting. As the whole journey was for Shoemaker only a vacation tour, he had not contacted any German geologists, and when he finally arrived in Copenhagen this fact was regarded as a diplomatic problem by a high-ranking USGS official. However, after some discussion, Shoemaker in his talk nevertheless announced the discovery of coesite at Meteor Crater as well as in the Ries Basin.[2]

Edward Chao gave a talk about *New Evidence for the Impact Origin of the Ries Basin, Bavaria, Germany*, at the 39th Annual Meeting of the German Mineralogical Society (DMG) 11 to 17 September 1961, a conference with some 350 delegates and 54 lectures. At this conference Chao explained the discovery of coesite (Figure 9.2) and the extreme heterogeneity of the suevite glass, which was unlike any volcanic glass (Gehlen 1961). He also presented data concerning bright round metallic spherules as inclusions in suevite glass. These metallic spherules had elevated nickel concentrations and appeared to be kamacite[3] (Preuss in AG Ries 1963: 11, 21; Preuss 1969a; Wagner 1963; Engelhardt and Hölder 1977: 61; Shoemaker and Chao 1961b). Edward Chao used his stay in Germany not only to give his talk in Tübingen, but also to see the Ries Basin personally and to take more samples. The work by Shoemaker and Chao was then finally published in October 1961 (Shoemaker and Chao 1961a).

The issue had been quickly picked up by the Tübingen Professor of Mineralogy Wolf von Engelhardt (1910–2008) (Figure 9.3). Engelhardt had studied science in Halle, Berlin and Göttingen between 1929 and 1935, with a focus on geology, mineralogy and chemistry. In 1935 he obtained his PhD under the supervision of the mineralogist and geochemist Victor Moritz Goldschmidt, who shortly afterwards had to flee Germany due to his Jewish descent (Chapter 7). From 1935 to 1938, Engelhardt was a scientific assistant to the mineralogist Carl Wilhelm Correns (1893–1980) at the mineralogical institute of the University of Rostock. In 1940, he *habilitated* there. During the Nazi time, Engelhardt became a member in the *Sicherheitsdienst*[4] (1942–45). Military service occupied him for some time between 1937 and 1942. In 1944, he became a professor in Göttingen, but after the War (possibly forced by denazification) he switched to a

---

[2]   Interview of Dr Eugene Shoemaker by Ronald Doel on 17 June 1987, Niels Bohr Library and Archives, American Institute of Physics, College Park, MD USA, <www.aip. org/history/ohilist/5082¬_3.html> accessed 25 February 2011.

[3]   Main component of iron meteorites. The mineral consists of some 90 per cent iron and ten per cent nickel (<http://webmineral.com/data/Kamacite.shtml> accessed 8 March 2013).

[4]   *Sicherheitsdienst* means Security Service. It was the intelligence agency of the SS and the Nazi Party. This organisation was the first Nazi Party intelligence organisation to be established and was often considered a 'sister organisation" to the Gestapo (<http:// en.wikipedia.org/wiki/Sicherheitsdienst> accessed 7 April 2011).

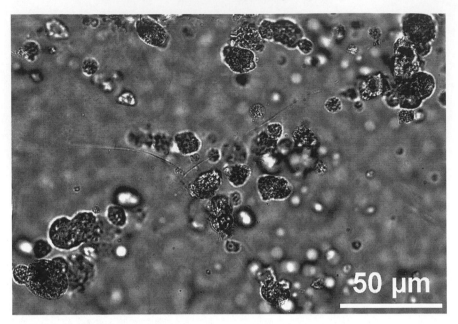

Figure 9.2   Coesite   from   Lake   Bosumtwi   (Ghana):   Thin-section
photomicrograph (in plane-polarised light) of coesite aggregates
(dark clusters) within diaplectic quartz glass (Sample LB-44B;
Suevite from the Bosumtwi crater, Ghana). (Credit: Ludovic
Ferrière, NHM Wien).

leading position in a petroleum company in Hannover (1948–57). Parallel to his
position in the industry, he lectured as an honorary professor in Göttingen from
1952 until 1957, when he became Professor of Mineralogy and Petrography
in Tübingen. After his official retirement in 1978, he continued lecturing on
extraterrestrial petrology and researching impact metamorphism together with
his PhD students.[5]

Engelhardt had been occupied for ten years with sedimentology for the oil
industry in northern Germany, and consequently his initial knowledge of the
Ries Basin, as he was well aware, was non-existent. But when he heard about the
new discovery of the high-pressure mineral as impact evidence, he was intrigued
and 'began immediately with collaborators and diploma and PhD students to
work on the one hand in the Ries and on the other experimentally on the effect

---

[5]   <http://de.wikipedia.org/wiki/Wolf_von_Engelhardt> accessed 25 July 2011;
Lang 1995; Arndt 1986; Pichler 2000.

Figure 9.3   Wolf von Engelhardt (standing), Professor of Mineralogy at the University of Tübingen, lecturing about Ries Crater geology in front of members of the Apollo 14 crew. (Photo courtesy of NASA).

of highest pressures on minerals. Thereby, we succeeded to further support the impact theory' (letter from Engelhardt to the author, 16 March 2000).[6]

Rudolf Hüttner, who in 1958 had finished his PhD on geological investigations in the Württemberg part of the Ries-foreland near Neresheim and Wittislingen (Engelhardt and Hölder 1977: 255), introduced Engelhardt to the Ries Basin in a three-day field trip, during which Engelhardt became completely fascinated by the suevite (Hüttner interview).

Coming from northern Germany, Engelhardt was not intimidated by any previous traditional rivalry between Württemberg and Bavaria. Thus, he readily

---

6   '... begann ich sofort mit Mitarbeitern, Diplomanden und Doktoranden einerseits im Ries und andererseits experimentell über die Wirkung von höchsten Drucken auf Mineralien zu arbeiten. Dabei gelangen weitere Bestätigungen der Impakttheorie' (letter from Engelhardt to the author, 16 March 2000).

agreed to cooperate with the geophysicists Gustav Angenheister (1917–91)[7] and Hermann Reich from the University of Munich, the Bavarian Geological Survey (under Helmut Vidal) and Alfred Bentz from the Federal Geological Survey to plan research drilling at the Ries Basin. The intention of this drilling project was to investigate the nature of some highly magnetic material that had been detected by geomagnetic studies and had hitherto been assumed to constitute a magmatic intrusion. Engelhardt and his students were to perform the petrographic investigations. The group officially became the Ries Working Group on 2 December 1961 (Vidal 1969: 9; see Chapter 11).

Impact in Tübingen

While Engelhardt thus embarked enthusiastically on the new field of research, there was no big 'hooray' among the traditional Ries community when the high-pressure mineral coesite was discovered at the Ries Crater and presented as substantial evidence of a meteorite impact. And so, on 17 December 1961, tumultuous scenes occurred at a geological meeting, the traditional Thomas-conference in Tübingen.[8] Wolf von Engelhardt had announced a talk about the impact nature of the Ries Basin as deduced from the new evidence that he had learned from Edward Chao and possibly verified since. Meanwhile, his geological colleague Georg Wagner, a most influential advocate of the Kranz–Löffler model of a central volcanic explosion, had tried to organise an opposition of Ries experts, in order to corner Engelhardt publicly. This Wagner

---

[7]   Gustav Angenheister studied geophysics in Göttingen, interrupted by military service. He obtained his doctoral degree in 1952. Between 1952 and 1957, he undertook geophysical prospecting work for the Geological Survey of Niedersachsen in Hannover. In 1957, he became Professor of Geophysics and Director of the Geophysical Institute at the University of Munich, as well as director of the university's geophysical observatory in Fürstenfeldbruck. He retired in 1983 (Freyberg 1974: 7; <http://de.wikipedia.org/wiki/Gustav_Angenheister_(1917–1991)> accessed 17 April 2012).

[8]   The Thomas-conference was an annual meeting of the *Verein für vaterländische Naturkunde* [loosely translatable as 'Society for Patriotic Natural Philosophy']. The society was founded in 1834 after a meeting of the Assembly of German scientists and physicians (see Chapter 7). In 1969, it was renamed *Gesellschaft für Naturkunde in Württemberg*; Society for Natural Philosophy in Württemberg (<http://www.ges-naturkde-wuertt.de/word/geschichte-m.html> accessed 25 July 2011). At a time when geology at the University of Tübingen was basically concerned with local affairs, this used to be a rather important conference for geologists in southern Germany; at least as vital as general assemblies of the German Geological Society and more important than any meeting of mineralogists or international congresses.

endeavoured to achieve with the help of the chairman of the conference session, Helmut Hölder.

Hölder (1915–2014) had obtained his PhD in 1939 from the University of Tübingen with geological investigations near Lauchheim, just outside the eastern edge of the Ries Basin. After military service, he became a research assistant in Tübingen from 1945 to 1950. He was appointed palaeontological curator in 1950 but also served as a lecturer from 1949. Hölder kept both positions until 1963, when he became Professor of Geology in Münster. Hölder also published on the history of geology.[9]

Hölder recalled the dramatic event of December 1961 in an interview:

> Shortly before this meeting, Prof. Georg Wagner, our most honoured colleague at the Tübingen institute, said to me, Mr Hölder, he said, Mr von Engelhardt will be speaking about his tiny nickel spherules. After he has finished his talk, you are going to give the first word to – that was an order, because I was the much younger one – then you give the first word to my friend Löffler. Löffler was a Ries-geologist, volcanologist,[10] highly respected, teacher at the Lyceum in Gmünd, and a friend of Georg Wagner. And so Wagner thought, he should speak first. You give the word to my friend Löffler and he will tell Engelhardt what is necessary, and then you give the word to me. And so it happened. Indeed, Engelhardt gave his talk about nickel spherules in suevite, and then I opened the discussion and said, as I was told, as desired by the old Georg Wagner: "Maybe Prof. Löffler has something to say concerning this talk, if you please". Then Mr Löffler stood up and said: "Yes, after I have heard this talk, I now think, that we shall need to think again". And that was a big blow for my Georg Wagner. I still see it before me, as he collapsed horrified. And then, of course, I said, ehm, and now, maybe Prof. Wagner wants to say something. Wagner, in shock, struggled to his feet and said: "I cannot follow here, I don't believe it. For 50 years, I have walked the Ries with my students and I have never seen a sign of that meteorite". Then, in the last row of the completely occupied lecture hall ..., Prof. Siedentopf[11] got up, who was then the astronomer in Tübingen, and said: "Colleague Wagner, from this meteorite

---

[9]   Engelhardt and Hölder 1977: 251, 260–61; <http://de.wikipedia.org/wiki/Helmut_Hölder> accessed 29 December 2011.

[10]   Löffler is here considered a volcanologist, because he was an expert on the Ries Basin 'volcano'.

[11]   Heinrich Friedrich Siedentopf (1906–63) was Professor of Astronomy at the University of Jena (<http://de.wikipedia.org/wiki/Heinrich_Friedrich_Siedentopf> accessed 27 March 2012) and afterwards in Tübingen. In his acknowledgments Heide (1934) thanked Siedentopf for constructive discussion of the impact problem, that is, Siedentopf already had some contact with the impact problem before 1961. Thus, he knew about Heide's paper, which demonstrated that large meteorites cannot be preserved upon impact (see Chapter 6).

you cannot have found anything, because there is a boundary level in the impact energy and if the force of the impact, because of the size of the meteorite, is larger than this boundary level, then it will melt and cannot become fossil". This was an immensely thrilling discussion then; a dramatic discussion. (Hölder interview)[12]

Another eye-witness, Rudolf Hüttner, also has lively recollections of this event:

It was dramatic, and most of all, it was important that Mr Löffler ... said, very carefully: "If this should be confirmed, then it would be an explanation". And ... Georg Wagner [was] very angry with Löffler that he now left the field without fighting. But Löffler had been the one who knew much more about the Ries than Georg Wagner. Georg Wagner himself did not work much on the Ries; he had written about the Buchberg pebbles and the work then practically had to be stopped, because it came to nothing. It had been a preliminary note, which he had published. And then, there is also a little paper where he, like Seemann, concerned himself with the rock-mechanical conditions [the local stress regime], that there is some pushing from the Alps continuing into the area and then that it comes to intense brecciation at the tip of the wedge as Seemann had thought,

---

[12]   'Kurz vor dieser Sitzung hat der Professor Georg Wagner, unser verehrter alter Kollege damals am Tübinger Institut, und er hat zu mir gesagt, also Herr Hölder, da spricht doch der Herr von Engelhard über seine Nickelkügelchen. Wenn der seinen Vortrag gehalten hat, dann geben Sie zunächst – war nun also eine Anweisung an mich, ich war der weit Jüngere – dann geben Sie zunächst meinem Freund Löffler das Wort. Löffler war Riesgeologe, Vulkanologe, hochverdient, Oberstudiendirektor in Gmünd und eine Freund von Georg Wagner. Und so hat nun Wagner gemeint, er soll mal zunächst sprechen, dann geben sie meinem Freund Löffler das Wort, der wird dem Engelhard das Nötige sagen und dann geben Sie mir das Wort. Und so geschah es dann auch. Doch, Herr von Engelhard hat seinen Vortrag über Nickelkügelchen im Suevit gehalten, dann habe ich die Diskussion eröffnet und habe gesagt, beinahe weisungsgemäß, dem alten Georg Wagner entsprechend, vielleicht hat nun Professor Löffler zu diesem Vortrag etwas zu sagen, darf ich Sie bitten. Dann stand Herr Löffler auf und sagte: "Ja, nachdem ich diesen Vortrag gehört habe, meine ich nun, dass wir doch wohl umdenken müssen". Und das war für meinen Georg Wagner ein schwerer Schlag. Ich sehe ihn heute noch, wie er erschreckt gleichsam in sich zusammen sank. und dann habe ich natürlich gesagt, ja und jetzt vielleicht will Professor Wagner sich äußern. Da stand Herr Wagner etwas mühsam und betroffen auf und sagte: Also ich kann da nicht mitmachen, ich glaube das nicht, ich bin 50 Jahre mit meinen Schülern durchs Ries gelaufen und ich habe nie was von diesem Meteoriten gesehen. Und dann meldete sich in der hintersten Reihe unseres vollbesetzten Hörsaals ... Prof Siedentopf, das war der damalige Astronom in Tübingen, und sagte: "Herr Kollege Wagner, von diesem Meteoriten können sie auch nichts gefunden haben, es gibt nämlich einen Schwellenwert der Einschlagsenergie und wenn die Kraft des Einschlags infolge der Größe des Meteoriten diesen Schwellenwert übersteigt, dann schmilzt der auf und er kann nicht fossil werden". Das war diese ungeheuer spannende Diskussion damals, dramatische Diskussion' (Hölder interview).

and that this cannot work; that this cannot be. These are basically the two original papers of Georg Wagner. Otherwise, he was always in the Ries as friend of Löffler and he was the big promoter of Löffler's theory. And now Löffler in his view has, so to speak, failed him and because of that, he had taken this a bit amiss. But Löffler had been the one who had known about the great difficulties that there are at the Ries, and so he nevertheless had seen it as a possibility that had to be investigated, this impact theory. (Hüttner interview)[13]

Georg Wagner was not alone in his opposition to the renewal of the impact theory. The arguments that were brought forward immediately after Engelhardt's talk were quite similar to those that had been encountered by Herbert Rohleder or Otto Stutzer (Chapter 3), most prominently the unlikelihood of a meteorite impact at a geologically 'predestined' locality and the complex history of Ries genesis, which seemed to require a wider time-frame. Nevertheless the general atmosphere was somehow different:

Wagner then probably said nothing more, but Professor Dehm then did speak and presented the known objections of the geologists. Dehm had done much research in the Ries and he said that the Ries Basin had a special predisposition for unusual events, especially of a volcanic nature. There are tectonic lines that cross each other there, there is this well-known Ries-Rise; it is a point that has a special geological history, with which the volcanism there, even volcanism of an unusual nature, can be understood. And therefore, it would have been quite extraordinarily strange, if – of all localities – the meteorite would hit right there of all places. This argument was quite convincing for us geologists, and it continued to be felt very

---

[13] 'Es war dramatisch und vor allen Dingen wichtig, dass der Herr Löffler ... gesagt [hat] sehr vorsichtig: 'Wenn sich das bestätigt, so wäre das eine Erklärung' nicht war und der ... Georg Wagner [war] dem Löffler sehr böse, dass er jetzt so kampflos das Feld räumt, aber der Löffler, der war eben derjenige der über das Ries sehr viel mehr wusste als der Georg Wagner, der Georg Wagner hat über das Ries selbst nicht viel gearbeitet da hat er über die Buchberggerölle geschrieben und die Arbeit musste ja dann praktisch aufhören, weil er eben nicht weiterkam es war eine vorläufige Mitteilung, die er da rausgegeben hat und dann gibt es noch eine kleine Arbeit, wo er sich wie Seemann auseinandersetzt, also mit gebirgsmechanischen Bedingungen, dass da von den Alpen her sich so ein Schub fortsetzt und es dann zur Vergrießung an der Spitze des Keils führt, so wie das der Seemann gedacht hatte, dass es so nicht klappen kann, dass es so nicht stimmen kann das sind eigentlich die zwei Originaläußerungen von Gerold Wagner, sonst war er immer als Freund von Löffler mit im Ries und war der große Propagandist von Löfflers Theorie. Und nun hat der Löffler da versagt sozusagen in seinen Augen und deswegen hat er ihm das auch ein bisschen übel genommen. Aber der Löffler war eben der der über die großen Schwierigkeiten wusste, die es im Ries eben gibt und so hat er es dann doch als eine Möglichkeit, die man eben prüfen muss gesehen diese Impakttheorie' (Hüttner interview).

much for quite some years in the discussion: a geologically predestined point. Why should the meteorite of all places hit right there?

Well, of course one can say, and it was later said too, if the meteorite hits somewhere on Earth, this somewhere can be just this place. ... If it hits somewhere, it can also impact at a locality with special tectonic preconditions, which have no genetic connections to it.

One can also say something quite different. One can say that the Ries area has been investigated geologically since old times, in such a way that of course many phenomena – geologically mapped – have been noticed, which probably exist at many other places too. Only those many other places, some of them are covered with different sediments and so on, in part they are not as much investigated. Everywhere the Earth is interesting, everywhere something special is about, and only because there was so much research in the Ries, we had this notion: "This is something *very* special!"

At any rate, today we are of the opinion that it may be an interesting spot between the Bohemian Massif and the Black Forest on the other side, within the Swabian-Bavarian Table-Land, but not so very special, ...

It had been a very dramatic meeting and it had been one of the most dramatic experiences of my career. I am still happy that I have experienced it so directly, even actively as chairman. It had been the first time in Württemberg that this meteorite was publicly talked about.[14] (Hölder interview)[15]

---

[14]   Obviously, Hölder – and possibly most other geologists at the Tübingen Institute for Geology and Palaeontology – had not attended the conference of their colleagues of the Tübingen Institute for Mineralogy and Petrology earlier in September, and thus had not been aware of Chao's talk on nickel-bearing metallic spherules in suevite (Preuss in AG Ries 1963: 11, 21).

[15]   'Wagner sagte wahrscheinlich dann nichts mehr, es hat dann aber gesprochen Professor Dehm und hat die damals bekannten Einwände der Geologen gebracht. Dehm hat ja sehr viel im Ries geforscht und hat gesagt, das Ries hat eine besondere Prädestination für besondere Ereignisse, gerade auch vulkanischer Art, es gibt dort sich kreuzende tektonische Linien, es gibt diese bekannte Riesschwelle, es ist ein Punkt der geologisch eine besondere Geschichte hat, mit der man den sich dort abspielenden Vulkanismus, auch Vulkanismus besonderer Art, verstehen kann. Und deshalb wäre es doch ganz außerordentlich merkwürdig, wenn ausgerechnet an diesem Punkt der Erde, der vulkanistisch prädestiniert war, wenn da ausgerechnet nun der Meteorit einschlagen sollte. Und das war ja auch für uns Geologen eigentlich recht überzeugend und das hat noch längere Jahre sehr stark in der Diskussion nachgewirkt, geologisch prädestinierter Punkt. Warum soll gerade ausgerechnet da der Meteorit einschlagen? // Nun kann man dazu natürlich sagen, und das wurde später auch gesagt, wenn der Meteorit irgendwo auf der Erde einschlägt, kann das Irgendwo gerade dieser Punkt sein. ... Wenn er irgendwo einschlägt, dann kann er auch da einschlagen, wo besondere

The other aspect, mentioned by Hölder here, was indeed one of the greatest factual obstacles for the acceptance of the Ries and Steinheim Basins as impact craters. The intense geological research at these puzzling localities had produced so many details that investigators completely lost the feel for balancing these data in a more general, pan-European or even global setting. They, as a German proverb says, could 'no longer see the forest because of all the trees' – and of course, the romantic or nationalistic notion of 'German Geology' (see Chapter 7) consciously or subconsciously reinforced this attitude by ranking local geological history higher than comparative investigations.

The argument that the Ries area had a special geological prehistory was not only raised by Dehm but also by Walter Carlé (1912–96)[16] in the discussion immediately after Engelhardt's talk (Hölder 1962: 20). Nevertheless, Carlé seemed to have been rather impressed by Engelhardt's presentation, the quality of which he acknowledged positively and open-mindedly:

tektonische Vorgegebenheiten sind, die damit gar nichts zu tun haben. // Man kann ja da was ganz anderes sagen. Man kann ja sagen, das Riesgebiet ist seit alter Zeit derartig geologisch erforscht, dass natürlich da viele Erscheinungen, geologisch aufgenommen, zur Kenntnis genommen worden sind, die es wahrscheinlich an vielen anderen Stellen auch gibt, bloß sind die vielen anderen Stellen, zum Teil sind sie bedeckt mit anderen Sedimenten und so weiter, zum Teil sind sie eben nicht so erforscht. Überall auf der Erde ist es interessant, überall ist was Besonderes los, und nur weil im Ries so viel geforscht wurde, kam ein Bild zustande: 'Das ist was *ganz* Besonderes!' Also jedenfalls heute sind wir ja der Meinung, dass ist zwar ein interessanter Punkt zwischen Böhmerwald und auf der anderen Seite Schwarzwald im Schwäbisch-bayerischen Schichtstufenland, aber es ist eben nicht so besonders, ... // Also das war eine sehr dramatische Sitzung, das war eines der dramatischsten Erlebnisse meiner Berufszeit, diese Sitzung. Ich bin heute noch glücklich darüber, dass ich das so unmittelbar miterlebt habe, sogar aktiv miterlebt habe bei der Diskussionsleitung. Es war zum ersten Mal in Württemberg, dass von diesem Meteoriten öffentlich geredet wurde'. (Hölder interview).

[16] Walter Carlé (1912–96) studied in Tübingen, Kiel and Berlin, initially with a focus on botany but later switched to geology. He was active in the Christian Boy Scouts up until his PhD, despite the fact that this engagement was frowned upon by the Nazis. He obtained his PhD from Berlin in 1938, and from 1938 until 1940 he worked in ore prospecting in Spain. From 1940 to 1945, he was a geologist for the National Geological Survey in Berlin, a job that took him to World War II as a surveyor from 1941 onwards. He returned badly wounded in 1945. He maintained contact with the Geological Institute in Tübingen and in 1946 received employment with the Württemberg Geological Survey in Stuttgart. From 1949 onwards, he also lectured at the University of Stuttgart and obtained his qualification for a professorship in 1953. He was appointed professor of geology without chair at the Technical University in Stuttgart in addition to his survey position. From 1970 to 1975, when he retired, he was the director of the Stuttgart branch of the Geological Survey of Baden-Württemberg (Simon 1996; Freyberg 1974: 26).

At present the meteor hypothesis seems to be supported so convincingly by a completely unexpected field of expertise that under the impression of these results, obtained in an exact[17] manner, one hardly dares to support the volcanic origin of the Ries. After the very beautiful argumentation of colleague Baron von Engelhardt it is however admissible to once more point to the, in several respects, special geological status of the location. (Carlé in Hölder 1962: 20)[18]

Carlé's remark was quite revealing. Carlé's statement that a meteorite impact was now supported by 'a completely unexpected field of expertise' highlighted that mineralogy had indeed been perceived as alien to conventional Ries research. Thus far, it had had no business with the Ries and nobody at the geological institutes had expected any reasonable input from that field. This was felt by others as well: 'The paradoxical case arose that a mineral, at first only visible microscopically, should be regarded as sufficient proof for the meteoritic origin of the vast Ries Basin' (Ekkehard Preuss in AG Ries 1963: 10–11).[19]

This attitude was all too easily turned into polemics, which occasionally, as we shall see later on, matured into plain chauvinism. It made it convincing to dismiss as outsiders the new impact researchers, who simply had no idea about the true issue at the Ries Basin: 'The interpretation of the Ries as a meteorite crater has been tried every now and then; but only by people who knew the Ries in a cursory manner. Thorough investigation again and again had to lead to dismissal [of this hypothesis]. This is also true for the proof by coesite' (Georg Wagner in Hölder 1962: 17).[20]

Not only did the pre-Ries geology seem predestined for the Ries event, but also the Ries structure itself (as a result of intense research) appeared to be immensely complex, requiring an equally complex geological history for its development:

---

[17]   Obtained by an exact science. That is by physical and chemical means, rather than descriptive means as in contemporary geology.

[18]   'Nun erscheint aber zur Zeit die Meteorhypothese von einer ganz unerwarteten Fachrichtung her so überzeugend gestützt zu sein, daß man unter dem Eindruck dieser exakt gewonnenen Ergebnisse kaum wagen kann, eine Lanze für die vulkanische Entstehung des Rieses zu brechen. Es sei mir aber doch gestattet, im Anschluss an die sehr schönen Darlegungen des Herrn Kollegen Freiherrn von Engelhardt noch einmal auf die in verschiedener Hinsicht besondere geologische Stellung des Raumes hinzuweisen' (Carlé in Hölder 1962: 20).

[19]   'Es entstand der paradoxe Fall, daß ein zuerst nur mikroskopisch sichtbares Mineral als ausreichender Beweis für die meteoritische Bildung des großen Rieskessels gelten soll'. (Ekkehard Preuss in AG Ries 1963: 10–11).

[20]   'Die Deutung des Rieses als Meteoritenkrater ist immer wieder versucht worden; aber nur von Leuten, die das Ries nur flüchtig kannten. Eingehende Untersuchungen mußten immer wieder zur Ablehnung führen. Das gilt auch für die Beweisführung mit Coesit' (Georg Wagner in Hölder 1962: 17).

Upon first glance the meteor hypothesis seems very intriguing, as it gives a simple explanation for the genesis of the Ries bowl and the chaotic deposition of the Ries debris in the Ries foreland. The procuring of the immense energy necessary for this revolution poses it no problems. However, in its impressive simplicity lies at the same time its greatest weakness. The formation of the Ries bowl and the Ries debris represents the main genetic phase, but it is only *one* phase in the complex Ries events. (Rudolf Hüttner in Hölder 1962: 24)[21]

Intense research had introduced a complex history of Ries events from the (volcanic) Ries explosion to subsequent suevite emplacement in several stages[22] (Ahrens 1929; Chapter 3) and continuing tectonic displacement. Such a sequence of events required time, geologists were convinced, whereas a meteorite strike needed just a matter of seconds.

This was the line of arguments that Reinhold Seemann had used to buffer his tectonic Ries theory, and he too aired it again on the occasion of the 1961 Thomas-conference:

> Now as before, I am still convinced that normal geological ideas, that is synorogenic, tectonic movements … working for a long time … are sufficient to explain all the chaos of debris in the Ries as well as its foreland, and that it is only due to our restricted overview or our impatience, if we fail to restrict ourselves and to succeed with these tools. (Seemann in Hölder 1962: 27)[23]

Hölder, who one year later published these arguments brought forward at the discussion of Engelhardt's talk or in its immediate aftermath, felt induced in a remarkably thoughtful paper to think about the philosophical aspects that influenced his colleagues and also himself. He keenly felt his former world-view shattered by the

---

[21] 'Die Meteorhypothese erscheint auf den ersten Blick sehr einnehmend, da sie eine einfache Erklärung für die Entstehung des Rieskessels und die chaotischen Lagerungsverhältnisse der Riestrümmermassen im Vorries gibt. Die Herleitung der ungeheuren, für diese Umwälzung benötigten Energie bereitet ihr keine Schwierigkeiten. In ihrer bestechenden Einfachheit liegt aber zugleich ihr Mangel. Denn die Bildung des Rieskessels und der Riestrümmermassen stellt zwar die Hauptphase, aber eben nur *eine* Phase in dem komplexen Riesgeschehen dar' (Rudolf Hüttner in Hölder 1962: 24).

[22] As it is now commonly interpreted due to reworking, slumping and so on, although even up to the most recent time dissenting opinions (Zimmerman 2003) are occasionally voiced, returning to a volcanic interpretation for suevite generation; strange as it may seem.

[23] 'Ich bin aber nach wie vor überzeugt, daß normale geologische Vorstellungen: nämlich synorogenetische tektonische Bewegungen … in einer langen Zeit sich abspielend … genügen, um den ganzen Wirrwarr der Trümmermassen im Ries und Vorries zu erklären, und daß es nur an unserem beschränkten Überblick bzw. an unserer Ungeduld liegt, wenn wir mit diesen Hilfsmitteln nicht aus- und durchkommen'. (Seemann in Hölder 1962: 27).

events, and it is the best eyewitness description of the paradigm shift that can be traced to the Thomas-conference of 1961:

> In this situation, the new or ... renewed meteorite hypothesis must not be dismissed right away. Yes, quite apart from its potential correctness, it is of special interest in the history of science. It is based on the discovery of the mineral coesite, which belongs only to the last decade. It is the problem of research that it has always to expect such unknown things. If it does not want to be crippled by this fact, so it must accept error – as well as the tolerance of it. A meteorite impact, which alone according to the present state of knowledge can have formed the mineral coesite as a high-pressure modification of quartz, for geological research is something like a slap in the face: because Earth history actually endeavours to demonstrate the geohistorical conditions for the occurrence of a geohistorical event, for the self-appearance of a piece of earth crust. The strike from outside into a geohistorically self-formed space, which the Ries certainly represents in several respects, can be random chance, demonstrating that the Earth is a fundamentally open system facing the cosmos, historically not closed within itself. (Hölder 1962: 16)[24]

This cosmic 'slap' was felt even more severely after the 'romantic/holistic' ideas of 'German Geology' (Chapter 7). Quite apart from its nationalistic/chauvinistic background, 'German Geology' had supported the notion of a local, self-sufficient Earth history that shaped and reshaped a given piece of land; where the sum of historical events determined the fate of this particular space, without the influence of more global processes. This idea, of course, was doomed to fail completely in the face of an intrusion by an unconnected outside event.

---

[24] 'In dieser Situation darf auch die neue bzw. ... neu aktivierte Meteoritenhypothese nicht von vornherein verworfen werden. Ja es kommt ihr, auch hier von ihrer Richtigkeit zunächst ganz abgesehen, wissenschaftsgeschichtlich sogar besonderes Interesse zu. Sie fußt auf der Entdeckung des Minerals Coesit, die erst dem letzten Jahrzehnt angehört. Die Forschung hat in ihrer Problematik stets mit solchen Unbekannten zu rechnen. Will sie sich angesichts dieser Tatsache nicht lähmen lassen, so muß sie den Irrtum – und die Toleranz ihm gegenüber – in Kauf nehmen. Ein Meteoriteneinschlag, der nach dem augenblicklichen Stand der Erkenntnis allein das Mineral Coesit als Hochdruckmodifikation des Quarzes erzeugt haben kann, ist für erdgeschichtliche Forschung so etwas wie ein Schlag ins Gesicht: Denn Erdgeschichte bemüht sich ja eben darum, die irdisch-historischen Voraussetzungen für den Eintritt eines erdgeschichtlichen Ereignisses, für das Eigenbild eines Stückes Erdrinde aufzuzeigen. Der Schlag von außen in einen erdhistorisch eigengeprägten Raum, wie ihn das Ries in mancher Hinsicht sicher darstellt, kann Zufall sein, zeigt die Erde als ein grundsätzlich offenes, dem Kosmos zugewandtes, geschichtlich nicht in sich selbst beschlossenes System' (Hölder 1962: 16).

Chapter 10
Dismissing Impact II

While Wolf von Engelhardt and his students embarked on the new type of Ries research, seeking cooperation especially with Edward Chao, who visited the craters for extended field campaigns, Georg Wagner organised the resistance against this intrusion from outside, which, as Wagner saw it, was only reviving an old, long-since-dismissed and obsolete hypothesis.

Georg Wagner (1885–1972; Figure 10.1) had studied to become a teacher of mathematics and science. His favourite subject soon became geology. He was out in the field whenever possible, and obtained a PhD in geology with a very well received thesis with a stratigraphic focus. In 1911 he became a teacher at a grammar school, and two years later, a professor at a teachers' seminary. Then World War I interfered with his career and he had to become a soldier. He returned heavily wounded and it took quite some time for him to recover. In 1918 he married Hildegard Munder, and they had two sons and two daughters. Wagner resumed his position as a grammar school teacher in Stuttgart. In his spare time, he continued with his scientific interests, and in 1928 he finished his *Habilitation*. He continued as a teacher but also gave lectures at the University of Tübingen. In the 1930s he undertook some journeys, for example, to Syria and to see the Italian volcanoes.[1] In 1943 the Wagners lost their eldest son to World War II, and shortly afterwards their family home was destroyed in an air raid on Stuttgart. The family fled to Tübingen. There, in 1946, Georg Wagner received an extraordinary professorship of geology, to replace Edwin Hennig in his geological duties, who had problems because of his Nazi involvement. After some lengthy toing-and-froing during the denazification of the University of Tübingen (see Chapter 7), Wagner himself was finally cleared of charges against himself in 1948. He retired from his professorship in 1953, but continued to give lectures, being present as the institute's patriarch. In 1963 and in 1964 he suffered severe injuries in two automobile accidents. In the summer of 1965 his wife died and in 1967 his second son, the geologist Gerold Wagner, was run down and killed by a truck while doing fieldwork at the French impact crater of Rochechouart (Carlé 1972; Zauner 2010: 975; letter from Ernst Kraus to Georg Wagner, 18 October 1965, UAT 605/343).

---

[1]   That is he probably took part in the field trip to Italy organised by the German Geological Society (see Bernauer 1939).

Figure 10.1 Georg Wagner was the most vocal critic of the impact theory in the 1960s. (Source: Carlé 1972: 35).

The opening paragraph of Georg Wagner's obituary gives a good idea of the status Wagner enjoyed in Tübingen and its surroundings:

> If a man already reaches legendary greatness during his own lifetime, his death comes as a surprise to the younger generation: "Had he been still alive?" it is asked. "Can he die?" is the admittedly illogical, unspoken question among the older generation, because for half a century, one was accustomed to his having been there, that he gave witness; that he fought. (Carlé 1972: 35)[2]

This highly respected patriarch of the geological institute in Tübingen would become the most obstinate, and at times, a very emotional critic of the new paradigm. From his numerous letters, written over the next decade (UAT 605/343), it is clear how very desperate Wagner felt about his Ries problem. He was completely enraged because the situation seemed so unfair. 150 years of painstaking, diligent geological fieldwork, and then along comes a mineralogist – a foreigner, who had never before seen the Ries – who picks up a few rock samples, and then claims to know the solution to the Ries problem, based on evidence that can only be seen in a microscope! Wagner was not prepared to accept this, and his arguments were not always factual but were sometimes even chauvinistic (see below; compare also Carlé 1987).

Immediately after the International Geological Congress of 1960, about which Engelhardt presumably reported, Georg Wagner had tried to accommodate the impact theory, being conscious of his insufficient astronomical, physical and mineralogical expertise:

> We geologists can hardly progress with the meteorite theory of the Ries, since we must treat with our methods an alien field. Nearly all recent examples are missing to us, on which we might draw for our understanding through comparison. We know that in the Ries everything was much more violent than in our volcanoes of today, and also that the temperatures were higher. But we cannot investigate a modern analogue and proceed by comparisons. ... We have no trace of the meteorite's considerable mass! Everything is supposed to have evaporated! Where should we start? ... It is fine that our mineralogists pursue this question in more detail. But we are still a long way from successful clarification! We still must collaborate for a very long time until we see more clearly. ... And now our limited astronomical-physical knowledge still must be improved,

---

2   'Wenn ein Mensch schon zu Lebzeiten legendäre Größe erreicht, so überrascht sein Tod die jüngeren Zeitgenossen; lebte er denn noch? wird gefragt. Kann er sterben? lautet die freilich unlogische, unausgesprochene Frage der Älteren, denn man hatte sich ein halbes Jahrhundert lang daran gewöhnt, daß er da war, daß er aussagte, daß er kämpfte' (Carlé 1972: 35).

because for us it represents too much new territory! (Private note entitled *Zum Riesproblem* [concerning the Ries problem] by Georg Wagner, 9 September 1960, UAT 605/343)[3]

About a year later, however, at the Thomas-conference on 17 December 1961, Wagner began to organise adverse measures. He subsequently wrote many letters to colleagues to mobilise geological opposition against the 'veni vidi vici' methods[4] of mineralogists. What Wagner in his rage – and at times despair and dismay – failed to grasp was that the model of a central explosion, which he had held to be true and which after the death of Walter Kranz he had defended together with his friend Löffler against alternative explanations (especially the 'local outbursts' favoured by the Munich school under Richard Dehm), had stood the test. Only the means of the explosion were different; the fieldwork had been sound:

> Well, I think Georg Wagner ... had especially taken offence ..., but I believe the son, Gerold, has already been more guarded and even checked his father a little; but Georg Wagner was really enraged and his conduct against the virgin new views was rather direct; and for Mr Löffler, it had been his life's work, – ... . For them it had been really a hard blow, although they needed not to be offended. I mean, this had been new with the mineralogy [data], and there is no reason to deny that. (Seibold interview)[5]

---

3   'Wir Geologen kommen mit der Meteoritentheorie des Rieses kaum weiter. Denn wir müssen uns auf ein unseren Methoden fremdes Feld begeben. Uns fehlen fast alle rezenten Beispiele, die wir zum Vergleich für unsere Vorstellungen heranziehen könnten. Wir wissen zwar, daß es beim Ries viel gewalttätiger zuging als bei unseren heutigen Vulkanen, daß auch die Temperaturen höher lagen. Aber wir können keine rezenten Parallelen untersuchen und durch Vergleiche weiter kommen. ... Vom Meteoriten haben wir keine Spur von der doch erheblichen Masse! Alles soll verdampft sein! Wo sollen wir dann anpacken. ... Es ist erfreulich, daß unsere Mineralogen eingehender dieser Frage nachgehen. Aber von guter Klärung sind wir noch sehr weit entfernt! Wir müssen noch sehr lange zusammenarbeiten bis wir klarer sehen. ... Und jetzt muss man unserem beschränkten atronomisch-physikalischen [sic] Kenntnissen noch aufhelfen. Denn für uns ist allzu viel Neuland!' (Private note entitled *Zum Riesproblem* by Georg Wagner, 9 September 1960, UAT 605/343).
4   Letter from Georg Wagner to Mr Härtlein, 31 March 1965, UAT 605/343; see below.
5   'Nun ich meine, Georg Wagner ... war ganz besonders beleidigt, ... Aber ich glaube der Sohn, der Gerold, war schon vorsichtig, und der hat sogar den Vater ein bisschen gebremst, aber der Georg Wagner, der war wirklich auf Hundert und hat sich also sehr direkt benommen gegen die feldneuen Ansichten; und Herr Löffler, das war sein Lebenswerk, – ... Für die war das also schon ein harter Schlag, wobei sie gar nicht beleidigt sein konnten, ich meine, dass war halt neu mit der Mineralogie, das braucht man ja gar nicht abzustreiten' (Seibold interview).

A letter from Helmut Hölder to Reinhold Seemann offers a deeper insight into Georg Wagner's personality and allows a better understanding of his reaction:

> I want to speak up for Wagner. Even though there is no doubt that occasionally he lacks the correct tone and tact. ... As I understand it, he is too much concerned with the issue as such; too much with the single question: True or false? – so that he cannot see the value of pure research and the 'digging-into', even if it be into an error. ... .

> With the 'true – false' alterative another aspect of Wagner's character is also connected, that in the course of his work he too rapidly makes up his mind for one or the other choice and occasionally fights for it with vehemence, when we would prefer weighing and openness. I see in this forceful, occasionally hurting manner not uncertainty but rather a too great and too rapidly won conviction. For me, there is absolutely no doubt that Wagner is completely convinced of the correctness of his explosion-theory opinion and that his vehemence in discussions stems not from uncertainty but from anger that others do not want to see it as he does. Because he (and in this case, probably with <u>too</u> much pedagogic-mindedness), strongly sees a work as complete, only if he manages to enlighten others by it. The danger rests sometimes just in this joy in pat answers and their propagation; the danger that he might err despite his certainly unusually good geological understanding, and that he becomes fixed in a specific solution. (Letter from Helmut Hölder to Reinhold Seemann, 14 February 1949, GA3/208)[6]

---

[6] 'Ich möchte ein Wort für Wagner einlegen. Es ist zwar kein Zweifel, dass er zuweilen den richtigen Ton und Takt vermissen lässt. ... Es geht ihm für meinen Begriff zu sehr um die Sache, zu sehr um die einzige Frage: Richtig oder falsch? – so dass er den Wert des reinen Forschens und Sich-Vertiefens, und sei es auch in einen Irrtum, nicht zu erkennen vermag. ... // Mit der Alternative 'Richtig – falsch' hängt es als weiterer Zug von Wagners Wesen zusammen, dass er sich während seines Arbeitens rasch für das eine oder andere entscheidet und zuweilen auch mit Heftigkeit dafür eintritt, wenn wir ein Abwägen und Offenlassen lieber sehen würden. Ich sehe in dieser heftigen, zuweilen verletzenden Art nicht Unsicherheit, vielmehr eine eher zu große und zu rasch gewonnene Sicherheit. Es besteht für mich gar kein Zweifel, dass Wagner von der Richtigkeit seiner sprengtheoretischen Auffassung völlig überzeugt ist und dass seine Heftigkeit in Diskussionen nicht aus Unsicherheit, sondern aus dem Ärger darüber kommt, dass andere das nicht auch einsehen wollen. Sieht er doch stark und in diesem Fall wohl <u>zu</u> pädagogisch veranlagt – eine Arbeit erst dann als erfüllt an, wenn er anderen dadurch ein Licht aufstecken kann. Gerade an dieser Freude an fertigen Lösungen und an ihrer Weitergabe liegt manchmal die Gefahr, dass er trotz seines sicher ungewöhnlich guten geologischen Blicks irrt und in eine Lösung festgefahren ist' (letter from Helmut Hölder to Reinhold Seemann, 14 February 1949, GA3/208).

Seemann, who as a proponent of the tectonic Ries hypothesis was certainly among the favourite targets of Georg Wagner's pedagogical anger, understandably cultivated a somewhat less relaxed attitude towards Wagner and complained that 'to stand in scientific contrast to G[eor]g Wagner as such is already a lèse-majesté' (letter from Reinhold Seemann to Helmut Hölder, 14 January 1954, GA 3/320).[7]

When in a geological journal of the University of Erlangen, edited by Bruno von Freyberg (1894–1981)[8], there appeared a brief and plain German abstract of Chao's talk at the DMG meeting (Gehlen 1961), Wagner complained to the editor, pointing out all of the arguments against the impact scenario. Freyberg replied diplomatically:

> When Mr v[on] Gehlen[9] brought me the paper, I countered him with about the same arguments that you too offer. I too am very sceptical concerning the new Ries hypothesis. But Mr v[on] Gehlen has only reported the Americ[an] opinion. That does not mean that he identified himself with the result. And even if he did, I sometimes publish opinions that I do not share. First of all, I do not suppress an opinion, even when it contradicts our own views; and secondly the discussion becomes thus livelier and the old view, if it be correct, will be established by

---

[7]   ' ... mit Gg. Wagner in wissenschaftlichem Gegensatz zu stehen, ist ansich schon ein crimen laesae maiestatis' (letter from Reinhold Seemann to Helmut Hölder, 14 January 1954, GA 3/320).

[8]   In 1914, Freyberg began to study science in Halle. Soon, however, he volunteered for World War I, and spent some time as a prisoner of war in Russia, but was exchanged because he was seriously wounded, having lost one arm. After his recovery, he continued to study geography, geology, palaeontology and philosophy at the Universities of Munich and Halle, where he became an assistant at the Mineralogical Institute in 1917. In 1919, he obtained his PhD and continued as an assistant at the Geological Institute. In 1922 he obtained his qualification for a professorship and in 1928 was appointed professor without chair at the University of Tübingen. From 1925 to 1930, he undertook several research trips to South America and in 1932–33 also lectured at the Technical University in Stuttgart. In 1933 he moved to Erlangen as full Professor for Geology and Mineralogy and director of the institute. In 1945 he was dismissed due to denazification and worked as a porter in a clothing factory. After denazification was officially concluded in 1953, Freyberg returned to his professorship in Erlangen. He retired in 1962 (<http://www.catalogus-professorum-halensis.de/freybergbrunovon.html> accessed 4 October 2011; Freyberg 1974: 45; <http://de.wikipedia.org/wiki/Bruno_von_Freyberg> accessed 4 October 2011).

[9]   Kurt von Gehlen (born 1927) obtained his PhD from Freiburg University. In 1953 he became an assistant at the Mineralogical Institute at Erlangen University, where he obtained his qualification for professorship in 1960. He remained as a lecturer until 1962. He spent the following year at the Geophysical Laboratory of the Carnegie Institution in Washington. In 1966 he became Professor of Petrology, Geochemistry and Mineral Resources at the University of Frankfurt am Main (Freyberg 1974: 49).

modern arguments. ... What is correct will again prevail. So it will also be in this matter, and therefore I consider all this with calmness. But I shall be very pleased, if you write the announced essay for my journal. As I hear, in short time the paper by the Americans will appear with all of their arguments. Maybe you will wait so long as to deal with them in their entirety? It will be especially pleasant to me if such a valued and critical expert as you will take a position accordingly. (Letter from Bruno von Freyberg to Georg Wagner, 10 December 1961, UAT 605/343)[10]

One year later, Kurt von Gehlen (1962) also produced a German abstract of Shoemaker and Chao's paper in the *Journal of Geophysical Research* (Shoemaker and Chao 1961a), summing up their arguments: Suevite was the only rock at the Ries Basin that might possibly be of volcanic origin. This, however, was ruled out by several points of evidence, for example, the extreme heterogeneity of the glasses, which obviously had not had enough time to assimilate each other, and the presence of coesite and other features that were evidence for a sudden shock accompanying suevite formation. Gehlen added of his own accord that the authors had neglected to mention the recent dissertation of Willi Ackermann, who had interpreted the strange glasses as eutectic melting that had suddenly been arrested by a (volcanogenic) explosion, which caused cooling of the formerly heated rock (Ackermann 1958; see also Chapter 3).

Gehlen's abstract was followed immediately by another abstract, written by the journal's editor, Bruno von Freyberg (1962), reviewing a paper by Richard Dehm dismissing the impact theory for the Ries Basin (Dehm 1962; see below).

Wagner's essay was published in 1963 under the title *'Ries problem: not yet solved!'* compiling all his factual arguments: Coesite grains and nickel-spherules, according to Wagner, were but tiny and had been found in suevite, which represented a volcanic melt that had been found in more than 150 outcrops, many of them recorded as individual vents, which would account only for one

---

[10] 'Als mir Herr v. Gehlen das Referat brachte, habe ich ihm etwa dieselben Argumente entgegengehalten, die Sie auch einwenden. Ich nehme die neue Ries-Hypothese auch sehr skeptisch auf. Aber Herr v. Gehlen hat ja die amerik. Auffassung nur referiert. Damit hat er sich doch mit dem Ergebnis nicht identifiziert. Und wenn er es täte: ich bringe manchmal Auffassungen, die ich nicht teile. Erstens unterdrücke ich keine Meinung, auch dann nicht, wenn sie unseren eigenen Auffassungen widerspricht; und zweitens wird dadurch die Diskussion belebt und die alte Auffassung, wenn sie richtig ist, mit modernen Argumenten nur begründet. ... Was richtig ist, setzt sich doch wieder durch. So wird es auch hier sein, und ich betrachte deshalb das alles mit Ruhe. Aber sehr freuen wird es mich, wenn Sie mir für meine Blätter den angekündigten Aufsatz schreiben. Wie ich höre, soll in Kürze die eigentliche Arbeit der Amerikaner mit allen Begründungen herauskommen. Vielleicht warten Sie solange, um gleich ganze Arbeit zu machen? Es wird mir besonders lieb sein, wenn ein so gediegener und kritischer Sachkenner wie Sie dazu Stellung nimmt' (letter from Bruno von Freyberg to Georg Wagner, 10 December 1961, UAT 605/343).

per cent of all the Ries debris. The map of suevite distribution by Shoemaker and Chao had been misleading, because so far no suevite had been discovered within the Ries itself, but only in its surroundings of up to 25 km; a distance which, as Wagner claimed, was beyond the reach of impact effects. Suevite was thought to have been produced after the explosion because it cut through and overlay the explosion debris. Also, suevite at the Aumühle quarry definitely showed two distinct suevite-producing events. No trace of alien matter had been found in the Ries debris[11], and a calculation of the mass balance also did not require it. Nor had a trace of the meteorite been found. The Ries area, on the other hand, was a geologically predestined point, lying neatly in a whole line of volcanoes. In contrast to that, there are no meteorites, which home in on such a special target (Wagner 1963).

As regards the physical aspects of the impact theory, Wagner felt out of his depth and so sought interdisciplinary advice, for example, from Wolfgang Gentner,[12] director of the Max Planck Institute for nuclear physics in Heidelberg. Gentner, like Freyberg, reacted with commendable diplomacy, providing no foothold for Wagner:

> For your friendly letter concerning the Ries problem, I thank you very much. For me as a physicist, it is of course very difficult and dangerous to mingle in geological discussions. Therefore, I do not want to try it. I only want to tell you that Mr Schüller,[13] here in the Mineralogical Institute, has a paper in preparation in which he again has assembled all the reasons for a meteorite impact at the Ries from the mineralogical point of view.[14]

---

[11]   Wagner here was neglecting Chao's, respectively Engelhardt's, nickel-spherules which are in fact condensates of the vaporised projectile.

[12]   Wolfgang Gentner (born 1906) obtained his doctoral degree from the University of Frankfurt am Main in 1931. From 1933 to 1935 he held a stipend from the Radium Institute of the Sorbonne University in Paris. From 1936 to 1946 he worked as a scientist at the Kaiser Wilhelm Institute [the predecessor of the Max Planck Gesellschaft] for Medical Research. In 1945 he also became professor without chair at the University of Heidelberg and in 1946 became Professor of Physics at the University of Freiburg im Breisgau. In 1959 he became director of the Max Planck Institute for Nuclear Physics and at the same time professor at the University of Heidelberg (Freyberg 1974: 50).

[13]   Arno Schüller (1908–64) obtained his doctoral degree and a geological diploma in 1934. He was an assistant at the Mineralogical Institute in Leipzig from 1934 to 1938. In 1938–39, he worked as a mining mineralogist, but then became a soldier. After his military service, he returned to his assistant position until 1945. In 1946–47, he was a member of the Technical Office of the Mining Academy in Freiberg. In 1947 he became a mineralogist at the Geological Survey in Berlin. From 1952 to 1959, he was Professor of Mineralogy and Petrography at *Freie Universität Berlin*. Then he moved to Heidelberg (Freyberg 1974: 143; Wagenbreth 1999: 151).

[14]   Schüller and Ottemann (1963); see Chapter 11.

I agree with you that clarification will probably soon come, now that so many geologists and mineralogists are in the process of tackling this problem. (Letter from Wolfgang Gentner to Georg Wagner, 2 October 1962, UAT 605/343)[15]

Such soft, non-committal and diplomatic answers, which provided no target, probably vexed Georg Wagner, but there was one person who was willing to support him.

Walter H. Bucher and the Sceptics of Tübingen

Walter Herman(n) Bucher (1888–1965; Figure 10.2) had been born in the United States to parents of Swiss descent, who shortly after Bucher's birth moved to Germany, where young Walter grew up. He studied geology and palaeontology and obtained his doctorate from the University of Heidelberg in 1911. Soon afterwards, he returned to the United States and became a lecturer at the University of Cincinnati. The year 1915 saw him as assistant professor and 1920 as associate professor. In 1924 he was appointed full professor of geology. In 1940 he moved to Columbia University, specialising in structural geology. From 1950 to 1953 he was president of the American Geophysical Union, whose Walter H. Bucher Medal is named after him. After retirement, he worked as a part-time consultant in a petroleum company in Houston.[16]

Bucher was especially interested in so-called crypto-volcanic explosion sites, localised disturbances without actual volcanic rocks, which later, and by others, would be reinterpreted as deeply eroded impact structures (see Mark 1995: 59–73, 148–54). He knew about the Ries and Steinheim Basins, the latter being the type-locality of 'crypto-volcanic explosions' since at least the early 1920s, when Walter Kranz had had contact with Bucher, and thanks to

---

[15] 'Für Ihren freundlichen Brief über das Ries-Problem danke ich Ihnen bestens. Es ist für mich als Physiker natürlich sehr schwierig und gefährlich, mich in geologische Diskussionen einzulassen. Ich will es deswegen auch garnicht versuchen. Ich möchte Sie nur darauf verweisen, daß Herr Schüller im hiesigen Mineralogischen Institut eine Arbeit in Vorbereitung hat und darin alle Gründe für eine Meteoriteneinschlag im Ries noch einmal von mineralogischer Seite aus zusammengestellt hat. // Ich bin mit Ihnen der Meinung, dass wahrscheinlich in Bälde eine Klärung kommen wird nachdem so viele Geologen und Mineralogen neuerdings dieses Problem in Arbeit genommen haben' (letter from W. Gentner to Georg Wagner, 2 October 1962, UAT 605/343).

[16] <http://en.wikipedia.org/wiki/Walter_Hermann_Bucher> accessed 5 April 2011; Mark 1995: 59; <http://sites.agu.org/honors/walter-herman-bucher-1888%e2%80%931965/> accessed 14 November 2011.

Figure 10.2   Walter Herman Bucher, impact critic from the USA, joined in the Ries debate in 1963. (Photo: American Geophysical Union (AGU), courtesy AIP Emilio Segre Visual Archives).

Bucher, likened the Steinheim Basin to structures in the United States such as the Serpent Mound Disturbance in Ohio (Kranz 1926c: 97).

Bucher had mapped Serpent Mound in 1920 and published a short paper the following year, pointing out the strict analogy to the Steinheim Basin, then the only known example of a crypto-volcanic structure (Mark 1995: 63). Bucher visited the Steinheim Basin (and Ries Basin) in 1937 to confirm the analogy by his own field work (Bucher 1963a). Bucher also described Jeptha Knob (Kentucky), Upheaval Dome (Utah), Decaturville (Missouri), Wells Creek (Tennessee) and the Kentland disturbance (Indiana), all in the United States (Mark 1995: 70), and meanwhile regarded as impact structures.

After World War II, Walter Bucher was the main person in the United States to oppose impact ideas on principles against Robert Dietz and others. Upon the invitation of Georg Wagner, Bucher visited the Ries and Steinheim Basins again in 1962:

> In America, [Bucher] had been concerned with these crypto-volcanic structures, which today are also regarded as impact structures, but at that time Mr Bucher thought that he was able to explain them as crypto-volcanic, somehow hidden. So he went on a field trip, and I accompanied them, Schorsch[17] Wagner, Gerold[18] and me and Mr Bucher. We were in the Ries and afterwards in the Steinheim Basin, and all in all, it was quite agreeable and at that time we too were not yet convinced of the impact theory. ... Bucher [was] a superior, calm, and then already elderly gentleman ... and he made a friendly and good impression on me; [he was] so competent in geology too. (Hüttner interview)[19]

According to Bucher (1963a), Richard Löffler was also present, and they spent 'four most valuable days in the Ries Basin'. Bucher tried to explain the presence of coesite by an excess of volcanic energy: 'Some of us here believe that the presence of coesite in such excess is due to the excess energy, delivered in the form of kinetic energy within the rising mass' (letter from Walter H. Bucher to Georg Wagner, 18 February 1963, UAT 605/343).

---

[17] That is Georg.

[18] Gerold Wagner, Georg's son, also a geologist and a close friend of the interviewee.

[19] '[Bucher] hat sich ja in Amerika mit diesen kryptovulkanischen Strukturen beschäftigt, die man ja heute auch als Impaktstrukturen ansieht, aber damals meinte der Herr Bucher die Sache eben auch kryptovulkanisch erklären zu können, verborgen irgendwie. Da hat er mal eine Exkursion gemacht, da war ich mit dabei mit Schorsch Wagner so und Gerold und ich und Herr Buchner. Wir waren im Ries und dann im Steinheimer Becken und es war an sich recht angenehm und da waren wir ja auch noch nicht überzeugt von der Impakttheorie. // ... Er [war] ein überlegter, ruhiger, damals schon älterer Herr ... und er hat mir einen freundlichen und guten Eindruck gemacht; auch so kompetent in der Geologie' (Hüttner interview).

With the same letter, Bucher asked for more samples of strongly deformed granite fragments to be sent to him. One year later, he published his findings in the *American Journal of Science* (Bucher 1963a) and a summarising version in *Nature* (Bucher 1963b).

Bucher admitted that an impact origin of the Ries Basin was possible and that this new theory would explain many 'puzzling factors', but he also cautioned by insisting on a descriptive, non-interpretative nomenclature. Although Bucher accepted coesite as evidence of shock, he still claimed, as all the German Ries geologists had done before him, that suevite occurred as vent fillings and thus had to be younger than the actual Ries-forming explosion. Bucher likened the suevite 'vents' to the diatremes of the Swabian Volcano and the Maar-volcanoes of the Eifel. A strong magnetic anomaly near the village of Wörnitzostheim was supposed to be a large vent, covered by Tertiary sediments.[20] Bucher pointed also to a thermal anomaly beneath the town of Nördlingen (Bucher 1963a).

Bucher's main argument, however, was the allegedly very special geological-geographical location of the Ries Basin:

> It would be difficult to find a more clearly defined position within the structural framework of southern Germany for a giant meteorite to hit. Pure chance could, of course, produce such an event. If this were a unique phenomenon, the argument involving chance would have little weight. But the meteorite hypothesis demands that the Ries giant had a small companion which is supposed to have produced the Steinheim Basin, and that it was placed neatly on the line connecting the Ries with the volcanic Urach area and, more remarkable still, almost exactly halfway between these two major structures. (Bucher 1963a)

In the much shorter paper in *Nature*, Bucher dwelt at length on this crucial (as he saw it) factor. Impact craters should be distributed randomly, but according to him, they were not:

> No doubt most, if not all, geochemical and geophysical (tectonic) features of individual cryptovolcanic structures could have been produced by meteorite impact. But were they? Discussion based on these features neglects a fundamental point: If meteorite impact created the structures, their incidence in time and space must be consistent with the hypothesis, that is, they must be distributed at random. This means that they must not bear any demonstrable relation to structural axes in the Earth's crust, nor to features the formation of which was accompanied by deep-seated volcanic activity.

---

[20]   At this locality research drilling was done in 1965 (see Chapter 11).

... I apply this test to the largest cryptovolcanic structures of Europe, America and Africa: the Ries Basin [24 km], the Wells Creek structure [12 km], and the Vredefort Dome [ca. 300 km; South Africa] ...

In all three cases, inspection of the regional setting reveals conclusively that each lies on a major regional structural axis aligned with one or several structures involving localized mafic and ultramafic igneous rocks. (Bucher 1963b)

And so again the special location of the Ries became the main factual argument against impact. Only with hindsight do we seem able to acknowledge that the unique features of the location had been an artefact of the intense research in that area: If you look long enough, every place becomes special, or as Robert Dietz answered to Bucher's argument: 'it would be difficult to find a spot on the tectonic map of the United States that was not associated with regional trends' (quoted after Mark 1995: 151), while Bevan French[21] later remarked with respect to all known impact structures at that time that '[a] few arbitrarily selected examples (Bucher 1963) are not statistically sufficient to establish the random or nonrandom distribution of a group of more than 50 structures' (French 1968a: 8–9). This remark made clear that impactists like French defined 'random distribution' differently from Bucher. Bucher assessed this quality from the relation of individual structures to local geology, while French had the entirety of known impact craters and their global distribution in mind.

Thus, a methodological problem became crucial in the new discussion. It is, however, no 'explanation' for an enigmatic event to point out a number of tectonic lineaments and other linear features to argue for a remarkable and specially predestined location. A theory must always explain the actual process that formed this allegedly unique structure. Wagner, however, worried about this question only within the impact scenario, while 'volcanic eruption' remained just another 'garbage-bin term' for which no actual, physically feasible eruption mechanism had been demonstrated.

While Shoemaker eventually managed to convince Bucher about the impact origin of Meteor Crater, Bucher remained sceptical as to the Ries Basin until his death in February 1965 (Mark 1995: 152), and so for some time Bucher provided Wagner with arguments. After Bucher's death, however, Georg Wagner, by then himself old and frail, failed to find another champion for his

---

[21]   The petrologist and planetary geologist Bevan M. French joined NASA's Goddard Space Flight Center in 1964. In 1972 he went to the US National Science Foundation as Program Director for geochemistry. In 1975 he returned to NASA to work in administration and science-management. French retired from NASA in 1994, and returned to scientific research as a research collaborator at the Smithsonian Institution. He was also a visiting professor at the University of Vienna in 1994, 1997 and 2001 (<http://nutley.bccls.org/nutleyhalloffame/hofbfrenchpg.htm> accessed 24 July 2012).

cause, although he continued to address likely candidates such as G. Christian Amstutz (1922–2005):

> The conduct of Shoemaker-Chao has angered us very much as 'Göbbels-propaganda;'[22] so we have clashed severely, while with Bucher most beautiful harmony reigned. We want to discuss [these matters] factually and scientifically, and not in the form of fun-fair propaganda. Bucher would have been our man. Will you now be that? There are too many unsolved questions. (Letter from Georg Wagner to G. Christian Amstutz, 20 June 1965, UAT 605/343)[23]

Swiss-born Amstutz studied engineering geology at the *Eidgenössische Technische Hochschule* (ETH) in Zurich and received his diploma in 1947. For the next two years he was research assistant at the ETH's Geophysical Institute. From 1949 to 1951, with the aid of research grants, he travelled several times to the United States. From 1951 to 1952 he was research assistant to Paul Niggli[24], again at the ETH Zurich. In 1952, he became geologist and petrologist of the Cerro de Passo Corporation in Peru. In October 1956 he was appointed associate professor of geology at the University of Missouri School of Mines and Metallurgy in Rolla. In 1964 he became full professor and director of the Institute for Mineralogy and Petrography of the University of Heidelberg (Park et al. 2008; Freyberg 1974: 7).

In 1964, Amstutz had argued somewhat polemically against an impactist interpretation of the Decaturville disturbance with the well-known argument of a special, pre-destined location: 'Anyone examining the polygonal and ring structures[25] and their geologic environment carefully can hardly fail to dismiss the assumption that a meteor would fall on the earth like a guided missile, and sense exactly where there is a regionally favorable place for punching a hole into the earth' (Amstutz 1964: 346–7).

Amstutz continued to sow doubt where none should have been, as an SiO_2 phase diagram including coesite had been published in 1960 (Boyd and England 1960), and stishovite, a considerably denser modification of SiO_2

---

[22]   Josef Goebbels had been the Nazi propaganda minister.

[23]   'Das Vorgehen von Shoemaker-Chao hat uns als 'Göbbelspropaganda' [sic] sehr geärgert; so sind wir scharf zusammengestoßen, während bei Bucher schönste Eintracht herrschte. Wir wollen uns sachlich, wissenschaftlich auseinandersetzen nicht in Jahrmarktpropaganda. Bucher wäre unser Mann gewesen. Werden nun Sie es werden? Denn der ungelösten Fragen sind zu viele' (letter from Georg Wagner to Amstutz, 20 June 1965, UAT 605/343).

[24]   Paul Niggli (1888–1953) lectured in Zurich from 1913 onwards. In 1915, he became professor without chair in Leipzig. In 1918, he switched to Tübingen, and returned to Switzerland in 1920 to become full Professor of Mineralogy and Petrography in Zurich (Freyberg 1974: 111).

[25]   Amstutz's term for cryptovolcanic structures.

than coesite, and thus requiring even higher pressure, had been discovered in meteorite craters, including the Ries crater in 1962 (Chao et al. 1962; Chao and Littler 1963; see Chapter 12): 'The physical criteria based on coesite and other polymorphs of SiO_2 are perhaps all premature as long as we know so little about their phase diagrams. It may well be possible that volcanic and tectonic processes produce coesite' (Amstutz 1964: 349).

And the presence of shatter cones was an argument that Amstutz deemed 'rather one-sided and superficial' (Amstutz 1964: 349), as:

> A detailed morphometric study [... ; of shatter cones] showed some of these cones are likely to be diagenetic cone-in-cone features; whereas others may have been produced by the breakage and pressure during tectonic movements. To the knowledge of this author no shatter cones have ever been found in meteorites themselves. This, of course, is rather striking evidence. (Amstutz 1964: 350)

Amstutz mentioned the possibility of interpreting cryptovolcanic structures as calderas or as volcanic ring-complexes, although he had to admit that lavas or dyke rocks were missing. To avoid this discussion he preferred to postulate mechanical deformation of the crust, which led to the observed faults and brecciation and which was supposed to sit at the cross-points of large fault systems; for which, however, no evidence was presented (Amstutz 1964: 353). Thus the paper chiefly remained destructive, that is only denying the new impact evidence without presenting an alternative. The author rested content with his opinion that:

> we do not need to invent a blow from outer space. Inherent crustal processes known elsewhere on the crust of the earth provide an entirely satisfactory analogue and explanation. ...

> The fact that scientists of our generation can depart so much from the scientific method of working with congruent analogues, or, in other words, the fact that the symmetry relationship between the polygonal structures and the regional geological pattern can be ignored, is alarming. ...

> [T]he science of mineral deposits is presently going through the same development as experienced by paleontology one hundred years ago: the belief into a magic creation from the outside is exchanged for a mature scientific interest in the evolution of ore forming processes in and with the host rock. The magic belief in a deus ex machina, a creationistic influence from some unknown external source is exchanged for a careful scientific interest and procedure.

What we now experience in regard to the revival of creationistic belief in impact from outer space is an alarming relapse into patterns which belong to the dark middle ages. After we have gone through a mass psychosis of beliefs and visions of flying saucers, which C. G. Jung (1959) has shown to be caused by deep fear of our time, the work with outer space causes now a wave of meteor-impact belief. It makes scientists so blind for sound basic geological assumptions that they build complex mathematical structures on top of assumptions which are pure geological and topological nonsense, and the mechanically derived conclusions, of course, must lead to more nonsense. (Amstutz 1964: 351–2)

A year later, Amstutz assembled his scepticism against Dietz's interpretation of shatter cones as shock and thus impact indicators in another paper, in which, in an over-generalising way, he lumped together all sorts of remotely similar-looking features, only to conclude that 'cones may be of many different origins' (Amstutz 1965: 1054).

Even though Amstutz was strongly opposed to the impactist cause, there is no evidence of further contact to Georg Wagner; instead he corresponded amicably with Wolf von Engelhardt (Engelhardt papers at GA 101), and Wagner eventually began to show signs of resignation, when more and more geologists, including his own son, the geologist Gerold Wagner, accepted the new ideas (see Chapters 11 and 12).

Although Georg Wagner was the most vocal and most organised of the impact sceptics in Tübingen, there were other people there just as unhappy about the latest developments in Ries research. Reinhold Seemann had not been at the Thomas-conference in 1961 (Hölder interview), but later added comments (printed in Hölder 1962), in which he criticised the new impact theory:

He then, somewhat with the stubbornness of old age, continually tried to elaborate his thoughts. Now he saw that with tectonics alone it would not work, but now he thought we could go far [towards a solution] concerning the chaos at the Ries with thrusting caused by gravity. ... And so he ... tried somewhat fantastically to enlarge his theory into the exogenic realm, however definitely not by meteoritic processes but by purely terrestrial means, that is by weathering and atmospheric phenomena and so on, by much movement due to gravity and slumping and so on to explain the turbulent picture of the Ries surroundings. And there I could no longer follow him. What he did then seemed to me too farfetched. (Hölder interview)[26]

---

[26] 'Er [hat] sich dann etwas altersstarr verhalten und hat nun immer weiter versucht diese Gedankengänge auszubauen. Jetzt hat er gesehen mit Tektonik allein geht es nicht, aber nun hat er gedacht, nun also dieses Durcheinander im Ries, da können sehr viel durch Schwerkraft sich vollziehende Übereinanderschiebungen und so weiter kommen. ... Also er hat ... etwas phantastisch versucht, seine Theorie durch Ausweitung in das exogene Geschehen, aber eben nicht meteoritischer Art, sondern rein irdischer Art durch also Verwitterung und

Others remembered faintly that in the 1930s impact theories had already been proposed, and now wished to re-read and re-evaluate these old papers but first had to hunt for the long-lost references. Walter Carlé, for example, contacted Edwin Hennig, who in his comment against Stutzer's proposal of an impact at Ries Crater and Steinheim Basin (Hennig 1936; see Chapter 3) had briefly mentioned Rohleder's and Kaljuvee's contribution:

> Yes, oh my goodness, where might I have picked up the theses of Rohleder and Kaljuwee [sic]? I am in the happy condition that I cannot lose my memory in old age, because I never had one. After more than a quarter of a century. – In vain, I sort through my braincase, but next week – today is Saturday – I shall browse the institute's library. It might of course have been in *Nature* or something similar, the issues of which I have no longer available. The name Kaljuwee I had to verify in my off-print to make sure it was quoted correctly by you. I faintly remember it was a Slav, but you are right: It sounds more Indian.

> Unfortunately, I cannot give you much hope, but at least I shall try my best.

> Suevite, in the new light, looks rather interesting. But I stick to my word: A meteorite would be an especially mean individual were it to dress up in carnival-fashion as a link in a long chain of tectonic events and hide right in the middle of volcanoes of all sorts. And where are the meteorites of the whole history of the Earth? Must they have already existed in former times?? (Letter from Edwin Hennig to Walter Carlé, 10 March 1962, GA 11/177)[27]

---

atmospärische Erscheinungen und so weiter, durch sehr viele Schwerkraftbewegungen und Rutschungen und so weiter, das turbulente Bild der Umgebung des Rieses zu deuten. Und da konnte ich ihm nun auch nicht mehr folgen. Das schien mir also zu weit hergeholt, was er da dann gemacht hat' (Hölder interview).

[27] 'Ja, du lieber Himmel, wo mag ich die Thesen von Rohleder u. Kaljuwee [sic] aufgegabelt haben? Ich bin in der glücklichen Lage im Alter das Gedächtnis nicht verlieren zu können, weil ich nie eins gehabt habe. Nach mehr als ¼ Jahrhundert ... Ich stöbere vergeblich in meinem Gehirnkasten, will aber – heut ist Samstag – in nächster Woche die Instituts-Bücherei unsicher machen. Es könnte natürlich auch in 'Nature' oder dergleichen gestanden haben, deren Jahrgänge mir nicht verfügbar sind. Den Namen Kaljuwee mußte ich erst in meinem Separat als richtig von Ihnen zitiert feststellen. Mir schwant eher etwas von einem Slawen, aber Sie haben recht: indisch klingt's eher. // Viel Hoffnung kann ich Ihnen leider nicht machen, aber mir wenigstens Mühe geben. // Der Suevit sieht im neuen Licht recht interessant aus. Aber ich bleibe dabei: Ein Meteorit wäre ein besonders gemeiner Vertreter, wenn er sich faschingsmäßig als Glied langanhaltender Tektonik mitten zwischen Vulkanen aller Art versteckt. Und wo bleiben die Meteoriten der gesamten Erdgeschichte? Muß es sie schon früher gegeben haben??' (Letter from Edwin Hennig to Walter Carlé, 10 March 1962, GA 11/177). For meteorites in former times see Hennig's comment in Chapter 3.

Scepticism and Apologetics in Munich

Scepticism was by no means restricted to Tübingen University. Munich too showed initial reluctance to accept the new situation, such as in the person of Ernst Carl Kraus, the chair of applied geology and ore geology at the Geological Institute of the Ludwig-Maximilian University in Munich (see Chapter 5):

> As to the meteorite interpretation of the Ries, I am like you very unbelieving: So far by no means are all of the pressure minerals known that originate in huge volcanic explosions! Of course, deep enough drillings are still missing at the Ries and the masses of suevite too are not explained by a meteorite – more so as it [the suevite] is missing in the few other, well-known true meteor craters. Should the hypothetical meteorite in the Ries have had the ambition to hit the ridge and the suevite there down below? (Letter from Ernst Carl Kraus to Georg Wagner, 18 October 1965, UAT 605/343)[28]

Another of Kraus' colleagues, Oskar Kuhn[29] (1908–90), in 1964 published the third newly revised and enlarged edition of his semi-popular *Geology of Bavaria*, in which according to the preface, dated July 1964, the Ries chapter gave much more detail than before. Kuhn wrote there:

---

[28] 'Bezüglich der Meteoriten-Deutung des Rieses bin ich mit Ihnen sehr ungläubig: Man kennt doch noch keineswegs alle Druck-Mineralien, die bei riesigen Vulkanexplosionen entstehen! Natürlich fehlen noch genügend tiefe Bohrungen im Ries, und die Massen des Suevits erklärt auch kein Meteorit – zumal er in den wenigen anderen, gut bekannten wirklichen Meteorkratern fehlt. Sollte der hypothetische Meteorit im Ries den Ehrgeiz gehabt haben das Schwellen-gebiet [sic] und den Suevit da unten zu treffen?' (Letter from Kraus to Georg Wagner, 18 October 1965, UAT 605/343).

[29] Oskar Kuhn obtained a palaeontological doctorate from the University of Munich in 1932 and continued to work there until 1938, when he moved to the University of Halle, working on the Tertiary fossils of the Geiseltal. In 1939 he qualified for a professorship at the University of Halle, becoming a lecturer. A conservative and devout Catholic, he saw evolution only possible within predetermined morphological classes. Consequently, he rejected the theory of common descent. 'After a political conflict with his mentor, Johannes Weigelt [1890–1948] over evolution, Kuhn's teaching certification was withdrawn ... in November 1941. He had to leave Halle and was immediately called up for wartime service in the Wehrmacht. In February 1942 he was released because of lung disease. (He had been a member of the SA from 1933 to 1936 but left for health reasons.)' (<http://en.wikipedia.org/wiki/Oskar_Kuhn> accessed 14 November 2011). In 1947 he became a professor at the Philosophical-Theological High School in Bamberg, but left after a short time (Freyberg 1974: 88; <http://en.wikipedia.org/wiki/Oskar_Kuhn> accessed 14 November 2011). For Weigelt see Grüttner 2004: 181–2 or <http://de.wikipedia.org/wiki/Johannes_Weigelt> accessed 14 November 2011.

The Ries explosion was 14.8 +/- 0.7 million years ago (potassium-argon age). The recently advanced opinion that the impact of a large meteorite (meteor crater) was the cause is to be rejected, even though it is supposed to have only acted as a catalyst for the magmatic processes. The Ries is characterized by an unusual proximity of the Variscan basement. The deep [seated] magma, however, played only a passive role in the explosion, magma itself did not flow out. In the Ries, volcanic bombs and ashes are widely distributed, the debris of the blasting flew as far as Augsburg, where a block of 150 kg weight was found. ...

The theory advanced since a few years ago that the Ries basin owes its origin to a huge meteorite is generally rejected. (Kuhn 1964: 105)[30]

Similar scepticism was displayed by [Karl] Werner Barthel, one of Dehm's PhD students, in a popular brochure, which was issued by a publisher located in the Ries Basin (Barthel 1964). Barthel (born 1928) had obtained his PhD from Munich University in 1955 and continued to work for Richard Dehm first as a scientific amanuensis and then as research assistant and eventually, from 1960 onwards, as curator at the Bavarian State Collection for Palaeontology and Stratigraphy. He eventually became Professor of Geology at the Technical University in Berlin (Freyberg 1974:11).

Dehm, who had had numerous PhD students mapping in the Ries and surrounding area, was himself unconvinced:

I remember a rather agitated [irritated?] conference shortly after the announcement of the meteorite theory by Shoemaker and Chao: Mr Preuss[31] at that time already

---

[30] 'Die Riesexplosion liegt 14,8 +- 0,7 Millionen Jahre (Kalium-Argon-Alter) zurück. Die neuerdings vertretene Auffassung, daß der Einschlag eines großen Meteoriten (Meteorkrater) vorliege, ist abzulehnen, wenn dieser auch nur auslösend auf die magmatischen Vorgänge gewirkt haben soll. Der Rieskessel ist durch abnorme Nähe des variszischen Untergrunds gekennzeichnet. Das Tiefenmagma spielte aber bei der Explosion nur eine passive Rolle, Magma selbst floß nicht aus. Im Ries sind vulkanische Bomben und Aschen weit verbreitet, die Trümmer der Sprengung flogen bis nach Augsburg, wo ein 3 Zentner schwerer Block gefunden wurde. ... Die seit wenigen Jahren vertretene Theorie, daß der Rieskessel seine Entstehung einem riesigen Meteor verdanke, wird allgemein abgelehnt' (Kuhn 1964: 105).

[31] Ekkehard Preuss (1908–92) studied mineralogy and geology in Jena from 1927 to 1935, when he obtained his doctoral degree. His supervisor was Fritz Heide. From 1936 to 1939, Preuss was an assistant at the Mineralogical Institute in Göttingen. From 1939 until 1945 he served in World War II as a soldier. From 1946 to 1948 he was a lecturer in Göttingen, and in 1949 he began to work as a scientist for an optics company [*Optische Werke Steinheil*] in Munich. In 1952 he became an honorary professor in addition to his industrial job. In 1960, he switched fully to the Technical University of Munich as a full professor. He

defended the new genetic hypothesis vigorously, while Mr Dehm – for whom after a decade-long labour under the prerequisite of the volcano theory the foundation [of his work] broke down – remained wisely guarded. (Letter from Wolf-Dieter Grimm to the author, Munich, 20 September 2001)[32]

But Dehm was considerably more reluctant than just 'wisely guarded'. Like Georg Wagner, he published a summary of all the arguments against an impact at Ries Crater in 1962. His general tone was less polemical but more arrogant. 'During the many years of concern with the geology of the Nördlinger Ries, we also put the *meteor interpretation* before our mind (Schröder & Dehm 1950, page 134); it seemed to us to postulate a wholly unthinkable chance event and therefore to be beyond reason' (Dehm 1962: 70).[33]

In 1950, however, despite the affirmation, Schröder and Dehm had not addressed the impact theory in any detail but mentioned its existence dismissively in just a single sentence (see Chapter 1), without further discussion or explanation, because it was 'generally recognized as wrong' (Schröder and Dehm, 1950: 134).

Dehm then brought forth the old argument of an extremely unlikely chance event right at the special, predestined locality of the Ries, which together with its smaller companion was situated right in the middle of a 'Southwest-German Mega-Plate', an isometric triangular structure with a side length of some 450 km bordered by the Rhine-Graben, the Bohemian Massif and the Alps (Figure 10.3). Ironically, this placed the Ries and Steinheim Basins as far away as possible from the major active tectonic zones, which bordered the Mega-Plate. The Ries Basin and the Steinheim Basin, which Dehm without further thought still interpreted as a volcano, were on a straight line with the diatreme field of the Swabian Volcano and the carbonatite volcano Kaiserstuhl in the Rhine-Graben, but in view of the distribution of other tertiary volcanoes in the German southwest – the Hegau Volcanoes, the Vogelsberg, the Rhön area and other smaller

---

retired in 1973 (Franke 1983: 291). The affiliation with the TU Munich from 1960 onwards might also have included, in the 1960s, a curatorship at the Research Institute for Applied Mineralogy in Regensburg (Freyberg 1974: 120). In 1934 Preuss published about elevated chromium and nickel contents in tektites (Franke 1983: 183).

[32] 'Ich erinnere mich an eine recht aufgeregte Konferenz ganz kurz nach Bekanntgabe der Meteoritentheorie durch Shoemaker & Chao: Herr Preuss vertrat schon damals energisch die neue Entstehungshypothese, während Herr Dehm – für den ja nach jahrzehntelangen Arbeiten unter der Voraussetzung der Vulkantheorie eine Grundlage zusammenbrach – sich klug abwartend verhielt' (letter from Wolf-Dieter Grimm to the author, in Munich on 20 September 2001).

[33] 'Bei der langjährigen Berührung mit der Geologie des Nördlinger Rieses haben wir uns (Schröder & Dehm 1950, S. 134) auch die *Meteordeutung* vorgelegt; sie schien uns einen ganz undenkbaren Zufallstreffer zu postulieren und daher abwegig zu sein' (Dehm 1962: 70).

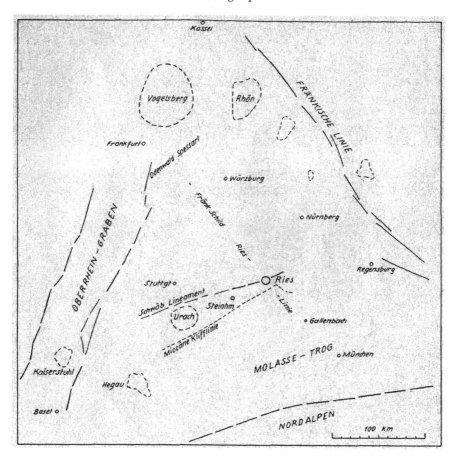

Figure 10.3   Sketch map by Richard Dehm showing the Ries Crater (*Ries*), the
Steinheim Basin (*Steinhm*), various structural 'lines' (the tectonic
graben structure of the Upper Rhine Valley (*Oberrheingraben*)
to the west, the tectonic western boundary of the Bohemian
Massif (*Fränkische Linie*) to the east, the northern margin of the
northern Alps (*Nordalpen*) to the south, the tectonic Swabian
Lineament (*Schwäbisches Lineament*), the northern coastline
(*Miozäne Klifflinie*) of the Mid-Tertiary Molasse Sea north of
the Alps (*Molasse-Trog*) and the *Fränkischer Schild Ries-Linie*)
and volcanically active areas in the Tertiary period (*Kaiserstuhl,
Hegau, Urach* [that is the Swabian Volcano], *Vogelsberg, Rhön* and
unnamed irregular areas surrounded by dashed lines). (Dehm
1962: Figure 2).

occurrences – there was nothing particularly special about this alignment. It was made to look special because on Dehm's map it was bordered by two – unrelated – lines, the Miocene cliff line in the south, which had no tectonic significance whatsoever, but was simply the coastline of a shallow Molasse Sea north of the Alps, and the Swabian Lineament, a highly hypothetical fault line running along the escarpment of the Swabian Alb. From seismic investigations, it was known that 'from the middle Black Forest to the east northeast, there runs a series of depressions below the Urach area [the area of the Swabian Volcano] and the Steinheim area to the 'bulge' of the Ries' (Dehm 1962: 80).[34] In this way the importance of the Swabian Lineament was bolstered with arguments that had no statistical significance: two of three structures were associated with a depression and the third with a bulge – so what? At the Ries Basin the line connecting the Swabian Volcano, the Steinheim Basin and the Ries Basin, was crossed by another line, which Dehm called the *Fränkischer Schild Ries-Linie*, running straight from Frankfurt to Munich, and according to Dehm 'a notable hinge since the Variscan orogeny and into the Cainozoic for see-saw-motions within the depositional area of southern Germany' (Dehm 1962: 81–2).[35] In other words, while elevation or subsidence occurred to the east or west of the line, nothing really happened at the line itself. Following Deecke (1925), Dehm also postulated obscure deep-seated structures, which explained the distributions of volcanoes: '[Deecke], by a surprisingly simple system of straight lines and circles on the map, is able to connect the volcanic and thermal areas of Middle Europe. He is providing thirty such relationships; in seven of them the Ries participates' (Dehm 1962: 82).[36]

Dehm's 'lines' and connections thus conspicuously resemble those of 'Ley-Lines' or various esoteric grids (see Knoblauch 1991). A current observer gains the impression that no matter where the meteorite might have struck, it would always have been at a special and therefore most unlikely location.

Consequently, Robert Dietz replied to Walter Bucher, when confronted with a similar line of argument, that 'There is a great geologic literature about lineaments and trends, but often the points correlated seem as random as the stars in the sky' (Dietz quoted after Bourgeois and Koppes 1998: 148), and

---

[34]   ' ... vom mittleren Schwarzwald zieht nach Ostnordosten eine Reihe von Einmuldungen unter dem Uracher und Steinheimer Gebiet zum 'Buckel'' des Rieses' (Dehm 1962: 80).

[35]   ' ... eine höchst bedeutsame, seit dem Varistikum immer wieder bis in das Känozoikum bemerkbare Achse für Schaukelbewegungen innerhalb des süddeutschen Sedimentationsraum' (Dehm 1962: 81–2).

[36]   '[Deecke] kann durch ein überraschend einfaches System von Geraden und Kreisen auf der Karte die Vulkan- und Thermengebiete Mitteleuropas miteinander in Verbindung bringen. Dreißig solcher Beziehungen führt er auf; bei sieben davon ist das Ries beteiligt' (Dehm 1962: 82).

Bevan French commented: 'The structures are not comparable; in some cases two cryptoexplosion structures establish a straight line (e.g., Ries Steinheim (Bucher, 1963[a]) ...), and other dissimilar structures are arbitrarily added along the trend' (French 1968a: 8–9).

Nevertheless, the argument kept on impressing even people who had otherwise reacted positively to the renewed impact idea. They had to reflect on the problem and find their own viewpoint:

> The spatial association with the geology of the Southwest-German Mega-Plate ... and with the temporal sequence of the volcanism [in that region] shows connections which seem to place the Ries firmly within this spatial-temporal frame. If one, however, accepts the Ries event to be an entirely unusual phenomenon, then I should like to view the location of the Ries to be not adequately prominent. Rather the Ries should be regarded within the larger frame, for example, of a continent. Then, however, its present geological locality seems to be entirely random, because it does not correspond to the uniqueness of the Ries. It is different, however, as to the spatial and temporal connection of the Ries with the Steinheim Basin and the volcanoes near Urach [the Swabian Volcano]. These too are unique (see Dehm). (Preuss 1964: 288)[37]

The impact theory was further discredited by association with its earlier propagator Kaljuvee: 'I do not know of any newer observations and reflections concerning meteor craters, which today would add more weight to the 'fanciful publication' of Kaljuwee [sic]' (Dehm 1962: 83).[38]

For Dehm, the Steinheim Basin was definitely a volcano, and he mentioned it always in connection with the vents of the 'Swabian Volcano' near the town of Urach (see Chapter 1): 'The Ries Basin, the Steinheim Basin and the volcano embryos of Urach are by no means normal, widely distributed phenomena often

---

[37] 'Die räumliche Einordnung in die Geologie der Südwestdeutschen Großscholle ... und in die zeitliche Abfolge des Vulkanismus zeigen Verknüpfungen, die das Ries fest in diesen räumlich-zeitlichen Rahmen zu stellen scheinen. Geht man dagegen davon aus, daß das Riesereignis eine völlig ungewöhnliche Erscheinung darstellt, dann möchte ich die Lage des Rieses nicht als entsprechend bevorzugt betrachten. Das Ries sollte vielmehr in den größeren Rahmen z.B. eines Kontinentes gesehen werden. Dann aber erscheint sein jetziger geologischer Platz rein zufällig, denn er entspricht nicht der Einzigartigkeit des Rieses. Anders ist es mit der räumlichen und zeitlichen Verknüpfung des Rieses mit dem Steinheimer Becken und den Uracher Vulkanen. Auch diese sind einzigartig (DEHM)' (Preuss 1964: 288).

[38] 'Davon, daß neuere Beobachtungen und Überlegungen zu Meteorkratern heute der 'phantasiereichen Veröffentlichung' Kaljuwees [sic] (Kranz 1937, S. 201) mehr Gewicht als damals verleihen könnten, ist mir nichts bekannt geworden' (Dehm 1962: 83).

present in the various rock formations; instead, each in itself is unique' (Dehm 1962: 84).[39]

Dehm then undertook calculations of the probability of a conjoining of volcanism and postulated impact, in which he chose to assume that the Tertiary volcanism in Germany was a tightly focused event in terms of time rather than a phenomenon that lasted for millions of years, and which in the dormant volcanism of the Laacher See area is still not extinct. But of course the more focused Dehm assumed the volcanic events to have been the less probable was it that an impact might have happened contemporarily:

> The high improbability of such a contemporaneity of independent factors is obvious; it speaks against the meteoric interpretation of the Ries. On the other hand, following the newest findings, the necessary energy can be provided neither by volcanic nor by other known means apart from a meteorite impact. This true dilemma seems to me unsolvable in the present state of Ries and meteor research; a neglect of facts certainly would be no solution. Rather, additional intensive and varied work will be necessary in the area of the Nördlinger Ries and possibly for comparable structures, as well as on the occurrence of high-pressure minerals (and similar) on Earth.
>
> As for the reasoning on the energy balance, it will be possible to make a comparison with the kimberlite pipes of Southern Africa. (Dehm 1962: 84)[40]

For Dehm, more was at stake than just the adoption of a new idea, as he had dedicated a large part of his career (considerably more than Georg Wagner) to the Ries Basin:

---

[39] 'Ries, Steinheimer Becken und Uracher Vulkan-Embryonen sind nicht etwa normale, regional weit verbreitete und während der Formationen oftmals auftretende Phänomene, sondern jedes ist für sich einmalig' (Dehm 1962: 84).

[40] 'Die hohe Unwahrscheinlichkeit solchen Zusammentreffens voneinander unabhängiger Fakten ist offenbar; sie spricht gegen die Meteordeutung des Rieses. Andererseits kann nach den neuen Erkenntnissen die erforderliche Energie weder von vulkanischen noch von sonstigen bekannten Vorgängen geliefert werden außer von einem Meteoreinschlag. Dieses echte Dilemma scheint mir beim gegenwärtigen Stand der Ries- und der Meteorforschung nicht lösbar; ein Beiseitelassen von Fakten wäre sicher keine Lösung. Es ist vielmehr weitere intensive und vielseitige Arbeit sowohl im Gebiet des Nördlinger Rieses und etwaiger vergleichbarer Gebilde als auch über die Vorkommen von Hochdruckmineralien u. ä. auf der Erde erforderlich. // Bei den Überlegungen zur Energiebilanz wird man die Kimberlitschlote Südafrikas zum Vergleich heranziehen können' (Dehm 1962: 84).

Dehm was certainly much hurt by it. ... this indeed is another case, similar to the case of Schorsch Wagner, in which the work of a life was made largely valueless by a new theory also fitting the facts and [being promoted] by people who had worked only for a short time [at the Ries crater]. I know from the institute [in Munich] that there had been many who worked on the Ries. There these maps were made; and [Dehm] must have felt it that way. It was a deep disappointment. The ground on which he had stood suddenly went brittle and fell away. There is also the human component – (Pichler interview, commenting on Dehm 1962)[41]

Dehm was particularly hurt by it being foreign scientists who now seemed to write the agenda:

Dehm said it would have been truly Engelhardt's task to solve the Ries riddle. He was annoyed because Americans came to tell the Germans what to do. Dehm has engaged himself in that pretty openly. ... Dehm admitted that the meteorite was the only possibility from an energetic [physical] point of view; nevertheless he refused to admit it internally. People were then so kind as not to quote [Dehm's paper of 1962] too often. (Volker Fahlbusch, 1934–2008, personal communication by phone to the author, 19 September 2001)[42]

Blaming Engelhardt was of course rather unfair, as he had not come to Tübingen before 1957 and had worked much of his previous career in sedimentary petrology. Under these circumstances it is remarkable and highly commendable that he embarked so quickly and efficiently into Ries research (Chapters 11 and 12).

Dehm's personal problem with a non-German intrusion into Ries research became apparent again in a paper on the history of Ries research that he published in 1969, after the impact origin had been accepted. It was a marvellous example of apologetics. It suggested that he had never objected to a central explosion,

---

[41] 'Dehm hat das sicher sehr getroffen. ... das ist tatsächlich ein weiterer Fall, ähnlich wie der Fall Schorsch Wagner, dass also eine Lebensarbeit weitgehend wertlos gemacht wird durch eine neue Theorie, die also auch den Tatsachen entspricht und durch Leute, die also ganz kurz nur gearbeitet haben. Ich weiß ja vom Institut, da gab es ja viele, die über das Ries gearbeitet haben, da wurden diese Karten gemacht und er muss das so empfunden haben. Das war eine große Enttäuschung. Der Boden, auf dem er gestanden hat, der plötzlich brüchig wird und wegsackt. Da ist auch die menschliche Komponente – ' (Pichler interview, commenting on Dehm 1962).

[42] 'Dehm hat gesagt, es wäre eigentlich Engelhardt's Aufgabe gewesen, das Riesrätsel zu lösen. Er regte sich auf, weil da Amerikaner kamen und den Deutschen sagten, wo es lang geht. Dehm hat sich da ziemlich aus dem Fenster gelehnt'. ... 'Dehm gab zu, dass der Meteorit energetisch die einzige Möglichkeit war, trotzdem hat er sich innerlich gewehrt. Die Leute waren dann so nett, sie [Dehm's paper of 1962] nicht zu oft zu zitieren' (Fahlbusch, personal communication by phone 19 September 2001).

and that even in the 'early period' of Ries research, from 1686 onwards, people had had a notion of the 'very special conditions of origin'. In the section on the period between 1920 and 1960 newer research appears – Hüttner and Wagner (1965b), Pohl (1965), Förstner (1967), Treibs (1965), Birzer (1969) – so that a casual reader might gain the impression that German geologists had already been working on impact issues when the two Americans appeared on stage. In Dehm's paper the two get just a brief mention for their discovery of coesite:

> In a quite unexpected manner, in 1961, Ries geology received some especially effective help from outside, when during comparative investigations of the effects of meteorite impacts and nuclear explosions the two Americans E. M. Shoemaker and E. C. T. Chao discovered in a suevite sample a high-pressure modification of quartz, coesite, which so far had only been found in meteor craters and in an underground nuclear explosion. (Dehm 1969: 29–30)[43]

Then the narrative was completely taken over again by the diligent, hard-working and efficient German scientists. The paper ended with the statement that in 1964 the geophysicist Preuss had given a complete formulation of the impact-theory, which in 1904 had already been uttered for the first time by E. Werner (Dehm 1969). And so the honour of the 'fatherland' remained unstained.

From Scepticism to Chauvinism

Dehm's apologetics point to another motive to dismiss the impact theory. There was reluctance to accept anything that came from outside the community of Ries experts; worse still if it came from outside Germany. When Shoemaker and Chao proposed the impact origin of the Ries Crater by way of their new evidence in 1960–61, it was but fifteen years after World War II and the Nazi regime. Things had definitely changed for the better and open patriotism was no longer fashionable and remembrance of former ultra-nationalism was a cause for collective shame. Yet while the new democracy prospered – even though a remarkable number of former Nazis were still about in politics and other influential positions – and thus found access to people's hearts – these same individuals still carried with them their personal history affected by a totalitarian system with ultra-nationalist propaganda, the effects of which

---

[43] 'In ganz unerwarteter Weise erhielt die Riesgeologie 1961 von außen eine besonders wirksame Hilfe, als im Zuge ihrer vergleichenden Untersuchungen der Wirkung von Meteoreinschlägen (*impacts*) und Atomsprengungen die beiden Amerikaner E.M. Shoemaker & E.C.T. Chao in einer Suevitprobe eine Hochdruckmodifikation des Quarzes, Coesit, entdeckten, die bisher nur in Meteorkratern und bei einer unterirdischen Atomsprengung gefunden worden war'. (Dehm 1969: 29–30).

sometimes still lingered (at least in the subconscious). In connection with the Ries Crater and Steinheim Basin, this past of a nationalist interpretation of 'German Geology' (see Chapter 7) surfaced, for example, in chauvinism against the two foreign scientists, Shoemaker and Chao, and especially against Edward Chao, the 'Chinese': 'Generations [of geologists] have walked there and made an effort and now the Americans come and additionally even such a Chinese' (Hüttner interview: quoting Georg Wagner from memory).[44] 'And my Professor Georg Wagner said: "My goodness, there comes this American and he even has a Chinese name! And he thinks he can intrude into our Ries business and say what it is. That is more than unlikely!"' (Hölder interview: quoting Georg Wagner from memory).[45]

Wagner literally raged against 'writing-mad Asian-Americans': '[T]rustworthy work needs time. With 'veni, vidi, vici' we do not want to compete. We are disgusted by this method!' (Letter from Georg Wagner to Mr Härtlein, 31 March 1965, UAT 605/343).[46]

For the first time it became evident that apart from the reasons for dismissal of the impact concept reviewed in former chapters, there was still another set of motives, which kept people like Georg Wagner or Richard Dehm from its acceptance, chauvinism being one of them. This particular motive was owed to the specific recent history of Germany and was thus a historical and political problem, which persisted in the present as a lingering social problem.

As has been shown, in Richard Dehm, German 'patriotism' found its way out in a distorted history of Ries research, which polished up the merits of German researchers while at the same time purposefully neglecting the input of foreign geoscientists.

Georg Wagner in contrast had a considerably more emotional personality and thus was not able to contain his intense feelings, disappointment and despair. He took the shift of paradigm after 1961 as a personal affront and so his arguments – justified by the initial 'offence' of a 'foreign intrusion' by Chao and Shoemaker, as Wagner seemingly felt – strayed from facts to polemics and even to chauvinism tinted with nationalism:

---

[44] 'Jetzt laufen doch schon Generationen da rum und bemühen sich und nun kommen die Amerikaner und noch dazu so ein Chinese' (Hüttner interview: quoting Georg Wagner from memory).

[45] 'Und mein Professor Georg Wagner hat gesagt: "Hach, da kommt so ein Amerikaner und hat auch noch einen chinesischen Namen! Und da meint er, er könne uns im Ries dreinreden und sagen, um was es sich da handelt. Das ist doch mehr als unwahrscheinlich!"' (Hölder interview).

[46] 'schreibwütige Asio-Amerikaner ... solide Arbeit braucht Zeit. Mit 'kam, sah, siegte' wollen wir nicht in den Wettkampf treten. Uns ekelt diese Methode an!' (Letter from Georg Wagner to Mr Härtlein, 31 March 1965, UAT 605/343).

That went as far as racism, so to speak. ... The thing with Chao was quite simply the emotional reaction of a deserving old geologist [Georg Wagner], who, as I have said, has been occupied with the Ries during his whole life. And there come two colleagues from a far-off country, take two samples, take them to the United States, look at them, find coesite, and state that this is the proof for the impact theory of the Ries. That of course was frustrating. ... One must imagine the frustration of a colleague who has worked for years when someone comes along and takes two samples, has them analysed, and the scientist's whole life's work and ideas collected life-long about the origin of a structure or a rock are rendered obsolete by that. That must be understood; and he [who did it] is of Chinese origin, is a Chinese immigrant ..., and that has tempted [Georg Wagner] to express [his opinion] publicly. It of course was talked about and indeed damaged his reputation. But it is understandable ..., especially if one [knows] the personality and emotionality of Georg Wagner, who had been an enthusiastic geologist and, when he was in the field, expressed enthusiasm. ... The personal emotionality is very important [for understanding Wagner]. (Pichler interview)[47]

Wagner's numerous letters, preserved at the University Library of Tübingen, indeed show how desperate he felt. He was completely beside himself: 150 years of patient and diligent geological fieldwork had not elucidated the Ries problem but had rather clouded it. And there comes a foreign mineralogist and claims to have the solution. Wagner at best could only consider this cheeky. 'It was in [Georg Wagner's] nature. He has known friends, good friends. As soon as one

---

[47]　'Das ging ja bis ins fremdenfeindliche eigentlich. ... Die Sache mit dem Chao, das waren ganz einfach emotionale Reaktionen eines verdienten alten Geologen, der ja, wie gesagt, sich ein ganzen Leben lang mit dem Ries beschäftigt hat. Und da kommen also zwei Kollegen aus einem fernen Land, nehmen zwei Proben, nehmen die mit rüber in die Vereinigten Staaten, schauen sie durch, finden Coesit und stellen fest, das ist der Beweis für die Impakttheorie des Rieses und das war natürlich frustrierend. ... Man muss sich den Frust vorstellen eines Kollegen oder einer Kollegin, die jahrzehntelang gearbeitet und nun kommt einer daher und nimmt zwei Proben auf, lässt sie analysieren und die ganze Lebensarbeit und die ganze Vorstellung ein Leben lang gesammelt über die Entstehung einer Struktur oder eines Gesteines sind damit hinfällig geworden. Das muss man verstehen; und der ist ja chinesischer Herkunft, ist also ein Einwanderer aus China, ... und das hat [Georg Wagner] natürlich verleitet das auch öffentlich rauszulassen. Das ist natürlich kursiert und hat seinem Ansehen schon geschadet. Aber verständlich ist es, ... vor allem wenn man die Persönlichkeit und die Emotionalität von dem Georg Wagner [kennt], der ja ein begeisterter Geologe war und dann auch richtig, wenn er im Gelände war, ausgebrochen ist in Begeisterung von sich aus. ... Die Veranlagung emotionaler Art ist ganz wichtig' (Pichler interview).

had obtained his favour, one had it for ever. And he also knew enemies ... and these then had it very difficult with him' (Hüttner interview).[48]

Unfortunately, Shoemaker and Chao, especially the latter, who often came to Tübingen, Munich and Nördlingen for fieldwork and conferences, were not among Wagner's friends:

> You must be patient for some time more! We cannot compete in a contest of time with the illustrated journals and writing-mad Asian-Americans. And thorough work needs time. This method sickens us! It is not as simple as Chao-Shoemaker and the newspaper writers have it. The proof minerals are a fraction of a millimetre large and can only be detected in thin section in a microscope by X-rays.[49] Up till now they have only been found under enormous pressures in atomic explosions. Whether they can originate otherwise, we do not yet know. If we want proofs, it is said that in such meteorite impacts all is evaporated. Rocks cannot be found at all! With traditional physics and chemistry and geology one cannot proceed because we can no longer imagine these new things. Spellbound, one looks at the high-pressure mineral, which is destined to solve all riddles. ... The more someone speaks of meteorite impacts, the less he has observed! In any case, so far everything is still uncertain. (Letter from Georg Wagner to Mr Härtlein, 31 March 1965, UAT 605/343)[50]

A geophysicist from Munich recalled: 'I can remember Georg Wagner. He once stood up at a conference and spoke of the Communist-Nazi methods of the

---

[48] 'Es lag in [Georg Wagner's] Wesensart. Er hat eben Freunde gekannt, gute Freunde. Sobald man bei ihm einen Stein im Brett hatte, so hat man den für immer gehabt. Und er hat auch Feinde gekannt ... und die hatten es dann sehr schwierig mit ihm' (Hüttner interview).

[49] Wagner here was mixing up two methods, thin-section microscopy and X-ray (powder) diffraction. Both mineralogical methods were obviously alien to him.

[50] 'Sie müssen sich noch etwas gedulden! Wir können nicht mit den Illustrierten und schreibwütigen Asio-Amerikaner in zeitlichen Wettkampf eintreten. Und solide Arbeit braucht Zeit. Mit 'kam, sah, siegte' wollen wir nicht in den Wettkampf treten. Uns ekelt diese Methode an! So einfach, wie Chao-Shoemaker und die Zeitungsschreiber es darstellen, geht es nicht. Die beweisenden Minerale sind Bruchteile eines Millimeters groß und können nur im Schliff im Mikroskop mit Röntgenstrahlen bestimmt werden. Bis jetzt hat man sie nur bei enormen Drucken bei Atomexplosionen gefunden. Ob sie nicht auch anders entstehen können, wissen wir noch nicht. Wollen wir Beweise, so heißt es, bei solchen Meteoriteneinschlägen verdampft alles. Gesteine kann man gar nicht finden! Mit normaler Physik und Chemie und Geologie kommt man nicht weiter, weil wir uns die neuen Dinge nicht mehr vorstellen können. Wie gebannt blickt man auf das Hochdruckmineral, das alle Rätsel lösen soll. ... Je mehr einer von Meteoriteneinschlägen spricht, desto weniger hat er beobachtet! Jedenfalls ist jetzt noch alles in der Schwebe' (letter from Georg Wagner to Mr Härtlein, 31 March 1965, UAT 605/343).

Americans, who simply come along, come into the Ries and take a few little stones' (Pohl interview).[51]

The development of the issue was ultimately easier to swallow after German scientists had participated in the new impactist Ries research:

> Yes, that was especially Wagner. On the other hand, after Mr von Engelhardt had presented his talk at the Thomas conference, [Wagner] said: "Of course if now a German scientist with German diligence approaches the matter, then it might already appear quite differently". He was of course a bit 'national', Schorsch Wagner, nationalistic, one might nearly say. Today, one would classify it as such. 'German Geology' was something very special to him. (Hüttner interview)[52]

Rivalry Between Geology and Mineralogy

Their foreign nationality was not the only factor delaying the acceptance of Shoemaker's and Chao's expertise. They were also classified as 'mineralogists' because of their methodology. The discussion of the Ries Crater and Steinheim Basin in the early 1960s was indeed hampered by differences in the methods used in geology and mineralogy, respectively. The conservatives were field-geologists arguing with field data; with large structures like faults, the superposition of rock types or the spatial and temporal relationship of the two craters to other geological structures in southern Germany, while the promoters of the impact scenario operated with microscopically small minerals and metamorphic structures, with chemical analyses and X-ray diffraction spectra. Is seems that there was no common language. Impactists initially tended to neglect field data as equivocal and thus not relevant to the discussion, while the volcanists often did not understand the meaning of mineralogical and chemical data or the physical modelling of impact processes. They doubted the expertise of mineralogists because their own expertise was insufficient for judging the mineralogists' data:

> Here the Americans want, because of heavy minerals (coesite, a quartz, which requires very high temperatures and pressures, but is no bigger than a grain of

---

51 'An Georg Wagner kann ich mich erinnern. Er ist mal aufgestanden bei einer Tagung und hat von den Kommunazimethoden der Amerikaner gesprochen, die da einfach kommen, ins Ries kommen und sich ein paar Steinchen holen' (Pohl interview).

52 'Ja, das ist speziell Wagner. Er hat sich andererseits wiederum, nachdem der Herr von Engelhart bei der Thomastagung diesen Vortrag gehalten hat, hat er gesagt: "Natürlich wenn jetzt ein deutscher Wissenschaftler mit deutscher Gründlichkeit da an die Sache herangeht, da sieht das vielleicht schon wieder anders aus". Er war natürlich schon ein bisschen national der Schorsch Wagner, nationalistisch könnte man fast sagen. Heute würde man das schon so sagen. 'Deutsche Geologie', das war ihm eben was ganz besonderes' (Hüttner interview).

salt and can only be detected in thin section by X-rays), to explain the whole Ries phenomenon as a meteorite impact that led to melting in the crystalline basement, so that on top of the ejected, penetrated rock, molten suevite was also thrown. The entire meteorite is said to have evaporated, because meteoritic rock has never been found in the Ries. Only the quartz in the melt fragments contains coesite, which by the way is also generated in atomic explosions. With infantile self-consciousness, all of this was trumpeted about, and all of the world's riddles were [supposedly] solved. In modern volcanic eruptions no such high-pressure and high-temperature minerals occur. One knows of them only for about ten to twenty years. And we poor foot-walkers cannot make use of it [or understand it]. But dozens of other questions remain unsolved. Especially that it was a well-aimed meteorite in temporal and local respects, of whose own material so far not a single piece of rock has yet been found. Yet on the other hand [there are] the molten masses, which are increasingly discovered in the whole Ries (in the vicinity of Wörnitzostheim more than 80m! We come into greatest conflict as regards time, space and material. A Swiss-American (professor in the Northeastern USA) [Walter H. Bucher] for five days was with us in the Ries, and has also seen the Arizona Crater, but the Coesite-Fair, he would not take part in. It is typical of infantile thinking, but not for complex geological thinking; Gerold[53] stands still undecided in this quarrel. My chemical-physical-astronomical thinking is at its end. We can not yet explain the extremely high pressures and temperatures, and the others fail with the jumbled succession of events. Will I live until this issue is elucidated? We must not lose the ground under our feet and wander in Cloud-Cuckoo-Land. (Letter from Georg Wagner to the Priehäußer family, 19 November 1965, UAT 605/343)[54]

---

[53]   Gerold Wagner, Georg's son, who had just finished his PhD at the University of Bonn with a dissertation on the Ries Basin.

[54]   'Hier wollen die Amerikaner auf Grunde von Schwermineralien (Coesit, ein Quarz, der sehr hohe Temperaturen und Drucke erfordert, aber nit [sic] größer ist als ein Salzkorn und nur im Dünnschliff mit Röntgenstrahlen nachgewiesen werden kann) das ganze Riesphänomen als einen Meteoriteneinschlag [sic], erklären, der zu Aufschmelzungen im Grundgebirge führte, so daß über ausgeworfenes durchschlagenes Gebirge noch aufgeschmolzener Suevit geworfen wurde. Der ganze Meteorit sei verdampft. Denn man hat im Ries noch nie Meteoritengestein gefunden, nur der Quarz der Fladen hat Coesit, der übrigens auch bei Atomexplosionen entsteht. Mit infantilem Selbstvertrauen wurde das alles ausgepsosaunt [sic] und alle Welträtsel waren gelöst. Nun kommen bei Vulkanausbrüchen heute keine solchen Hochdruck- und Hochtemperaturmineralien vor – Diese Werte werden heute nur bei Atomexplosionen erreicht. Man kennt sie erst seit etwa 10–20 Jahren. Und wir armen Fußwanderer können nichts damit anfangen. Aber es bleiben Dutzende anderer Fragen ungelöst. Vor allem, es war ein zeitlich und örtlich gut gezielter Meteorit, von dessen Eigenmaterial noch nicht ein Steinchen gefunden worden ist, dagegen Aufschmelzmassen, die im ganzen Ries immer mehr entdeckt werden (bei Wörnitzostheim über 80m! Wir

Even today there is sometimes an attempt to gloss over the scepticism of the early 1960s. Coesite was later also found in eclogitic rocks in the Alps, and thus, so Rudolf Trümpy[55] claimed, it had simply been prudent to remain sceptical, and only with the discovery of stishovite, another high-pressure modification of silica, was the issue solved properly.[56] What Trümpy failed to stress sufficiently is the fact that in the Ries coesite is found in prodigious amounts in rocks right on the Earth's surface, while the eclogites, which are high-pressure metamorphic rocks from deep within the Earth's mantle, contain but little coesite and only as inclusions within garnet crystals, which served as pressurised vessels to prevent the coesite inclusions from recrystallising to plain quartz during the ascent of the eclogite. These fine details of the mineralogical argument failed to impress the geologists:

> Well, the discovery then of coesite was of course sensational. Meanwhile, we know that coesite can also be generated under terrestrial conditions. We have high-pressure rocks ... in the western Alps that contain coesite, mostly as inclusions in garnet. These were rocks that had been exposed to very high pressures in depths of around 100 km and which were then rapidly elevated to the surface. Thus coesite is no unequivocal 'index fossil' for impacts. But then many other criteria were added, which connected the Ries with confirmed impact craters.

> Well, at the beginning, I was somewhat sceptical, because I was always sceptical against fashions of the time; yet on the other hand, the location and time situation of the Ries was somewhat suspicious. The Ries originated nearly contemporaneously with the Swabian volcanics, in the surroundings of [the town] Urach; and these Swabian volcanics, they are certainly something very strange. It must have been a highly explosive volcanism, which basically produced

---

kommen mit Zeit, Raum und Material in größte Konflikte. Ein Schweizer Amerikaner (Prof. in NO-USA) war mit uns 5 Tage im Ries, hatte auch den Arizonakrater gesehen, aber den Coesitrummel mache er nicht mit. Er ist typisch für infantiles Denken, aber nicht für das komplexe geologische Denken, Gerold steht noch unentschieden in diesem Streit. Mein chemisch-physikalisch-astronomisches Denken ist am Ende. Wir können noch nicht die unerhört hohen Drucke und Temperaturen erklären, und die anderen scheitern an der bunten Abfolge der Ereignisse. Ob ich es noch erlebe, bis man das hindurchsieht? Wir dürfen eben nicht den Boden unter den Füßen verlieren und im Wolkenkukuksheim [sic] wandern'. (Letter from Georg Wagner to the Priehäußer family, 19 November 1965, UAT 605/343).

[55] Rudolf Trümpy (1921–2009) obtained his doctorate degree in 1947 and then worked at the University of Lausanne. In 1953, he became Professor of Geology at the ETH Zürich with a main focus on alpine geology (Bernoulli 2009).

[56] Stishovite was discovered in 1961 (Chao et al. 1962) and its presence in Ries suevite had been known since 1962 (Preuss in AG Ries 1963: 10).

only hot air and rock debris. The Ries lies on this line of undoubted volcanic rocks, which eventually continue into the Hegau,[57] although the volcanism there is of a different kind.

Therefore it seemed strange to me that the Ries should lie specifically on a line of the Miocene volcanism, and so I thought, not publicly but secretly, of some sort of highly explosive volcanism unknown to us, perhaps of the same type as Santorini. On an excursion to the Ries I then let myself be convinced that the features there were much better reconciled with an impact than with a somewhat mysterious volcanism for which we do not know any good analogy, present or historic, maybe with the exception of the explosions of Santorini and similar ones. (Trümpy interview)[58]

When Georg Wagner dispatched four heavy boxes of samples to Walter H. Bucher, he informed him in an accompanying letter:

---

[57]   A group of Tertiary volcanoes near Lake Constance.

[58]   'Ja, nun die Entdeckung des Coesits war damals natürlich sensationell. Unterdessen wissen wir dass Coesit auch unter irdischen Bedingungen entstehen kann. Wir haben Hochdruckgesteine ... in den Westalpen, die Coesit enthalten, meistens als Einschluss in Granat. Das waren Gesteine, die sehr hohen Drucken, in Tiefen um die 100 km. ausgesetzt waren und die dann sehr schnell an die Oberfläche befördert wurden. Also Coesit ist nicht ein eindeutiges Leitfossil für Impakte. Es kamen dann aber viele andere Kriterien dazu, die das Ries mit sicheren Impaktkratern verband. Nun, ich war am Anfang etwas skeptisch. Einerseits weil mir Zeitmoden etwas skeptisch waren und andererseits eben die raumzeitliche Situation des Ries etwa verdächtig war. Das Ries ist ja fast gleichzeitig entstanden wie die Schwäbischen Vulkanite etwa in der Umgebung von Urach. und diese Schwäbischen Vulkanite, die sind ja etwas sehr Merkwürdiges, das muss ein hochexplosiver Vulkanismus gewesen sein, der eigentlich nur heiße Luft und Gesteinsbrocken förderte. Das Ries liegt auf der Linie dieser nicht angezweifelten Vulkanite, die dann schließlich bis in den Hegau weitergehen, ob schon dort der Vulkanismus von etwas anderer Art ist. Es schien mir also merkwürdig, dass das Ries ausgerechnet auf einer Linie eines miozänen Vulkanismus liegen sollte und ich habe dann nicht öffentlich, aber heimlich, an eine uns nicht bekannte Art von hochexplosivem Vulkanismus, vielleicht vom Typus Santorin, gedacht. Auf einer Exkursion ins Ries ließ ich mich dann überzeugen, dass die dortigen Erscheinungen eben viel besser mit einem Impakt vereinbar waren, als mit einem doch etwas geheimnisvollen Vulkanismus, für den wir keine guten Analoga, jetzige oder geschichtliche Analoga, kennen, vielleicht abgesehen von Explosionen von Santorin und dergleichen' (Trümpy interview).

Recently, [Kavasch][59] had several visits from two mineralogists, Prof. Dr Preuß[60] from Regensburg and an American from Philadelphia (Venn or similar);[61] both were convinced meteorite people! The causative meteorite is supposed to have been split; the larger [piece] caused the Ries, the smaller flew as far as Steinheim! It seems to me that the horizon of mineralogists and astronomers is much tighter than ours!

Do publish your opinions on the Arizona crater too, which have impressed me especially, because with this the card-house might collapse.

I enclose my summary of our findings, asking for your criticism and suggestions for improvements and additions, because the muteness of the geologists is seen as a victory of the mineralogists! (Letter from Georg Wagner to Walter H. Bucher, 27 August 1962, UAT 605/343)[62]

After the death of Walter H. Bucher in 1965, Georg Wagner increasingly lost his confidence as a geologist, resorting to resignation instead of understanding:

At present, one is reluctant to write anything about the Ries, because now the mineralogists have their say. The especially high temperatures and pressures that are required by them, we cannot yet explain. But the mineralogists cannot explain to us how there came to be such extensive melting of suevite. Their meteorite

---

[59]   The amateur geologist Julius Kavasch (1920–78) was a schoolteacher in Nördlingen (<http://adsabs.harvard.edu/full/2005M%26PS...40.1555P> accessed 14 November 2012).

[60]   Ekkehard Preuss.

[61]   This is most possibly Vladimir Vand[t] (1911–68), who was born in Russia. He obtained his first doctorate degree in 1937 from the Charles University in Prague. From 1940 onwards, he held an industrial position in the UK; from 1950 onwards, he was at the University of Glasgow, where he obtained a second PhD in 1954. Subsequently, he became Professor of Physics and later Professor of Crystallography at Pennsylvania State University (Freyberg 1974: 161). He published an early review paper on the two German Craters and their relation to tektites (Vand 1963).

[62]   '[Kavasch] hatte in letzter Zeit wiederholt Besuch von Mineralogen, Prof. Dr. Preuß, Regensburg und ein Amerikaner aus Philadelphia (Venn oder ähnlich). beide eingeschworene Meteoritenleute! Der erzeugende Meteorit habe sich geteilt; das größere habe das Ries erzeugt, das kleinere sei bis Steinheim weiter geflogen! Mir scheint daß der Mineralogen- und Astronomen-Horizont viel enger ist als der unsrige! // Veröffentliche Sie nicht auch Ihre Ansichten über den Arizonakrater, die mich besonders beeindruckt haben. Denn mit diesen könnte das ganze Kartenhaus zusammenbrechen. // Ich legen Ihnen meine Zusammenfassung unsere [sic] Befunde bei mit der Bitte um Kritik, Besserungs- und Ergänzungsvorschläge. Denn das Schweigen der Geologen wird als Sieg der Mineralogen aufgefaßt!' (Letter from Georg Wagner to Walter H. Bucher, 27 August 1962, UAT 605/343).

would have to be much bigger than they say. And the suevite dykes they can only interpret with difficulty. And up to this day no trace of a meteorite has been found. So we remain waiting and sceptical; see only gigantic explosions but can hardly imagine something correct! ... We geologists always base our information on observable events, which offer examples to us. At Steinheim I have not yet found anything volcanic or meteoritic. So I have no new weapon apart from my disbelief. This alone, however, is not yet sufficient. So one prefers to wait before one subscribes. (Letter from Georg Wagner to Mr Schedler, 2 February 1966, UAT 605/343)[63]

The psychological aspect of a rivalry between geologists and mineralogists was less trivial than might be assumed at first impression. In Germany at that time, there had been a distinct division of labour and interest between mineralogy and geology. There were actually two cultures in geoscience, which historically went back to two different conceptions in geology: natural history that is the descriptive and taxonomic sciences, reconstructing the history of the globe, and natural philosophy that is developing general laws and principles, which was process oriented (Laudan 1982; Rudwick 2005: 84–115). While modern geology represents a synthesis of these two conceptions, 'at certain periods in history the emphasis has moved decisively from historical to physical geology and back again' (Laudan 1982: 9), and during the early twentieth century and until the early 1960s, geology in Germany was decidedly more at home in the historical, stratigraphic realm. Geology dealt with a temporal, historical frame derived from stratigraphy and was thereby closely related to palaeontology. This emphasis had clearly been strengthened by the notion of 'German Geology' during the Nazi era (Chapter 7).

Process-oriented research in Germany was, on the other hand, more at home in mineralogy, which considered physical and chemical processes such as metamorphism, and whose data often did not come from Earth history but from experimental, lab-based activities (compare Laudan 1982: 8). The early

---

[63] 'Man hütet sich zur Zeit, etwas über das Ries zu schreiben. denn jetzt haben die Mineralogen das Wort. Die besonders hohen Temperaturen und Drucke, die diese verlangen, können wir noch nicht erklären. Aber die Mineralogen können uns nicht erklären, wie es zu so ausgedehnten Aufschmelzungen von Suevit kan [sic]. Ihr Meteorit müßte weit größer gewesen sein, als sie sagten. Und die Suevitgänge sind von ihnen schwer zu deuten. Und bis heute ist keine Spur eines Meteoriten gefunden worden. So stehen wir abwartend und skeptisch da, sehen bloß gewaltige Explosionen, aber können uns kaum etwas Richtiges vorstellen! ... Wir Geologen gehen eben immer aus von beobachteten Ereignissen, die uns Beispiele geben. bei Steinheim habe ich noch nichts Vulkanisches oder Meteoritisches gefunden. So habe ich keine neue Waffen als den Unglauben. Der genügt aber allein noch nicht. So wartet man lieber ab, ehe man sich festlegt' (letter from Georg Wagner to Schedler, 2 February 1966, UAT 605/343).

1960s for Ries research was a time 'when the [new] techniques [became] so fruitful that the traditional concept of a particular science [geology] [was] left languishing, its theories shown to be unacceptable, and a new concept gain[ed] dominance in the community' (Laudan 1982: 12).

The division of labour between geology and mineralogy occurred at the conscious level; Georg Wagner, for example, repeatedly argued for his resentment against mineralogists with this 'cultural' distinction:

> In my opinion, the mineralogists work too much with decimal places and not enough with space and time! And the young ones adore the new fashion with delight, while we ask: How? Where? Why? How long? (Letter from Georg Wagner to Mr Loner, 31 July 1967, UAT 605/343)[64]

> The new Ries people are mainly mineralogists, not enough Earth historians! They are concerned with their theory and hardly with the explanation of the present landscape. (Letter from Georg Wagner to Friedrich G.J. Birzer, 13 July 1968, UAT 605/343)[65]

> A foreign mineralogist[66] even calculated how many tons of coesite must be present in Long Daniel![67] With that, the mineralogists were content, but not the geologists, who are concerned with the history of the Earth and its morphology! (Undated note by Georg Wagner, UAT 605/343)[68]

It was by no means Wagner alone who felt this way. Richard Dehm also uttered similar thoughts when he reflected on his task as a geologist, apologetically denigrating the moment of impact in the view of millions of years of geological history:

---

[64]   'Die Mineralogen arbeiten mir zu sehr in Zehnerpotenzen, zu wenig ins [sic] Raum und Zeit! Und die Jungen beten den neuen Schlager entzückt an, während wir fragen: Wie? Wo? Warum? Wielange?' (Letter from Georg Wagner to Loner, 31 July 1967, UAT 605/343).

[65]   'Die neuen 'Rieser' sind vorwiegend Mineralogen, zu wenig Erdgeschichter! Es geht ihnen um ihre Theorie, wenig um die Erklärung der heutigen Landschaft' (letter from Georg Wagner to Friedrich G.J. Birzer, 13 July 1968, UAT 605/343).

[66]   Edward Chao.

[67]   The *Lange Daniel* is the tower of the Nördlingen church, which is built from suevite.

[68]   'Ein ausländischer Mineraloge rechnete sogar aus, wie viel Tönnen [sic] Coesit in langen Daniel erhalten seien! Damit waren dann die Mineralogen zufrieden, aber nicht die Geologen, denen es um die Geschichte der Erde und ihrer Formen geht!' (Undated note by Georg Wagner, UAT 605/343).

In Ries research the main question had been: How did the crater originate? Now we know: in an event of a mere second! The task of the geologist, however, is to appreciate the whole of history, the millions of years before the second of catastrophe as well as the millions of years that followed it. This was managed by special geological mapping: For the Ries this had been done especially from Munich, by the university institute there. (Dehm in Barsig and Kavasch 1983: 47)[69]

While geology was seemingly shaken in its confidence, mineralogy on the other hand gained in prestige:

When in 1964 I [Hans Pichler] came to this house [the Mineralogical Institute in Tübingen], the last resistance of the traditionalists had just been broken; the resistance [of the traditionalists] who tenaciously clung to the inherited view of a Ries created terrestrially or even volcanically. Georg ('Schorsch') Wagner and the old, mud-caked ideas of Ries genesis against the unassailable mineralogical proofs of a meteorite impact. "Mineralogy has its uses after all!" I wrote into my diary then. (Pichler 2000)[70]

For several years, Georg Wagner seemed to have struggled in establishing a new feeling for the role of geology in Ries research. Several note sheets, rather similar in general content and mostly undated, among Wagner's papers at UAT, bear witness to this mental process:

Concerning the Ries problem, many questions still have to be solved, because it is not only a physical, astronomical, mineralogical but a geological-historical [problem]. We are not interested in temperatures and pressures but in the course of the events too. During the sixteen million years [since the Ries event] much has changed, was eroded; we are very distant from the event and have to be attentive to that.

---

[69] 'Bei der Riesforschung war die Hauptfrage gewesen: Wie ist der Krater entstanden? Jetzt wissen wir es: ein Sekunden-Ereignis! Aufgabe des Geologen ist es aber die ganze Geschichte zu erfassen, die Jahrmillionen vor der Sekunde der Katastrophe, ebenso die Jahrmillionen, die ihr folgten. Dies leistet die geologische Spezialkartierung: sie ist im Ries insbesondere von München aus durch das Universitätsinstitut betrieben worden' (Dehm in Barsig and Kavasch 1983: 47).

[70] 'Als ich 1964 als Assistent in dieses Haus kam, wurde gerade der letzte Widerstand der Traditionalisten, die zäh an der ererbten Scholle eines irdisch, gar vulkanisch entstandenen Rieses festhielten, gebrochen. Georg – Schorsch – Wagner und die alten, verkrusteten Vorstellungen der Ries-Entstehung gegen die hieb- und stichfesten mineralogischen Beweise eines Meteoriten-Einschlags. "Die Mineralogie ist doch zu etwas gut!" schrieb ich damals in mein Tagebuch' (Pichler 2000).

Much has disintegrated, was eroded or covered [by sediment]. It is nearly ridiculous to argue with the many tons of coesite that have been built into the [church tower of] Long Daniel in Nördlingen. We are not pleased with propaganda, which is too well-known to us. We want clean proofs! Unfortunately, in the whole of the Ries no alien rock was found. The basement rock, which is amply present in the ejecta too, is not dissimilar to that between the Bohemian Massif and the Black Forest. And meteorite iron, as in the Arizona Crater, has not been found. Only in a microscopically [small] ore grain was some nickel allegedly also found. This however proves nothing, since even in the Black Forest nickel iron does occur. Also, a meteorite can be so hot that it evaporates! How the enormous event was caused, geologists must leave to others. Their task is to find out when it occurred and how our landscape then looked and how it was changed. The great volcanic period of Southern Germany was over, the deposition of the Miocene Sea too. We see in the Upper Miocene freshwater limestone a guiding time horizon [a stratigraphic marker], which for us is also some sort of level to which we relate the heavily affected Upper Miocene network of streams, which we will have to reconstruct. Equally important for us is the surface on which were deposited the Ries ejecta, which also have enormously reshaped the water courses; because old valleys were completely filled in, rivers dammed. The water divide between Northwest and Southeast was much affected and mostly altered. The erosion of the escarpments [of the Swabian Alb and further to the north] was slowed down. Cut off valleys, especially at the escarpment of the Upper Triassic, again sent their water towards the Rhine, because the water divide of the cut-off valleys lay deeper than the lakes, which have been dammed in the vicinity of the Ries; for up to 50 m! The upper river Jagst was bypassed towards the Rhine! A horrible destruction of plant and animal life occurred – but bone beds and coal seams are missing. The cover of the polymictic breccia was perhaps not sufficiently regarded. One thought not enough of life and morphology, too much of decimal points! (Note by Georg Wagner, 5 October 1968, UAT 605/343)[71]

---

[71] 'Zum Riesproblem müssen noch viel Fragen geklärt werden; denn es ist nicht nur ein Physikalisches, Astronomisches, mineralogisches sondern ein ein [sic] geologisches-geschichtliches. Uns interessieren nicht die Temperaturen und Drucke sondern auch die Vorgänge. In den 16 Millionen Jahren ist viel verändert, abgetragen worden; wir sind sehr weit vom Geschehen entfernt und müssen das berücksichtigen. // Viel hat sich zersetzt, ist abgetragen oder zugedeckt worden. Es ist fast lächerlich, mit den vielen Tonnen Coesit zu kommen, die im langen Daniel in Nördlingen verbaut wurden. Wir haben keine Freude an uns zu bekannter Propaganda. Wir wollen saubere Beweise! Leider hat man am ganzen Ries kein Fremdgestein gefunden. Das Grundgebirge, das reich, auch in Ausgeworfenem vertreten ist, weicht nicht von dem zwischen Bayrischem und Schwarzwald ab. Und Meteoreisen, wie im Arizona-Krater, hat man nicht gefunden. Nur in einem mikroskopischen Erzkorn soll auch etwas Nickel gefunden worden sein. Das beweist aber nichts, da im Schwarzwald sogar Nickeleisen vorkommt. Auch kann ein Meteorit so heiß sein, daß er verdampft! Wie das

Geologists like Georg Wagner regarded themselves as 'Earth-historians'. They were mainly stratigraphers. For them the superposition of different strata at least subconsciously implied ample time, and thus it seemed inconceivable that a substantial stratigraphic section of various distinct rock units could have been deposited in just a few minutes. Rocks were treated as representing the passing of time and not as testimony to processes (Figure 10.4), and hence the geologists had severe difficulties in reconciling themselves with the idea of an impact.

The German mineralogists as well as the geophysicists, on the other hand, who after 1961 ardently jumped onto the American bandwagon, seizing the opportunity to do research on impact geology and establish their own research programmes about impact processes, had not routinely been concerned with the relationship between time and rock sequences. They were used to dealing with (ahistorical) processes, which they sought to deduce from the rock. Thus, they were less disturbed by the idea that just a few seconds to minutes were sufficient to create the basic outline of a geological structure like the Ries crater. Consequently, those who by profession already before the crucial years of 1960–61 had been concerned with geological processes in a general way rather than with unravelling Earth history from local rock sequences, found it easier to accept the new paradigm.

Rapprochement between the two cultures became possible, where mineralogists (or those perceived as such) showed that they themselves cherished geological fieldwork and were not quite ignorant of it; and where geologists finally got up to test their old, dogmatic assumptions and thus were forced to come to new interpretations of old field data (Chapter 11).

---

gewaltige Ereignis ausgelöst wurde, müssen die Geologen anderen überlassen. Ihre Aufgabe ist es zu fassen, wann es erfolgte und wie damals unsere Landoberfläche aussah und wie sie umgestaltet wurde. Die große vulkanische Periode Süddeutschlands war vorbei, auch die Ablagerung des Miozänmeeres. Wie sehn [sic] in den obermiozänen Süßwasserkalken eine führende Zeitmarke, für uns auch eine Art Nivellement in dem schwer betroffenen obermiozänen Flußnetz, das wir erst wieder rekonstruieren müssen. Ebenso wichtig ist uns die Auflagerungsfläche des Riessprengschuttes, welcher auch die ganze oberflächliche Entwässerung gewaltig umgestaltete. Denn alte Täler waren völlig verschüttet, Flußläufe abgedämmt. Die Wasserscheide zwischen NW und SO war stark betroffen, wurde weitgehend umgewandelt. Das Rückschreiten der Stufenränder wurde gehemmt. Geköpfte Täler, vor allem am Keuperstufenrand, flossen wieder nach dem Rheine aus. Denn die Wasserscheiden der geköpften Täler lagen tiefer als die ums Ries gestauten Seen, um bis zu 50m! Die obere Jagst wurde zum Rheine abgelenkt! Eine grauenhafte Vernichtung von pflanzlichem und tierischem Leben fand statt – Aber Knochenfelder und Kohlenlager fehlen. Die Auflagerung der Bunten Bresche wurde vielleicht zu wenig beachtet. Man dachte zu wenig an Leben und Formen, zu viel an Potenzen!' (Note by Georg Wagner, 5 October 1968, UAT 605/343).

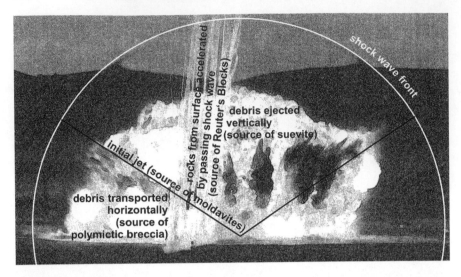

Figure 10.4   Sketch of a test explosion of 500 tons of TNT. The developing
explosion plume consists of several sections, which lead to distinct
deposits that can also be recognised in the much larger Ries impact
crater. (Graphics: Kölbl-Ebert).

Edward Chao[72] surprised the Ries experts by actually spending three months
in the Ries, doing field-work and considering large quarries just as interesting
as his tiny and exotic minerals. The astonished guide – the young geologist
Rudolf Hüttner from the Geological Survey in Baden-Württemberg – was most
impressed, because Chao not only considered all these various dimensions to
be of equal importance, but also had the methodological expertise for them.
Hüttner on these trips learned to use a simple hand lens, by which means Chao
discovered a polish on the grains in the explosion breccia that resulted from the
transport mechanism (Hüttner interview). Chao later published his findings in
a German journal (Chao 1977a) – an intelligent piece of diplomacy:

> He not only operated his machines in America and analysed his rocks, but he was
> also here and has lived in the Ries. We sat together on long evenings. For some

---

[72]   While it seems that for Shoemaker the Ries was just a brief and not very important
episode in his life, notable only for the fact that he and his collaborator succeeded in
showing that large impact craters also existed on Earth (interview of Dr Eugene Shoemaker
by Ronald Doel on 17 June 1987, Niels Bohr Library and Archives, American Institute of
Physics, College Park, MD USA, <www.aip.org/history/ohilist/5082¬_3.html> accessed
25 February 2011), Edward Chao stayed interested in continuing Ries research, including it
in his global itinerary to review possible impact structures as part of his work for the USGS
Astrogeology Group.

time, I showed him around and collaborated with him. My travel costs – then it was very difficult to get a travel grant from the Survey – and so the travel money for some time was paid by the American embassy, but at first a member of the embassy did inquire ... whether it was not dangerous to climb down into that crater. ...

[Chao] then also published a paper on the Ries in the journal *Geologisches Jahrbuch*, where he summarized his analyses, and where he also presented some observations, which were new. ... At that time he always took the hand lens and with that you can see much more, and thus he also saw the fine striae on the stones within the polymictic breccia. He called it 'polish' – polished surfaces with these fine scratches on them. This had not been seen until then. Chao was the first to see them, by application of the hand lens; and then he always had his Ries model as a final goal in mind. ... His colleagues at NASA ... said: "Yes, we cannot calculate this". [So] he simply left off the physical things and believed that he would be able to explain [the process] solely from the geological clues. (Hüttner interview)[73]

Thus, a rather simple instrument (in the form of a hand lens) opened a new world of observational data to the classical geologist of the day, who until then, it seems, had only a hammer and walking-boots.

But those now were clearly no longer sufficient. Field evidence had to be revisited; and thus a number of tacit assumptions regarding data, which seemed to fit or even to point to the volcano, were now put to the test for the first time ever.

---

[73] 'Der hat also nicht nur seine Apparate bedient in Amerika und seine Steine analysiert, sondern er war auch hier und hat im Ries gewohnt. Wir haben dann lange Abende zusammen gesessen. Ich habe ihm auch eine Zeitlang so einiges gezeigt und mit ihm zusammen gearbeitet. Meine Reisekosten – das war damals vom Amt aus sehr schwierig, die Reisekostenfrage – und da wurden die Reisekosten eine Zeitlang auch von der amerikanischen Botschaft gezahlt und dann hat aber erst mal eine Botschaftsangehörige nachgefragt ... ob es denn nicht gefährlich sei in diesen Krater hinabzusteigen. // ... [Chao] hat ja dann auch eine Arbeit über das Ries veröffentlicht im Geologischen Jahrbuch – wo er dann seine Analysen zusammengefasst hat – und da sind einige Beobachtungen da, die schon neu sind. ... Er hat damals auch schon ganz konsequent die Lupe genommen und da sieht man dann einfach mehr, und da hat er dann auch die feinen Striemen auf Steinen in der Bunten Brekzie [gesehen]. *Polish* hat er das genannt, so polierte Flächen und dann diese feinen Striemen drauf. Das war bisher nicht gesehen worden. Das hat der Chao eigentlich als erster erkannt durch das konsequente Anschauen mit der Lupe; und dann hat er natürlich auch immer wieder sein Riesmodell als Endziel vor Augen gehabt. ... seine Kollegen bei der NASA ..., die sagten: "Ja, das können wir eben nicht rechnen". [Da hat] er die physikalischen Dinge einfach außer Acht gelassen und hat die Sache rein so aufgrund der geologischen Indizien geglaubt erklären zu können' (Hüttner interview).

Chapter 11
Testing an Old Theory

While there was still quite a vocal resistance to the renewal of the impact theory among the older generation of geologists, some of the younger ones prepared to test the volcanic and impact theories against new field evidence, overcoming a methodological fallacy that had infested Ries research since its beginning.

An early attempt at testing an old presumption was undertaken by two mineralogists from the Mineralogical Institute of the University of Heidelberg, Arno Schüller (Chapter 10) and Joachim Ottemann (born 1914). These authors were intrigued by the success of Edward Chao in demonstrating the connection between coesite and impact craters, and likewise saw his discovery of nickel-iron spherules as an unassailable argument for the impact origin of the Ries crater – and thus suevite:

> Nevertheless, and up to the most recent time, people could not free themselves from the idea that suevites were the 'true and only' volcanic rocks of the Ries Basin, so that even today the majority of geologists argue for a volcanic origin of the Ries Basin despite the facts that Chao has meanwhile published. (Schüller and Ottemann 1963: 3)[1]

In this context, the two mineralogists set out to test the old paradigm of suevite as a volcanic rock. But what they found fell utterly short of the predicted outcome:

> Following our investigations, the suevites are not volcanic glasses based on their mineral contents, the form of the minerals (structure) and their texture. Namely, missing in suevite glass are all crystals and all idiomorphic early crystallisates, which are an essential character of the true volcanic glasses. ... That it is truly crushed rock material is incontestably proven in that all idiomorphic crystal forms are missing in suevite glass, and all grains are crossed by irregular black cracks and fissures ... Within such mineral fragments, areas develop that become gradually optically isotropic, because they were molten to form glass. Thus, it is

---

[1] 'Trotzdem konnte man sich bis in neueste Zeit von der Vorstellung nicht frei machen, daß die Suevite die 'echten und einzigen' vulkanischen Gesteine des Rieskessel darstellen, so daß also auch heute noch die Mehrzahl der Geologen trotz der von Chao inzwischen veröffentlichten Tatsachen eine vulkanische Entstehung des Rieskessels vertreten' (Schüller and Ottemann 1963: 3).

proven that the glass originates from rock and mineral fragments. In volcanic rocks the case is the other way round, crystals grow from the glass. (Schüller and Ottemann 1963: 11–12)[2]

The authors also noted the unusually high density of the glasses and their inhomogeneity, which also readily distinguished them from ignimbrites, which often contain mineral fragments in addition to idiomorphic crystals:

> The Ries of Nördlingen, according to the available facts, is of meteoritic and not of volcanic origin. In our opinion, the reasons why the Ries of Nördlingen was a subject of disputed hypotheses for an entire century (Dorn [1950]) are that efforts to understand its genesis were made solely through morphological analysis. However, all scientific disciplines have experienced that a morphological analysis is unable to provide unequivocal genetic results. (Schüller and Ottemann 1963: 20)[3]

In other words, it was not enough to state only that a given structure 'looks like' a volcano or an impact crater. Additional evidence, based on the scientific principle of prediction and testing, was required.

The geologists too, meanwhile, had grasped that problem. Gerold Heinrich Wagner (1928–67; Figure 11.1), Georg Wagner's son, began his geological studies in Tübingen, but his father did not want him to work on his PhD in Tübingen. This was because Georg Wagner harboured the dream that Gerold should one day become his successor in Tübingen, in private letters calling

---

2   'In Bezug auf Mineralbestand, Mineralform (Struktur) und Gefüge (Textur) sind die Suevite auf Grund unserer vergleichenden Untersuchungen keine vulkanischen Gläser. Es fehlen nämlich alle Kristalle und alle idiomorphen Frühausscheidungen im Suevitglas, die ein wesentliches Kennzeichen der echten vulkanischen Gläsern sind. ... Dass es sich wirklich um zerstoßenes Gesteinsmaterial handelt, wird einwandfrei dadurch bewiesen, daß alle ideomorphen Kristallformen im Suevitglas fehlen, daß alle Körner von schwarzen unregelmäßigen Sprüngen und Rissen durchzogen sind ... Innerhalb solcher Mineralbruchstücke entwickeln sich Felder, die allmählich optisch isotrop werden, weil sie zu Glas aufgeschmolzen wurden. Dadurch wird bewiesen, daß das Glas aus Gesteins- und Mineraltrümmern hervorgeht. Bei vulkanischen Gesteinen ist das Umgekehrte der Fall, die Kristalle wachsen aus dem Glas' (Schüller and Ottemann 1963: 11–12).

3   'Das Nördlinger Ries ist gemäß dem vorliegenden Tatbestand meteoritischen und nicht vulkanischen Ursprungs. Die Gründe dafür, weshalb das Nördlinger Ries ein ganzes Jahrhundert (Dorn 1948 [sic]) Gegenstand umstrittener Hypothesen war, liegen u. E. daran, daß man allein aufgrund der morphologischen Analyse seine Entstehung erkennen wollte; es ist aber in allen naturwissenschaftlichen Disziplinen die Erfahrung gesammelt worden, daß eine morphologische Analyse keine eindeutigen genetischen Resultate zu liefern imstande ist' (Schüller and Ottemann 1963: 20).

Figure 11.1   The geologist Gerold Heinrich Wagner, son of Georg Wagner. (Source: Carlé 1987, detail from Figure 6).

his geological chair '[his] well-provided hereditary farm' (letter from Georg Wagner to Leonhard Rückert, 29 September 1969, UAT 605/343).[4] However, the new university policy after the war was to prefer candidates from outside for professorships, to obtain fresh views and avoid the former intellectual 'incest':

> Therefore I sent him to Bonn, to Hans Cloos, our best geologist, with whom he was much pleased. But [Cloos] died just at that time, when he wanted to give him the topic [of his thesis]. It took more than a year until he received a topic from Cloos' successor Brinkmann; *small-scale tectonics in the Ries*. Cloos had excluded Ries topics on principle, because he thought them too difficult and because they took far too much time. When I confronted Brinkmann with my reluctance concerning this topic, he replied: That's why he is your son; he should be fit for such a task! (Undated note by Georg Wagner, UAT 605/343)[5]

Roland Brinkmann (1898–1995) had studied geology at the University of Freiburg, where he received his doctorate in 1921. He became an assistant to Hans Stille (1876–1966) in Göttingen, where he also qualified for professorship in 1923. In 1929 Brinkmann became a professor without chair at the same university. In 1933 he became Professor of Geology and Palaeontology in Hamburg. In May 1937 he was expelled from membership of the Nazi party (NSDAP) because of his critical attitude, and left Germany. After working as an ore geologist for a German company in Spain and Portugal, Brinkmann was ordered to Krakow (Poland) when his company was merged with the Geological Survey of Germany. He returned to Germany in 1944 and was appointed Professor of Geology and Palaeontology in Rostock (GDR) in 1946. In 1949 he was arrested and extradited to Poland, together with other people who had held high offices during the German occupation of Poland. Until 1951 Brinkmann was in detention pending trial, but in the following court-martial he was cleared of all charges and rehabilitated. In 1952, Brinkmann became Cloos's successor in Bonn. After his retirement in 1963, Brinkmann was given the task

---

4   'mein wohl vorbereiteter Erbhof' (letter from Georg Wagner to Leonhard Rückert, 29 September 1969, UAT 605/343).

5   'Darum schickte ich ihn nach Bonn zu Hans Cloos, unserem besten Geologen, an dem er auch große Freunde hatte. Doch starb dieser gerade zu der Zeit, wo er ihm das Thema geben wollte. Es dauerte über ein Jahr, bis er von dessen Nachfolger Brinkmann eine Arbeit erhielt. Kleintektonik im Ries. Cloos hatte grundsätzlich Riesarbeiten, weil zu schwer, abgelehnt, da sie viel zu viel Zeit erfordere. Als ich Brinkmann meine Bedenken gegen diese Arbeit vortrug, entgegnete er: Dazu ist er ja Ihr Sohn; der soll dieser Aufgabe gewachsen sein!' (Undated note by Georg Wagner, UAT 605/343).

of establishing a new geological institute at the university in Izmir (Turkey). In 1973, he finally returned to Germany.[6]

In contrast to Richard Dehm in Munich (who, in PhD projects he supervised, always cut the Ries into small individual mapping projects where Ries rocks were only part of the local stratigraphy), Brinkmann now for the first time commissioned a structural approach to the Ries Basin. Thus, when Gerold Wagner finished his PhD with Brinkmann in 1957, he was essentially the only person with a detailed overview over the entire Ries Basin and its foreland, and so his father had high hopes for him to defend the anti-impactist view. But as it turned out, the young man was not so easily led: 'Gerold has completed his doctorate in Bonn on the Ries. The dissertation was very good and he now stands in the Ries battle: Volcanism or meteorite. ... Gerold still stands undecided in this quarrel' (letter from Georg Wagner to the Priehäußer family, 19 November 1965, UAT 605/343).[7]

When Gerold published the results of his PhD project (Wagner 1964), he plainly stated his facts and observations, but refrained from interpretation in one or the other paradigm. Instead, he fairly admitted that 'set-up and arrangement of the investigated breccias surrounding the Ries may be interpreted within the meteorite theory just as well' (Wagner 1964: 588). However, he cautioned against an unquestioning adoption of the new paradigm, using the old argument of a special location, and referred to the relevant literature, of which Bucher (1963a) was the most recent (Wagner 1964: 588).

Gerold Wagner had also managed to show that the striation underneath the Ries breccia, wherever accessible, documented movement of the debris outwards radially from the Ries Basin centre. Thus, it became evident that the cause of the Ries structure was indeed some sort of central explosion, be it volcanic or of impact origin, but definitely not due to 'local outbursts', the explanation favoured by the Munich school (Hüttner interview).

Gerold Wagner's second Ries paper (published on 1 July 1965) again cultivated a carefully neutral attitude,[8] and thus, in the light of contemporary mineralogical research, conveyed a somewhat conservative impression. The

---

[6]   Freyberg 1974: 21; <http://de.wikipedia.org/wiki/Roland_Brinkmann> accessed 4 October 2011.

[7]   'Gerold hat in Bonn über das Ries promoviert. Die Arbeit war sehr gut, und jetzt steht er im Rieskampf: Vulkanismus oder Meteorit. ... Gerold steht noch unentschieden in diesem Streit'. (Letter from Georg Wagner to the Priehäußer family, 19 November 1965, UAT 605/343).

[8]   Gerold Wagner cited Shoemaker and Chao (1961a), Weiskirchner (1962), Schüller and Ottemann (1963) and Preuss (1964) as representatives of the pro-impact camp; and Dehm (1962), Buchner (1963) and Mosebach (1964) as impact deniers (Gerold Wagner 1965: 200). The name conspicuously absent on the latter list is that of Gerold's father, Georg Wagner. It may be speculated, perhaps, that because Gerold, inclining more to the

stated purpose of the paper was to review the question of whether suevite was delivered from the central crater or by local 'outburst' from a number of eruption points (Gerold Wagner 1965: 199), that is, it was explicitly not about deciding on the cause of a central explosion. Nevertheless, Gerold commented that the impact hypothesis rested on the occurrence of coesite, of iron spherules containing nickel and traces of stishovite:

> that is some substances in the suevite tuffs surrounding the Ries that occur only in traces, and in parts are only to be detected by x-ray methods. The representatives of this theory have so far avoided all relevant geological arguments. Then again, geologists are also rather baffled by the occurrence of these minerals, because the volcanic explosion theory had not previously had to reckon with such high pressures (20 respectively 100 kbar). Therefore, the major counter-arguments are not mineralogical but tectonic-regional: the temporal and spacial alignment with the volcanism of southern Germany and the palaeo-geographical and structurally predestined location of the Ries explosion (Dehm 1962, Buchner 1963). Thus, today, only the fact of a large central explosion, with all its effects, is confirmed; not, however, its cause. (Gerold Wagner 1965: 200)[9]

Gerold Wagner was still unable to position himself in one or the other paradigm, because the geological evidence he found seemed equivocal. In the summer of 1964 he again examined the quarries of Aumühle, where he found three distinguishable suevite beds, and of Otting, where there also were at least two distinct layers of suevite on top of each other (Gerold Wagner 1965: 203–4). On the other hand, the attempt to drill for a postulated volcanic vent at Otting failed. Fifty drill holes on a surface of 400 x 500 m showed no evidence of the vent. Instead, the thickness of the suevite was always less than 24 metres (Gerold Wagner 1965: 214). Gerold Wagner concluded:

---

former group, was reluctant to explicitly associate his father with the faction he expected to lose in the end.

[9]   '... also einige nur spurenhaft vorkommende, z.T. nur röntgenographisch nachweisbare Substanzen in den Suevittuffen der Riesumgebung. Die Vertreter dieser Theorie haben bisher alle wesentlichen geologischen Argumente beiseite gelassen. Umgekehrt stehen aber auch die Geologen dem Vorkommen dieser Mineralien im Riesgebiet ziemlich ratlos gegenüber, denn mit so hohen Drucken (20 bzw. 100 kbar) hatte die vulkanische Sprengtheorie bisher nicht zu rechnen. Die wesentlichen Gegenargumente sind deshalb bezeichnenderweise auch keine mineralogischen, sondern tektonisch-regionale: zeitliche und räumliche Einordnung in den süddeutschen Vulkanismus, paläogeographisch und strukturell vorgezeichnete Stelle der Riesexplosion (Dehm 1962, Buchner 1963). So ist heute lediglich die Tatsache einer großen zentralen Explosion mit all ihren Wirkungen gesichert, noch nicht aber deren Ursache' (Gerold Wagner 1965: 200).

Although steep contacts and suevite tuff dykes at some places in the Ries surroundings point to an eruption from independent vents, no such vents have been directly validated so far. The first attempt to drill a deep continuation of the tuff has failed (Hüttner and Wagner 1965[a]).

The glassy, flat bombs within the suevite tuff show a noticeable alignment. ... A structural investigation has shown that, at least for the tuffs of Otting, a transport from the Ries [centre] is quite possible.

Despite these observations, there are still severe reservations against a general ejection of the suevite tuffs from the Ries [centre]. For instance, at Aumühle there are several partly bedded tuffs that require several eruptions, and – because of the inclusions of tuffites [that is reworked tuffs] – eruptions of a longer duration. (Wagner 1965: 220)[10]

On 19 June 1965, some two weeks before Gerold Wagner's paper was published, a manuscript was submitted by Friedrich [Fred] P. Hörz, who was one of the first two doctoral students of Wolf von Engelhardt working on Ries suevite. Hörz obtained his PhD from the University of Tübingen in 1965. He then became a postdoctoral researcher at the NASA Ames Research Center in California and the Lunar Science Institute, working on shock metamorphism. He joined NASA in 1970 and became Mission Science Trainer of the Apollo 16 crew as well as a member of the Apollo Preliminary Examination Team. Hörz initiated NASA's Experimental Impact Laboratory, which he then managed. He was also a Visiting Scientist at the Max Planck Institute for Nuclear Physics in Heidelberg, Germany.[11]

Hörz's 1965 paper presented results from his PhD dissertation. Some of his methods were astonishingly geological for a mineralogical study, but far from conservative. Hörz described and classified the so-called *Flädle* that is glassy, ballistic bombs within the suevite. He convincingly showed that, unlike the

---

[10] 'Obwohl steile Kontakte und gangförmige Lagerung der Suevittuffe an einigen Stelle im Vorries für eine Förderung aus selbständigen Schloten sprechen, ließ sich bis jetzt keine solche Förderstelle direkt nachweisen. Der erste Versuch eine Tiefenfortsetzung der Tuffe zu erbohren, ist fehlgeschlagen (Hüttner & Wagner 1965). // Die im Suevittuff eingeschlossenen Glasfladen zeigen eine deutliche Einregelung. ... Ihre gefügekundliche Untersuchung hat ergeben, daß wenigstens bei den Tuffen von Otting ein Transport aus dem Ries durchaus möglich ist. // Trotz dieser Beobachtungen bestehen aber noch ernste Bedenken gegen einen einheitlichen Auswurf der Suevittuffe aus dem Ries. So liegen bei der Aumühle mehrere z.T. geschichtete Tuffe vor, die eine mehrmalige und, wegen der Einschlüsse von Tuffiten, eine länger anhaltende Fördertätigkeit erfordern' (Wagner 1965: 220).

[11] <http://ares.jsc.nasa.gov/People/horzfred.html> accessed 16 January 2012; Engelhardt in AG Ries 1963: 28.

previous assumption – also made by Gerold Wagner (1965: 220) – these bombs were not plastically deformed when they hit the ground. Instead, their characteristic form was acquired aerodynamically during ejection from the crater (Hörz 1965: 631–2). Hörz then argued that, if the distinction between a vent and a basin facies made by the old volcanists (see Chapter 3) was valid, he would then expect the general orientation of the flattened *Flädle* to be distinct: horizontal in the postulated basin facies and vertical or at least chaotic in the vent facies. However, it turned out that the postulated vent and basin facies were indistinguishable (Hörz 1965: 633). Instead, Hörz found that the elongated *Flädle* indicated a radial transport from the Ries centre outwards, thus confirming independently what Gerold Wagner had deduced from the radial striation underneath the Ries breccia (Hörz 1965: 637). For this and various other reasons, Hörz argued that the suevite outcrops were remnants of a once continuous ejecta blanket derived from a common (central) source (Hörz 1965: 638).

Hörz also undertook heating experiments to homogenise the suevite glasses, in order to collect data on their cooling times. Upon heating the inhomogeneities disappeared as expected, but the formerly transparent glasses also became black, obviously due to oxidisation of iron from FeO to Fe_2O_3 and Fe_3O_4. This pointed to a genesis of the glasses under reducing or oxygen-free conditions,[12] something that was not to be expected under normal volcanic conditions (Hörz 1965: 639–42). Hörz also showed that the elevated nickel and cobalt contents of the suevite glass could not have been derived from recycled basement rocks, whose concentration in these elements was definitely too low. Instead, they were derived from the meteoritic projectile (Hörz 1965: 648).

Already a rough calculation showed that the pressure required for formation of the high-pressure minerals would never be achievable by volcanic forces:

> The Ries catastrophe is depicted as a true, shallow explosion. The focus of detonation was presumably 300–400 m deep within the crystalline basement. Together with an overlying cover of sediments of some 600 m thickness, this adds up to a rock parcel of 1000 m thickness as a brace for the explosion.

> A simple calculation shows how weak this brace was. Assuming a mean density of d = 2.5 g/cm^3, we find that a column of 1000 m length and 1 cm^2 basal area weighs 250 kg. A pressure of little more than 0.25 kbar would already lift the undamaged board of the [Swabian/Franconian] Alb, thereby fracturing it.

---

[12]   Suevite glasses formed within the incandescent cloud of ejecta, which were expelled into the vacuum – or 'hole' in the atmosphere – which had been formed by the descending projectile. The atmosphere, pushed aside by the impact shock wave, was too slow to collapse back into the void to oxidise the glasses before they had cooled down too much for chemical reaction.

For [several] years the 'formation fracture' method has been used by the petroleum industry to increase recovery from oil fields. In this process, a special drilling fluid is pumped into the oil deposit until eventually the entire overlying rock-cover is lifted. ... For a depth of 1000 m, a maximum pressure of 0.25 kbar is sufficient. For the Mesozoic [rocks] of southern Germany, according to the experience of the German oil industry, a pressure of 0.2 to 0.3 kbar per 1000 m depth is required (friendly communication by Dr E.W. Straub, Elwerath Company).

These calculations enforce the conclusion that the Ries can only have originated from a cosmic event. The cover of the *Alb* was much too weak to withstand the high pressures – originating, according to the volcanic theory, within the Earth's interior – that are required by the occurrence of the high-pressure phases. (Hörz 1965: 653–4)[13]

Hörz fervently argued against the notion – still common, even among impact promoters – that suevite was a late volcanic product, erupting after the actual Ries catastrophe:

It is said the volcanic products [the suevites] were extruded from the unknown magma chamber only through the loose breccia masses that fill the explosion crater. Thereby, however, they would have to overcome hardly any resistance. Under such conditions the formation of stishovite, coesite, isotropic quartz and feldspar, as well as the high-pressure glasses, is impossible.

---

[13] 'Die Rieskatastrophe wird als echte Flachsprengung dargestellt. Der Explosionsherd lag vermutlich 300–400 m tief im Grundgebirge. Bei einer überlagernden Sedimentdecke von rund 600 m ergibt sich als Widerlager für die Explosion ein Gesteinspaket von 1000 m Mächtigkeit. // Eine einfache Rechnung zeigt, wie schwach dieses Widerlager war. Nimmt man eine durchschnittliche Dichte von $d = 2,5$ gr/cm^3 an, so ergibt sich, daß eine 1000 m lange Säule von 1 cm^2 Grundfläche 250 kg wiegt. Schon ein Druck von wenig mehr als 0,25 kbar würde die unversehrte Albtafel hochheben und dabei zerbrechen. // Seit Jahren wird in der Erdölindustrie das 'Formation-Fracture'-Verfahren angewandt, um die Ausbeute einer Erdöllagerstätte zu erhöhen. Dabei wird eine Spezialspülung in den Erdölträger gepumpt, bis sich schließlich die ganzen überlagernden Deckschichten abheben. ... [Es] genügt für eine Teufe von 1000m ein Druck von maximal 0,25 kbar. Für das süddeutsche Mesozoikum ergibt sich nach Erfahrungen der deutschen Erdölindustrie ein Druck von 0,2–0,3 kbar pro 1000 m Teufe (freundlichen Mitteilung von Dr. E.W. Straub, Fa. Elwerath). // Diese Berechnungen zwingen zu dem Schluß, daß das Ries nur durch ein kosmisches Ereignis entstanden sein kann. Die Albtafel war viel zu schwach, um den – nach der Vulkantheorie aus dem Erdinneren stammenden – hohen Drucken widerstehen zu können, wie sie nach dem Vorkommen der Hochdruckphasen geherrscht haben müssen' (Hörz 1965: 653–4).

The existence of the high-pressure phases proves the direct association of the suevites with the Ries catastrophe. Only the main explosion could have provided the pressure required for their formation. After their formation, the melt must have been rapidly cooled; otherwise they would not have been preserved. (Hörz 1965: 655)[14]

The allegedly multiphase deposition of suevite – as Gerold Wagner had again just confirmed for the Otting and Aumühle quarries – was easily debunked by Hörz with a mineralogical argument: 'In the '2nd suevite phase' of Hainsfarth, mentioned by Bucher (1963[a]), Löffler and G[eorg?] Wagner (personal communications), coesite was found as well. Its occurrence shows that this 'tuff' too must belong directly to the Ries catastrophe. A later phase of suevite should not contain any high-pressure indicators' (Hörz 1965: 655).[15]

Thus for Hörz, there was no longer any doubt possible: 'These clues force the conclusion that the Ries Basin can only have originated from a cosmic event. Pressures and temperatures of the Ries catastrophe are too high for a terrestrial event; not to mention the fact that the energy balance cannot be explained adequately by a subcrustal process' (Hörz 1965: 657).[16]

---

[14] 'Sie sollen ihre Förderprodukte vom ungekannten Magmaherd lediglich durch die den Explosionskrater ausfüllenden, lockeren Trümmermassen gepreßt haben. Dabei hätten sie aber so gut wie keinen Widerstand zu überwinden gehabt. Unter solchen Voraussetzungen ist die Bildung von Stishovit, Coesit, isotropisiertem Quarz und Feldspat sowie der Hochdruckgläser ausgeschlossen. // Die Existenz der Hochdruckphasen beweist die unmittelbare Zugehörigkeit der Suevite zur Rieskatastrophe. Nur die Hauptexplosion konnte die zu ihrer Entstehung notwendigen Drucke liefern. Nach ihrer Bildung muß die Schmelze rasch abgekühlt und ausgeworfen worden sein, sonst hätten sie nicht konserviert werden können' (Hörz 1965: 655).

[15] 'In der von Bucher (1963[a]), Löffler und G. Wagner (mündliche Mitteilungen) erwähnten '2. Suevitphase' von Hainsfarth wurde ebenfalls Coesit nachgewiesen. Sein Auftreten zeigt, daß auch dieser 'Tuff' unmittelbar zur Rieskatastrophe gehören muß. Eine spätere Suevitphase dürfte keinerlei Hochdruckindikatoren enthalten' (Hörz 1965: 655). In another argument, Hörz was mistaken: He argued that suevite was not welded like ignimbrites; a feature that at that time was obviously considered to be due to the presence of a water vapour phase during (volcanic) eruption. Since welding was missing in the suevite, suevite could not have been generated by a phreatic eruption (Hörz 1965: 656). The observation of phreatic eruptions such as the now-classic Surtsey eruption of 1963–64 has shown that phreatic processes and welding exclude each other. The lack of welding is therefore no clue against phreatic volcanism. Meanwhile, lenses of welded suevite have been recognised in the Ries (for example at Polsingen).

[16] 'Dieses Beweismittel zwingt zu dem Schluß, daß der Rieskessel nur durch ein kosmisches Ereignis entstanden sein kann. Drucke und Temperaturen der Rieskatastrophe sind für ein terrestrisches Ereignis zu hoch, ganz abgesehen davon, daß die Energiebilanz durch einen subkrustalen Vorgang nicht befriedigend geklärt werden kann' (Hörz 1965: 657).

No Volcanic Vent at *Altenbürg* Quarry

Gerold Wagner also did not leave things in a doubtful position. He conscientiously began to assemble and test the geological arguments for the different genetic models. Meanwhile, he was working for the Federal Geological Survey in Hannover, but around 1963 he was transferred to the Geological Survey of Baden-Württemberg in Freiburg, where he joined forces with his friend Rudolf Hüttner who had been in Freiburg since 1959 (Hüttner interview).

Hüttner had studied geology in Tübingen and obtained his PhD in 1958 with a dissertation on the geology of an area of Württemberg just outside the Ries Basin, on the map sheets of Neresheim and Wittislingen. His PhD supervisor was Helmut Hölder. From 1959 onwards, he worked as a geologist for the Geological Survey of Baden-Württemberg (Engelhardt and Hölder 1977: 255; Hölder interview).

In the quarry of *Altenbürg* some 5 km southwest of Nördlingen (Figure 11.2), there was suevite sandwiched between massive limestone in a steep contact on both sides. Consequently, it was assumed that here suevite was filling a volcanic vent. In the summer of 1963, the two friends obtained access to a small drill rig and workers, and they made a dense array of boreholes in the quarry:

> Then over lunchtime we always sat together drinking tea and the topic was always the Ries. We talked about these questions. And there was this open point, that is the suevite; the suevite, which had been interpreted as volcanic tuff, completely without doubt. It was simply a volcanic tuff, and as such it should at least at some localities be contained in a [volcanic] vent. And for this the quarry of *Altenbürg* is the most renowned, where in two walls the vertical contact can practically be seen. And if it was deposited in vents, then this would be a clue that it had truly been volcanism. And therefore we, Gerold and I, undertook these drillings.

> It happened that the survey at that time had some more money for outcrop purposes, and so a drill-rig was rented for a month or so, and then this drilling equipment, including labourers, was provided to various colleagues. And thus we too drilled for some days in the Ries, and so we drilled intensively in this quarry of *Altenbürg*, and nowhere did we find any extension [of the suevite] to below. Everywhere, it was gone after five, six metres, and that for us was basically the proof. (Hüttner interview)[17]

---

[17] 'Dann saßen wir so jeden Mittag beim Tee zusammen und das Thema war immer das Ries –wir haben uns eben über diese Fragen unterhalten. Und da war also ein offener Punkt, nämlich der Suevit; der Suevit, der ja als vulkanischer Tuff angesehen worden ist, völlig zweifelsfrei. Es war halt ein vulkanischer Tuff und der sollte ja stellenweise wenigstens in Röhren lagern. Und da ist ja der Steinbruch Altenbürg der bekannteste, wo man in zwei Wänden praktisch den vertikalen Kontakt sieht. Und wenn der also in Röhren gelagert ist,

As it turned out, the Jurassic limestone in steep contact with suevite was not the wall of a volcanic vent. Instead, the suevite everywhere was underlain by chaotic explosion breccia, and the limestone proved to be only a couple of gigantic, detached, house-sized blocks within the polymictic Ries debris. The two friends published these most interesting discoveries in 1965 (Hüttner and Wagner 1965a).

> The [suevite] deposit at Heerhof [some 6 km west of Nördlingen] too (which due to its rich content of *Fladen* ('bombs') had been regarded as a typical vent facies) is, according to the result of 19 drillings (1964/65), a comparatively horizontal-lying erosion remnant of a once more widely distributed [ejecta] blanket (suevite thickness up to 15 m) on top of the polymictic breccia masses. (Hüttner 1969: 189)[18]

And so Gerold Wagner and Rudolf Hüttner managed to support the impact scenario by removing a crucial obstacle against it, with the aid of purely geological fieldwork.

Georg Wagner, the father, was not too happy with the general open-mindedness of his son. In his doctorate dissertation, Gerold had mentioned that the meteorite theory too could be reconciled with the observed features – a remark that the father took badly. Everything was fine with Gerold's thesis, but he should not have written this, his father remonstrated. However, when the dissertation was finally published, Gerold Wagner was able to keep his statements without changes, adding only a remark concerning the new evidence of coesite (Carlé 1987: 93; Hüttner interview).

At the various Ries conferences and workshops that took place in the decade following the discovery of coesite at the Ries Crater, Gerold Wagner was a welcome guest. Thus, in 1963 for example, he was specially invited (together with Helmut Hölder), upon rather short notice, to represent the view of the

---

dann wäre das ein Indiz, dass es eben doch Vulkanismus ist. Und deswegen haben wir dann, der Gerold und ich, Bohrungen durchgeführt. Es war gerade so, dass das Landesamt – hatte da mal etwas mehr Mittel für Aufschlusszwecke und dann wurde ein Bohrgerät einen Monat lang, so etwa war es, gemietet und dann bekamen verschiedene Kollegen dieses Bohrgerät mit Mannschaft zur Verfügung gestellt. Und da haben wir dann auch einige Tage im Ries gebohrt und da haben wir diesen Steinbruch Altenbürg intensiv abgebohrt und haben halt nirgendwo eine Fortsetzung nach unten gefunden. Überall war es nach fünf, sechs Metern aus und das war für uns eigentlich der Beweispunkt' (Hüttner interview).

[18] 'Auch das Vorkommen am Heerhof (R 36 02 900, H 54 15 000), das wegen seines reichen Inhalts an Fladen ('Bomben') als typische Förderstelle galt, ist nach dem Ergebnis von 19 Bohrungen (1964/65) ein verhältnismäßig flach gelagerter Erosionsrest einer einst weiter verbreiteten Decke (Suevitmächtigkeit bis 15 m), die auf Bunten Trümmermassen lagert' (Hüttner 1969: 189).

Figure 11.2   In the quarry of *Altenbürg* a steep contact between suevite and
a huge allochthonous limestone block can be seen. A second
such contact with another limestone block is to the left outside
the image. Formerly, this feature was interpreted as a tuff-filled
volcanic vent in contact with autochthonous country rock.
(Photo: Kölbl-Ebert).

Württemberg geologists at a DFG[19] workshop in Munich. Gerold actually showed up and brought with him off-prints of Georg Wagner's 'battle-cry' against the impact theory (Wagner 1963) and the documentation of the 1960 Thomas-conference including the critical discussion (letter from Paul Schmidt-Thomé to Georg Wagner, 5 April 1963, UAT 605/343).

Notwithstanding the open-mindedness documented by his publications, at conferences Gerold Wagner acted the role of *advocatus diaboli*; possibly to please his father.[20] 'Gerold Wagner … still voiced the arguments against the impact theory. … At least, he wanted to explore for once all the arguments that possibly pointed to volcanism' (Pohl interview).[21] In 1966, at a Symposium for Meteorites, Tektites and Impact Craters in Nördlingen, Gerold Wagner (to the faint puzzlement of international guests) seemed to have been the only one who still actively argued for the volcanic theory (Signer interview):

> There was of course von Engelhardt, Stöffler – those were the German party – and then Ed Chao, Shoemaker – they were the USA party – pro-impact-crater, and then there was this man from Freiburg [Gerold Wagner] and a whole bunch of, numerous let's say, European geologists, whom I[22] did not know by name, who simply wanted to know what it now is; who were disbelieving in the one as well as in the other [impact origin or volcanic origin of the Ries Crater]. And that was interesting indeed, [to be a member of] a separate party, we 'Americans' including von Engelhardt and Stöffler, who by way of the Moon, by the investigation or near-future investigation of the Moon, were convinced of these impacts. (Signer interview)[23]

---

[19] *Deutsche Forschungsgemeinschaft*: a federal organisation providing research grants.

[20] Jean Pohl, a geophysicist at The University of Munich, at any rate received the impression that the father (Georg Wagner) had been 'somewhat tyrannical or authoritarian'. ('Der Vater muss jedenfalls ein wenig tyrannisch gewesen sein oder autoritär, sage ich mal' [Pohl interview]).

[21] 'Gerold Wagner … hat immer noch die Argumente gegen die Impakttheorie gebracht. … Zumindest wollte er alle Argumente, die eventuell für Vulkanismus zusprechen, einmal ausloten' (Pohl interview).

[22] The interviewee is the physicist Peter Signer, who had been a professor at the Institute of Crystallography and Petrography of the ETH Zurich. From 1965 onwards, he – together with two colleagues – had been responsible for the solar wind experiment of the Apollo missions (<http://www.nzz.ch/nachrichten/forschung_und_technik/was_apollo_der_wissenschaft_brachte_1.3043313.html> accessed 16 January 2012).

[23] 'Da war natürlich von Engelhardt, Stöffler – das waren die deutschen Vertreter – und dann Ed Chao, Shoemaker – das waren die USA-Vertreter – für den Einschlagskrater und dann gab es eben diesen Freiburger und ein Haufen, zahlreiche sagen wir mal, europäische Geologen, die ich eben nicht mit Namen kannte, die einfach wissen wollten, was ist es jetzt, [die] ungläubig waren fürs eine und fürs andere. Und das war schon interessant, eben auf der einen Seite, ich sage jetzt mal wir Amerikaner, die eben durch den Mond, die Erforschung

Gerold Wagner's Tragic Death at Rochechouart

Intensive research on the geological occurrence of high-pressure minerals led to the discovery of coesite at more and more of the known or postulated impact craters, such as, for example, Lake Bosumtwi (Preuss in AG Ries 1963: 18). But coesite also appeared outside such known sites of meteorite impact.

In 1966, at the same symposium in Nördlingen, François [Ferencz] Kraut (1907–83, subdirector at the Paris Natural History Museum) gave a talk on certain rocks around Rochechouart, a town in southern France. In this town, the church as well as the castle were built from a local suevite. The region had initially been interpreted as an old volcanic area and was known to Kraut from his first experiences in geological mapping in the 1930s. The geological neophyte had at that time found glasses, cleavable quartz and other unusual features; all in all, a rock that he could not compare with anything else (Kraut 1969).

Born in Hungary, Kraut studied mathematics, physics and mineralogy in Budapest and Leoben (Austria). In 1929, he and his family immigrated to France, and he continued his studies in Paris. In 1933, he became a mineralogist at the *Muséum national d'histoire naturelle*. In 1939, he joined the French army and spent the years 1940 to 1945 as a prisoner of war in Germany. After World War II, he returned to the Paris Natural History Museum and in 1963 became its subdirector, until his retirement in 1972.[24]

When Georg Wagner heard rumours of Kraut's discoveries at Rochechouart, he rejoiced: 'In the mid-1960s, Georg Wagner pointed me to some news. In volcanic deposits in France, coesite, shatter cones and planar elements had been found without a crater being visible anywhere near or far. Therefore, all these indicators are disqualified as proofs of an impact' (Carlé 1987: 94).[25]

In 1967, Gerold Wagner went to investigate the site and to acquire samples. He became friends with François Kraut, discussing the geological problem. Kraut also sought contact with American colleagues, to learn more about impact issues. In July 1968 Kraut guided numerous geoscientists from Germany, France and the USA on a field trip to Rochechouart and subsequently interpreted this site as a heavily eroded impact crater (Kraut 1969).

Gerold Wagner, however, did not take part in the field trip, as he had not returned alive from the last of his own field campaigns to Rochechouart: 'On his last trip Gerold was still investigating samples of a potential meteorite impact, and had brought

---

oder die bevorstehende Erforschung des Mondes an diese Einschläge glaubten, inklusive von Engelhardt und Stöffler' (Signer interview).

[24]   <http://fr.wikipedia.org/wiki/ François_Kraut> accessed 16 January 2012.

[25]   'In der Mitte der 60er Jahre wies mich Georg Wagner auf eine Neuigkeit hin. Man habe in Frankreich in vulkanischen Ablagerungen Coesit, shatter cones und planare Elemente gefunden, ohne daß weit und breit ein Krater sichtbar wäre. Also scheiden alle diese Anzeichen als Beweismittel für einen Impact aus'. (Carlé 1987: 94).

back samples, when on the Rhône bridge of St. Esprit he was rammed, hopeless and innocent, by a French lorry driver. His notebooks, however, were saved' (letter from Georg Wagner to Gustav Angenheister, 3 June 1967, UAT 605/343).[26]

The tragic accident, destroying such a hopeful young life, caused Gerold's father Georg Wagner to break down completely. He himself had suffered from two severe traffic accidents, in 1963 and 1964:

> My time has nearly run out. The murdering motors have cost me nearly two years in hospital, my dear, true spouse and my last son. And this at a time when so much of my harvest was ready to be cut. And the gap between University and the Geological Surveys is so difficult to bridge. ... I always try to bridge the gap, but I only succeeded at times. For this, my son Gerold was set up by me and he was on a good path for it, when he too became a victim of the murdering motors. His death has hurt me terribly. So many hopeful beginnings, simply severed. And where the closest cooperation would be necessary, greatest discord reigns ...

> Curator [W], on whom I set great hopes. Also on [Rudolf] Hüttner! Who is also familiar with the Ries, which we must not leave to the mineralogists completely! (Letter from Georg Wagner to Walter Carlé, 10 October 1968, GA unnumbered)[27]

Gerold, who had attended all the Ries meetings and field trips, who had undertaken his own geological investigations to test for one or the other theory, had become more and more convinced of the impact theory himself, although he poked at every little piece of doubtful evidence; and by his influence his father, Georg Wagner, had become less blatant in his dismissal of the impact theory (Hüttner interview). This

---

[26] 'Auch auf seiner letzten Fahrt hat Gerold noch Proben von einem möglichen Meteoriteneinschlag unterucht [sic] und Proben mit gebracht, als er auf der Rhonebrücke von St. Esprit von einem französischen Lastwagenfahrer hoffnungslos und schuldlos gerammt wurde. Seine Tagebücher wurden aber gerettet' (letter from Georg Wagner to Gustav Angenheister, 3 June 1967, UAT 605/343).

[27] 'Meine Zeit ist bald abgelaufen. Die mordenden Motore haben mich 2 Jahre Klinik gekostet, die liebe treue Gefährtin und den letzten Sohn. Und das zu einer Zeit, wo so viel meiner Ernte schnittreif war. Und die Kluft zwischen Universität und Geol. Landesanstalten ist schwer zu überbrücken. ... Ich versuchte immer, die Kluft zu überbrücken; aber es ist mir nur zeitweise gelungen. Mein Sohn Gerold war von mir dafür eingesetzt worden und war auf gutem Weg dazu, als auch er ein Opfer der mordenden Motore wurde. Sein Tod hat mich furchtbar getroffen. So viel hoffnungsvolle anläufe [sic] einfach abgehackt. Und wo die engste Zusammenarbeit notwendig wäre, herrscht größter gegensatz [sic] ... Kustos [W], auf den ich große Hoffnungen setze. Ebenso auf Rudi Hüttner! der sich auch im Ries auskennt, das wir nicht ganz den Mineralogen überlassen dürfen!' (Letter from Georg Wagner to Walter Carlé, 10 October 1968, GA unnumbered).

attitude still shines through in letters written by Georg Wagner shortly after Gerold's death:

> Through my last son Gerold, I continued to be somewhat informed about your field of expertise. Even though I am nearly 82 years old, more than 60 years of scientific interest have connected me so closely to the problem [of the Ries] that I cannot leave it alone. ... And I have spent several months together with geologists in the Ries, so that I do not want to lose contact. Especially also with the Swiss Dr Bucher (USA) we have still been in the Ries. And with the mineralogist Dr Bernauer I still participated in his great volcanological excursion to Italy,[28] to find comparisons. Because what we learned from highly proficient geologists, chemists, physicists is no longer sufficient here with such huge dimensions of temperature, pressure and velocity, so that unfortunately all familiarity is missing for us. At that although in the meteoritic theory so many problems of space, time, stratigraphy, masses remain completely unsolved. That is why I have so sharply rebuked Chao with his tons of coesite in the Long Daniel of Nördlingen. His base [of evidence] was too slim. A Ries friend of many years, R. Löffler, has turned off into the other camp, though. I, however, am still someone questioning! (Letter from Georg Wagner to Gustav Angenheister, 3 June 1967, UAT 605/343)[29]

---

[28] This may have been the field trip of the German Geological Society on which Ferdinand Bernauer (1892–1945) reported in 1939. There were thirty participants. The purpose of the journey was 'to get to know the active volcanoes of Italy, principally in order to provide the possibility of comparison with our German conditions, where due to sparse outcrops and missing context, and so on, the interpretation is often so difficult' ['die tätigen Vulkane Italiens kennen zu lernen, letzten Endes um eine Vergleichsmöglichkeit für unsere deutschen Verhältnisse zu geben, bei denen die Deutung wegen der spärlichen Aufschlüsse, des fehlenden Zusammenhanges usw. oft so schwierig ist'.] (Bernauer 1939: 450). Special topics had included the mechanics of lava movement and the structures formed by it, as well as the formation of volcanogenic ore deposits. The excursion visited Mount Vesuvius, which was active at that time and obliged with an actual lava flow, the Campi Flegrei with Solfatara and Monte Nuovo, Mount Etna and the Aeolian Islands with Lipari, Stromboli and Volcano. They also visited Rome, Pompeii, and several cities as well as antique sites in Sicily (Bernauer 1939). Bernauer had studied mineralogy and geology in Heidelberg. From 1919 onwards, he worked for the Geological Survey in Berlin. In 1928 he became a professor without chair of mineralogy and petrography at the Technical University in Berlin. He developed filters for the polarisation microscope, which after 1935 were produced by the Zeiss Company and marketed under the name 'Bernator'. From 1930 onwards, Bernauer also became interested in volcanology, and travelled to Iceland and the Aeolian Islands (Italy). For the ZDGG, he wrote a regular news column on worldwide volcanic activity. He died a few days after the end of World War II from a wound inflicted by a grenade fragment (Quiring 1955b; <http://de.wikipedia.org/wiki/Ferdinand_Bernauer> accessed 27 March 2012; ZDGG).

[29] 'Durch meinen letzten Sohn Gerold war ich etwas auf dem Laufenden geblieben mit Ihrem Arbeitsfeld. Wenn ich auch bald 82 Jahre alt werde, so haben doch über 60 Jahre wissenschaftlicher Berührung mich so eng mit dem Problem verbunden, daß ich nicht davon

After Gerold's death, however, it seems the only way the old man could bear the terrible loss was to get up again, returning, as he saw it, to battle. Thus, Georg Wagner suffered from a relapse and consequently became more and more isolated in his bitter resistance against the new views of the Ries Basin:

> The Ries experts did value [Gerold] very much, even though he was not as sceptical against the novelty as his experienced father. ... But now he is mute. Perhaps I will still find notes from him! Instead of saying: Reserve may rest! I now must go back into the harness, and there I would have liked your company in north-eastern Württemberg. Also, for informal talk and discussion! Then to the question: What was the influence of the Ries explosion in the area of the rivers Altmühl, Wörnitz, Tauber and Jagst! ... For this you are especially welcome to me! Because the many cubic kilometres that were blasted from the Ries by the explosion, up to 50 km off, have altered the surroundings of the Ries considerably. These 'traces', a thick, widespread blanket, is of no interest whatsoever to the coesite people! Therefore, you see: There are problems enough! And you are younger and still more enterprising than I am. Therefore, on to battle! (Letter from Georg Wagner to Mr Loner, 31 July 1967, UAT 605/343)[30]

This 'battle', however, took an enormous toll on the old man, not only because of the workload but also because of loneliness and the damage he was about to inflict to his own, once so exalted, reputation:

lassen kann. ... Und ich habe mehrere Monate mit Geologen im Ries verbracht, so daß ich die Verbindung nicht verlieren möchte. Besonders auch mit dem Schweizer Dr Bucher USA waren wir noch im Ries. Und mit dem Mineralogen Dr Bernauer habe ich noch seine große Vulkanexkursion nach Italien mitgemacht, um Vergleiche zu finden. Denn was wir bei tüchtigen Geologen, Chemikern, Physikern gelernt haben, reicht eben hier bei so hohen Dimensionen von Temperatur, Drucken und Geschwindigkeit nicht mehr aus, so daß uns leider alle Anschauung fehlt. Dabei bleiben eben bei der Meteoritentheorie so viele Probleme von Raum, Zeit, Lagerung, Massen völlig ungelöst. So habe ich Chao mit seinen Tonnen Coesit im Langen Daniel von Nördlingen scharf widersprochen. Seine Basis war zu schmal. Ein langjähriger Riesfreund R. Löffler ist zwar ins andere Lager abgeschwenkt. Ich bin aber noch Fragender!' (Letter from Georg Wagner to Gustav Angenheister, 3 June 1967, UAT 605/343).

[30] 'Die Rieskenner haben ihn [Gerold] sehr geschätzt, auch wenn er dem Neuen nicht so skeptisch gegenüberstand wie sein abgebrühterer Vater. ... Aber jetzt ist er eben stumm. Vielleicht finde ich noch notizen [sic] von ihm! Anstatt zu sagen: Reserve hat Ruh! nuß [sic] ich jetzt wieder ins Geschirr und da hätte ich Dich gerne in Ordostwürttemberg [sic] dabei gehabt. Auch zu gemütlicher Aussprache und Diskussion! Dann zur Frage: Wie hat sich die Riesexplosion im Altmühl-, Wörnitz, Tauber und Jagstgebiet ausgewirkt! ... Da bist Du mir besonders willkommen! Denn die vielen Kubikkilometr [sic], die aus dem Ries bei der Explosion hinausflogen, bis 50 km weit, haben doe [sic] das Vorries sehr stark umgestaltet. Diese 'Spuren', ein [sic] mächtige weite Schleierdecke, locken die Coesiter nicht im geringsten! Du siehst also: Probleme genug! Und Du bist jünger und noch unternehmender als ich. Also auf in den Kampf!' (Letter from Georg Wagner to Mr Loner, 31 July 1967, UAT 605/343).

But my work is not much use these days. I grow tired too quickly, and I am missing my Gerold too much. In the Ries the mineralogists chase their decimal powers too much and temperature and pressure, which have never been introduced to us. And the scenario is too much neglected. The effect on the landscape, when the Ries Lake reached 500 m above sea level and 50 km³ were blasted into the air! Reuter's blocks had a vast dimension that dramatically altered the whole river network of southern Germany ... It is an exceeding pity that my Gerold lives no longer. For him, that would have been now an area in which he would have been able to create wildly. ... Dehm is too much of a palaeontologist, and in his Ries patchwork carpet too much was only lightly touched on without leading thoughts. Gerold could have started now brilliantly as a young and fiery spirit! Freyberg[31] did do excellent work, but we must be glad if he still presents a summary publication of his area in the northeast.

And Birzer[32] is not enough of a 'fire-brain'! Paul Wurster[33] is too much focused on Bonn. So I see from the mountain of Nebo[34] the land in which milk and honey flow, and can no longer enter! ... Our mineralogists are exhausted [or literally: slain] by decimal powers of pressure and temperature, also by the propaganda. And now the Limes[35] is breached and we have no young fighters! From the Nebo, I saw the whole of the Jordan Graben and the Land of Promise full of problems too. ... But both sons are dead and my best students have been valued more in the north than by the Swabians, who cannot overlook the congenital defect![36] (Letter from Georg Wagner to Helmut Hölder, 30 September 1968, GA3/223)[37]

---

[31]   Bruno von Freyberg (see Chapter 10).

[32]   Friedrich Georg Joseph Birzer (born 1908) obtained his doctorate in 1932 in Erlangen, when he became an assistant at the Geological Institute. In 1939 he obtained a qualification for professorship. In 1945 he became a professor without chair, and in 1970, full Professor of Applied Geology at the University of Erlangen (Freyberg 1974: 17).

[33]   Paul Wurster (born 1926), after his PhD, became an assistant in Hamburg, and after his *Habilitation*, lectured at the University of Munich. In 1968 he became Professor of Geology at the University of Bonn (Freyberg 1974: 174).

[34]   See the Bible: Deuteronomy 34, 1–7.

[35]   An antique fortification along the frontier of the Roman Empire in what is today southern Germany; much like Hadrian's Wall but less well preserved.

[36]   Georg Wagner is here complaining about the new university policy in Germany, which after the war stipulated that candidates for professorships must come from outside to obtain fresh views. This policy had induced Wagner to send his son off to obtain his PhD in Bonn (see above), and it now barred Wagner's PhD students from a career in Tübingen.

[37]   'Nur meine Arbeit ist nicht mehr weit her. Ich ermüde zu rasch, und mein Gerold fehlt mir zu sehr. Im Ries reiten die Mineralogen zu sehr ihre Zehnerpotenzen ind [sic] Temperatur und Druck, die uns nie vorgestellt worden sind. Und das Geschehen kommt viel zu kurz. Die Auswirkung auf die Landschaft, wenn der Riessee bis 500m NN stand und 50km³ durch die Luft flogen! Die Reutterschen [sic] Blöcke hatten eine gewaltige Größenordnung, die das ganze süddeutsche Flußnetz gewaltige umgestaltet hat ... Es ist jammerschade, daß

Rudolf Hüttner tried to soothe his dead friend's father, and explained to him the overwhelming amount of positive evidence for an impact from space at the Ries Crater, and also the negative tests that had falsified the old theory of a volcanic Ries origin. He also pointed out that the new theory still required a central explosion and that, therefore, it was not as far away from the ideas that Georg Wagner had held all along:

> Even though the meteorite experts cannot yet at present answer all questions without doubt, they have erected an extensive edifice of ideas that conforms to the numerous newly discovered mineralogical facts. In contrast to this development, the volcanic interpretation has lost its footing more and more (suevite is not volcanic rock, it is not produced through vents; there is no limestone from thermal springs but algal limestone). Additionally, the newly discovered mineralogical facts (high-pressure modifications of silica; so-called planar elements in quartz and similar features in feldspars as well; vitrification of feldspars without melting, and many more things) are unexplained by known volcanic processes. They can, however, be produced by shock wave experiments. For the volcanic explanation of the Ries there would remain but one single clue: the timing of the upper-Miocene volcanism. But against the great weight of the counter-clues, this alibi is not striking. ...

> Dear Mr Wagner, this is the situation today. I think we should not gnash our teeth because of it, but be happy about the – not lightly gained – progress. The meteorite hypothesis, by the way, has brought about that the Bavarians have buried their 'local outbursts' without further ado. The central explosion – this is, what you, Löffler, Kranz and eventually also we young people, especially Gerold, have fought for – is everywhere accepted. And an additional bonus the meteorite hypothesis has brought: The Ries is now no longer a nearly unique event, but falls into line with a larger number of crater structures in which the same characteristic alterations of minerals can be detected. In recent years a hitherto-unknown form

---

mein Gerold nicht mehr lebt. Das wäre jetzt für ihn eine [sic] Aufgabengebiet, in dem er wild schaffen könnte. ... Dehm ist zu sehr Paläontologe, und in seinem Riesfleckerlteppich wurde zu viel nur gestupft, ohne führenden Gedanken. Gerold hätte jetzt als junger Feuergeist glänzend einsetzen können! Freyberg hat zwar ausgezeichnet gearbeitet. Aber wir müssen froh sein, wenn er noch seine [sic] Gebiet im NO abschließend darstellt. // Und Birzer ist zu wenig 'Feuerkopf'! Paul Wurster ist zu sehr in Bonn festgelegt. So seh ich vom Berge Nebo das Land, darinnen Milch und Honig fließt und kann nicht mehr hinein! ... Unsere Mineralogen sind erschlagen von Zehnerpotenzen von Druck und Temperatur, auch von der Propaganda. Und jetzt ist der Limes durchbrochen und uns fehlt es an den jungen Kämpfern! Vom Nebo aus sah ich den ganzen Jordangraben und das Gelobte Land, auch voller Probleme. ... Aber beide Söhne sind tot, und meine besten Schüler hat man im Norden höher eingeschätzt als bei den Schwaben, die nicht über den Geburtsfehler hinwegkommen können!' (Letter from Georg Wagner to Helmut Hölder, 30 September 1968, GA3/223).

of rock alteration has been traced (shock wave metamorphism). (Letter from Rudolf Hüttner to Georg Wagner, 29 August 1969, UAT 605/343)[38]

Georg Wagner, however, would not listen. He was unable to understand why nobody was prepared to fight against the mineralogical usurpation of the Ries Basin. For him, the problem was no longer about the genesis of a geological structure, but about the future of geology as an independent field of expertise; about the future of a science that once had produced clear and instructive images of the history of a landscape, and that now, he seemed to have thought, had been lost in a volley of abstract and incomprehensible numbers.

Georg Wagner had been the son of a farmer (Carlé 1972), and in the use of his metaphors he stuck to his agricultural upbringing, which might also have been responsible for his urge to leave not just money, manners and education to his offspring (or knowledge, experience and craftsmanship to his students), but also a well-provided *field* of expertise from which to go on. This spiritual conduct of life might also be the key to understanding his local patriotism, which showed him exceedingly bound to his native soil:

---

[38] 'Wenn auch die Meteoriten-Fachleute heutzutage noch nicht alle Fragen zweifelsfrei beantworten können, so haben sie doch ein umfangreiches Gedankengebäude errichtet, das mit den vielen neuentdeckten mineralogischen Tatsachen in Einklang steht. Dagegen wurde der vulkanischen Deutung des Rieses immer mehr der Boden entzogen (Suevit kein vulkanisches Gestein, nicht in Schloten gefördert, keine 'Sprudelkalke', sondern Algenkalke). Zudem finden die neuentdeckten mineralogischen Tatsachen (Hochdruckmodifikationen der Kieselsäure, sog. Planare Elemente in Quarz und ähnl. Bildungen auch in Feldspäten, Verglasung der Feldspäte ohne Aufschmelzung u.v.a.) durch bekannte vulkanische Prozesse keine Erklärung. Sie können jedoch weitgehend durch Stoßwellenexperimente erzeugt werden. Der vulkanischen Erklärung des Rieses blieb nur ein einziger Beweispunkt: die Zeit des obermiozänen Vulkanismus. Aber gegenüber der großen Last der Gegenbeweise ist dieses Alibi nicht schlagend. ... // Lieber Herr Wagner, so ist heute die Lage. Ich denke, wir sollten deswegen nicht mit den Zähnen knirschen, sondern uns über den – keineswegs leichtfertig gewonnenen – Fortschritt freuen. Hat es doch die Meteoritenhypothese fertiggebracht, daß die Bayern ihre 'örtlichen Aufbrüche' sang- und klanglos begraben haben. Die zentrale Sprengung, das ist ja das, wofür Sie, Löffler, Kranz und schließlich auch wir Jungen, insbesondere Gerold, gekämpft haben, ist allseits anerkannt. Und noch einen Gewinn hat die Meteoritenhypothese gebracht: Das Ries steht jetzt nicht mehr als fast singuläres Ereignis da, sondern reiht sich ein in eine größere Zahl von Kraterstrukturen, in denen man dieselben charakteristischen Mineralumwandlungen feststellen kann. Man ist damit in den letzten Jahren einer bisher unbekannten Form der Gesteins Umwandlung (Stoßwellenmetamorphose) auf die Spur gekommen'. (Letter from Rudolf Hüttner to Georg Wagner, 29 August 1969, UAT 605/343).

Now my [daughter] still has a gifted son of fourteen years; whether he will become a geologist I do not yet know. Alas, many a hereditary farm has become desolate before!

And my best students are at Universities in northern Germany. But my own field of expertise is hardly farmed by anybody.

It pains me especially that the Ries scenario, the vast changing of the southern German landscape, is hardly promoted due to all that focus on meteoritics; that this very fertile field, which is the true geology itself, is not regarded enough. (Letter from Georg Wagner to unknown friends, 27 November 1969, UAT 605/343)[39]

Wagner was in his eighty-fifth year, and when he eventually became less vocal it was by the resignation of an old man who had outlived his own fame, and not by acceptance. Geology had changed and moved on, and had left him behind. It was no longer enough to be an exceptionally gifted field geologist.

The Problem of Visualisation

Georg Wagner was, however, by no means completely off in his assessment that nothing much purely geological or 'Earth-historical' was done by geologists involved with the new Ries research. Reviewing the relevant volumes of the journal *Geologica Bavarica*,[40] it becomes apparent that, apart from very few, albeit notable, exceptions, there had been basically three different approaches to the Ries problem in the 1960s among local Ries researchers. There are many articles by geophysicists who produced mainly descriptive papers with the aim of answering once and for all the question of whether the Ries crater was due to an impact or volcanism (see BGL 1969). A simple *yes* or *no* seemed to suffice. There was no specific attempt to develop an impact scenario from geophysical data. Then there were the mineralogists around Wolf von Engelhardt, who did great

---

[39] 'Nun hat meine [Tochter] noch einen begabten Sohn von 14 Jahren; ob er Geologe wird, weiß ich noch nicht. A[ch] schon mancher Erbhof ist verödet! // Und meine tüchtigsten Schülrt [sic] sind auf norddeutschen Hochsculen [sic]. Aber mein eigenes Arbeitsfeld wird nur noch jaum [sic] beackert. // Besonders schmerzt mich, daß das Riesgeschehen, die gewaltige Veränderung der süddeutschen Landschaft vor lauter Meteoritenkunde kaum gefördert wird, daß dieses sehr fruchtbare Feld, das eigentliche Geologie ist, zu wenig beachtet wird' (letter from Georg Wagner to unknown friends, 27 November 1969, UAT 605/343).

[40] Edited by the Bavarian Geological Survey and containing the proceedings of the Ries meetings related to the scientific drilling programme (see below): the volumes 61, 72, 75 and 76.

work on details of rock and mineral metamorphosis caused by the shock wave (see Chapter 12), deducing general effects of the impact process. Much of their research was published in international journals anyway, and aimed at general scenarios that could be used for other craters in a global and even planetary setting, rather than restricting themselves to the local problem. Finally, there were the classical stratigraphers of the Munich school, producing conservative papers (Schmidt-Kaler 1969a, b; Haunschild 1969; Bolten and Müller 1969; Hollaus 1969)[41] about pre-impact or post-impact stratigraphical details, but seemingly unconcerned by the sudden and alien impact event. So the actual geologists again worked stratigraphically rather than structurally, and tended to neglect the actual impact scenario and what it meant for the Miocene landscape.

The geology chairs in Munich, despite the traditional idea (see Chapter 3) that most of the Ries belonged to Bavaria (not just territorially by also for research purposes), did not show much inclination to embark on geological Ries research. The chair of general and applied geology at the University of Munich, Albert Maucher,[42] was only prepared to let his assistant Wolf-Dieter Grimm[43] investigate 'the problem of Reuter's Blocks', and whether these limestone blocks within the molasse sediments might not be transported by an obscure volcanism within the molasse basin, rather than being Ries ejecta as Reuter had postulated in 1925 (see Chapter 1), albeit from the exploded Ries Volcano as envisioned by Walter Kranz (Maucher in AG Ries 1963: 31).

---

[41]   Hellmut Haunschild (born 1928) received his doctoral degree from the University of Erlangen in 1954. Up to 1955, he worked as a geological helpmate in Erlangen, and then became a geologist at the Bavarian Geological Survey. Hermann Schmidt-Kaler (born 1933) obtained a geological PhD in 1961, also in Erlangen, and then worked for the Federal Geological Survey in Hannover. In 1965, he switched to the Bavarian Geological Survey (Freyberg 1974: 62, 138).

[42]   Albert Maucher (born 1907) obtained a PhD in geological and mineralogical engineering from the Technical University of Munich in 1932. Two years later, he became an assistant at the Institute for Ore Research at the Technical University in Berlin-Charlottenburg. He obtained his *Habilitation* there in 1936. He went to work in Ankara (Turkey) as an ore geologist for one year, afterwards becoming an assistant and eventually lecturer at the University of Göttingen. From 1939 to 1944, he was in military service. In 1944, he was appointed professor without chair at the *Reichsuniversität* Strasbourg, and in 1946 he became Professor of General and Applied Geology at the University of Munich (Freyberg 1974: 101).

[43]   Wolf-Dieter Grimm (born 1928) obtained his doctorate from the University of Munich in 1953. Until 1957, he worked as a geologist for the oil industry, when he returned to university first with a scholarship and then working as an assistant. He obtained his *Habilitation* in 1964, and obtained a staff position at the Institute of Applied Geology at the Free University of Berlin. He was eventually appointed professor at the Institute of General and Applied Geology at the University of Munich (Freyberg 1974: 55).

Maucher's colleague at the Technical University of Munich, Paul Schmidt-Thomé,[44] was likewise not really interested, and declared at the first meeting of the Ries Research Working Group that he did not plan to apply for grant money concerning the Ries. His added comment suggested that he had not given much thought to the issue: 'The age relations to be expected between the forms of bedding of individual structural features ... and the larger structures of the Ries should give clues to the age of the Ries volcanism in relation to the Ries catastrophe. Thus, clues in favour or against individual Ries hypotheses are to be expected' (Schmidt-Thomé in AG Ries 1963: 34).[45] Like Dehm, Schmidt-Thomé obviously still regarded suevite as separate from the actual Ries explosion and formed by volcanism, whatever a postulated impact might or might not have done before suevite emplacement.

Ironically, among the very few people who did work exactly in the direction which Georg Wagner demanded was Edward Chao, whom unfortunately Wagner heartily disliked (and, due to the animosity and the language problem, probably never even spoke to personally). But even Chao, as a member of the USGS/NASA astrogeology group, was interested in the Ries mainly as one well-preserved example of a structure likely to be encountered on the Moon, and thus restricted his geological research to the actual impact event plus subsequent modification of the crater due to gravity and elastic rebounding. He did not investigate what weathering and erosion, altered river courses and the eventual formation of a Ries lake meant for the development of the Miocene landscape.

Rudolf Hüttner presented some geological work at conferences and also held alive the memory of his friend Gerold Wagner's achievement, but without being able to establish a working group of his own (as a member of the Baden-Württemberg Geological Survey he was not employed for such a thing), there was nothing much he could do. He did at least supervise the completion of the Württemberg part of the Geological Map of the Ries 1:50,000 (Gall et al. 1977).

Georg Wagner, however, had always been the prototype of a *Heimat*-geologist, a local expert, who – since he had trained as a teacher – always saw his task not only in geological research, but also in education; and he produced

---

[44]   Paul Schmidt-Thomé (born 1911) obtained his PhD from the University of Bonn and worked as an assistant in Hamburg until 1938, when he became a mining geologist in Spain. In 1941 he returned to Germany as an assistant in Berlin, where he obtained his *Habilitation* in 1943. Between 1941 and 1943 he was in military service. In 1946 he became a geologist at the Bavarian Geological Survey, and from 1953 onwards he was Professor of Geology at the Technical University of Munich (Freyberg 1974: 138).

[45]   'Die zu erwartende Altersbeziehungen zwischen Lagerungsformen einzelner Bauelemente ... und den Großstrukturen des Rieses dürften Anhaltspunkte für die Alterszuordnung des Riesvulkanismus zur Rieskatastrophe liefern. Damit sind die Hinweise für oder gegen einzelne Ries-Hypothesen zu erwarten'. (Paul Schmidt-Thomé in AG Ries 1963: 34).

3D drawings of geological features and landscapes as well as maps for museums, schools and popular books. He had additionally written several geological textbooks for use in schools:

> Out of principle, it must be said that our Tübingen is not just an ordinary university but also a *Landesuniversität*,[46] and our People take an interest in its work and take pleasure in any research concerning our *Heimat*; hence the great interest in the country for geology. And it is also worth promoting this tradition. (Letter from Georg Wagner to the Rector of the University of Tübingen, 19 December 1963, GA 20/031)[47]

From this impetus also stemmed Wagner's well-known textbook *Einführung in die Erd- und Landschaftsgeschichte mit besonderer Berücksichtigung Süddeutschlands* [Introduction to Earth history and history of landscape with special attendance to southern Germany], which he published between 1931 and 1960 in three continually revised editions. The book was not aimed only at students of geology, but also addressed teachers and interested laypersons, and thus sat on many a bookshelf in Württemberg. It contained much of what Wagner saw as his legacy.

Now, however, possibly for the first time in his life, Georg Wagner was at a loss for how he could explain the new ideas about the Ries Crater to others:

> It has been very exhausting and 'learnèd' [at the Annual Meeting of the *Oberrheinischer Geologischer Verein* in Nördlingen]. Most speakers would have required a practical training year as a schoolteacher. [There was] rarely a presentation attractive for ordinary mortals. Worse for me, the seeker [of truth]. In the thick blue book[48] too much knowledge [is] stored. But [it is] hardly usable directly! Everybody is astonished, but nobody dares to pass it [the knowledge] on. ... One is nearly crushed by the amount and intellectual heft of the data, but can hardly translate it into something illustrative. (Letter from Georg Wagner to his colleague Schneider, 9 May 1970, UAT 605/343)[49]

---

[46] Main university of the province (of Baden-Württemberg).

[47] 'Prinzipiell muß gesagt werden, daß unser Tübingen nicht nur gewöhnliche Universität sondern auch Landesuniversität ist, an deren Arbeit auch unser Volk Anteil nimmt, das sich über jedliche [sic] Erforschung unserer Heimat freut; daher auch das große Interesse im Lande für Geologie. Und diese Tradition ist auch der Pflege wert' (letter from Georg Wagner to the Rector of the University of Tübingen, 19 December 1963, GA 20/031).

[48] The thick blue book is *Geologica Bavarica*, volume 61 (1969), which contained the proceedings of a Ries conference featuring the results of a scientific drilling campaign at Wörnitzostheim (see below).

[49] 'Es war sehr anstrengend und 'gelahrt'. Die meisten Redner hätten ein Referendarjahr durchmachen müssen. Selten eine anziehende Darstellung für gewöhnliche Sterbliche. Umso

His was the tragedy of a passionate teacher who in full consciousness no longer felt able to understand or explain a subject to others. The aesthetic problem of visualising an impact increasingly preoccupied Wagner, and he felt frustrated when at conferences and on field trips (even those of the *Oberrheinischer Geologischer Verein*, which attracted many geological laypersons), he could no longer follow the physical and mineralogical arguments:

> In the Ries it was very nice, interesting and exhausting. More than 250 fieldtrip participants. During a blizzard too, and across country! It reached my upper limits. Also, much talk above people's heads.[50] There you could have gone fishing and catch but few things! Pedagogics is very much neglected! What is offered is but rarely connected to ideas of space and time! People like me must concentrate like hell, to get something still understandable! (Letter from Georg Wagner to his colleague Schneider, 9 April 1970, UAT 605/343)[51]

Anschauung, or in a more modern term, *Anschaulichkeit* (visualisation, graphicness), had occupied a prominent role in the pedagogic ideas of the Swiss pedagogue and educational reformer Johann Heinrich Pestalozzi (1746–1827), and thus in Germany had become fundamental for general education and schools. Local nature (the familiar), within the reach of personal experience, was the starting point for pupils within the school subject of *Heimatkunde* (Chapter 7; Nyhart 2009), from whence their views and minds were broadened until a global perspective was reached. The pedagogic plan was thought to reflect the intellectual ripening of the schoolchildren, and thus a thorough understanding of local affairs was mandatory for teachers (Peter Schimkat, pers. comm. 23 September 2011).

Thus, the teacher within Georg Wagner desperately needed a scenario, a story he could relate vividly – not abstract numbers beyond which he could not see – and he enlisted the help of one of his nephews to hunt for good descriptions

---

schlimmer für mich, den Suchenden. Auch im dicken blauen Buch viel Wissen gespeichert. Aber kaum direkt verwendbar! Alles staunt, aber niemand wagt es weiter zu geben. ... Man wird von Fülle und Höhe des Stoffes schier erdrückt, aber kann es kaum ins Anschauliche übersetzen'. (Letter from Georg Wagner to his colleague Schneider, 9 May 1970, UAT 605/343).

[50] That is too many incomprehensible technical terms and complicated matters were discussed, which Wagner expected most of the participants not to understand.

[51] 'Im Ries war es sehr schön, anregend und aufreibend. Über 250 Exkursionsteilnehmer. Auch bei Schneesturm und querbeet! Es ging an meine obere Grenze. Viel auch über die Köpfe weg. Da hätten Sie fischen können und wenig gefangen! Die Pädagogik wird noch recht klein geschrieben! Das Gebotene ist wenig mit Raum- und Zeitvorstellungen verknüpft! Unsereins muß noch höllisch aufpassen, daß man noch etwas Verständliches erwischt!' (Letter from Georg Wagner to his colleague Schneider, 9 April 1970, UAT 605/343).

of a meteorite impact (letter from the nephew to his uncle Georg Wagner, 23 February 1968, UAT 605/343).

The problem, of course, became virulent as soon as Georg Wagner's resistance against the impact as such began to give way. He felt blind, just when his task was to describe a new image. This deeply felt emotion may have lain at the very root of his verbal attacks against mineralogists and their methods.

If only Wagner had talked to Chao! If we follow Hüttner's recollections, he would have been just the person to help Wagner along. Instead, Georg Wagner reacted to Chao's own pedagogical efforts, namely calculating the total amount of coesite present in the Nördlingen church tower, as if it were blasphemy. At least at the subconscious level, it might have added to his anger that here was someone using a different concept of *Anschaulichkeit* that was alien to Georg Wagner, thereby trespassing on his pedagogical grounds.

Thus, Wagner bitterly complained about the missing *Anschaulichkeit* of the impact model, which no teacher would be able to teach, while his fellow geologists were all not pedagogues, a fact that Wagner mourned again and again. The event must have been so enormous that he could not imagine it.

There is, of course, a certain irony in this observation, since the Ries event would have been just as enormous and unimaginable if it had been caused by volcanism instead of by an impact. Wagner, however, did not reflect on this paradox. Apparently, he had never shied from explaining the devastating Ries catastrophe in the past as long as he thought it volcanic in nature, and he had never before worried to such an extent about missing geological data concerning the transformation of the landscape by the event, although this problem was exactly the same within the old as well as the new paradigm.

The Ries Working Group

On 25 March 1963 a working group meeting took place in Munich. It was to be the first general meeting of a major research cooperation comprising a considerable number of individual researchers and small working groups, an interdisciplinary affair so far unheard of in its scope in the entire history of Ries research. The Ries Working Group was not a formal organisation, but a loosely knitted group of people with a common interest (Angenheister in AG Ries 1963: 4–5). This group planned to address the following questions, issues and aims:

> Is the Ries a Tertiary volcanic crater, which was formed by a main explosion and possibly subsequent volcanic activity, or is the Ries a meteorite crater? In this second case as well, a secondary volcanism as effect of the meteorite impact might be conceivable.

The following phenomena must be investigated to solve the Ries problem: Location and surroundings, especially also the farther surroundings of the Ries; the morphology of the Ries; the geological units – which have suffered, so to speak, from the Ries catastrophe – the fractures; the emplacement of the movable slivers; the masses of debris in the Ries basin; the suevites and the crystalline rocks; petrographical peculiarities, especially the high pressure minerals; the geophysical data; and last but not least, all events that took place after the Ries catastrophe.

The aim of our research should be the complete description of cause and consequence of the Ries catastrophe. (Angenheister in AG Ries 1963: 4)[52]

This group had applied for a major research grant from the DFG, and was now invited to defend their proposal. In a series of talks, various interested parties also presented their view of the status quo. The minutes of the meeting were copied and distributed among the participants (AG Ries 1963). The research was eventually supported by the DFG within its special grant focus *Upper Earth Mantle*. This (upon first glance, illogical) assignment had the advantage that more money for seismic purposes was available, and that additional institutions and university institutes became informed of and interested in the project, which then were able to provide the most modern equipment and methods for geological/geophysical exploration as well as petrographic/mineralogical investigation (Vidal 1969: 9).

Gerold Wagner, who had been invited to participate at rather short notice (see above), was not the only one still defending the volcanic view. Richard Dehm too was reluctant to accept the new data:

I remember of course Mr Dehm. He was always very cautious with his statements. He always brought forth all sorts of scruples. He himself mapped for 20, 30 years

---

[52] 'Ist das Ries ein tertiärer Vulkankrater, der durch eine Hauptexplosion und möglicherweise durch eine nachfolgende vulkanische Tätigkeit gebildet worden ist, oder ist das Ries ein Meteoritenkrater? Auch in diesem zweiten Fall könnte an einen sekundären Vulkanismus als Folge des Meteor-Einschlages gedacht werden. // Folgende Phänomene müssen zur Lösung des Riesproblemes untersucht werden: Ort und Umgebung, insbesondere auch die weitere Umgebung des Rieses, die Form des Rieses, die geologischen Einheiten – die gewissermaßen die Leidtragenden der Rieskatastrophe waren – die Frakturen, die Lagerung der beweglichen Schollen, die Trümmermassen im Rieskessel, die Suevite und kristallinen Gesteine, petrographischen Besonderheiten, insbesondere die Hochdruck-Minerale, die geophysikalischen Daten und nicht zuletzt alle Ereignisse, die nach der Rieskatastrophe eingetreten sind. // Das Ziel unserer Untersuchung sollte die vollständige Beschreibung von Ursache und Wirkung der Rieskatastrophe sein' (Angenheister in AG Ries 1963: 4).

in the Ries and had also any number of people, who mapped in the Ries Basin. And thus he always uttered many scruples. (Pohl interview)[53]

Richard Dehm now admitted that there was no truly volcanic rock in the Ries: 'but a pseudo-volcanic [rock existed], the suevite consisting of brecciated and partially molten components of crystalline rocks, containing spindle bombs and lapilli and unusual minerals, among them high-pressure modifications of quartz; it is younger than the Ries breccias' (Dehm in AG Ries 1963: 6).[54]

Like many of his contemporaries (for example Schmidt-Thomé in AG Ries 1963: 34), Dehm still considered the suevite to be younger than the actual explosive Ries event that must have produced the polymictic Ries breccia, because the suevite rested above the breccia. There was still no realisation that suevite simply represented a different part of the explosion plume (see Figure 10.4.) and was thus distinct from the debris-like polymictic breccia below, not by the factor of time, but through a different process within the very same event.

This lack of understanding of suevite even influenced and handicapped people otherwise proactive in the new impact crater research. One of these was Ekkehard Preuss, who as a student had already become acquainted with the idea of large impacts through his PhD supervisor Fritz Heide (see Chapter 6), and so had reacted with open-minded curiosity to Shoemaker's announcement of coesite in Copenhagen and Chao's lecture in Tübingen (Preuss 1969a: 12). In his review talk to the Ries Working Group, Preuss would not definitely preclude that suevite was erupted in several phases from volcanic vents somehow triggered by the impact event (Preuss in AG Ries 1963: 19), although he himself preferred to interpret suevite as being generated by the impact explosion itself (Preuss in AG Ries: 14, 16), and he still considered the Steinheim Basin to be of volcanic origin just like the diatremes of the 'Swabian Volcano'. He speculated whether the earthquake from the impact might not also have caused eruptions at the 'unique formations' of the Swabian diatremes and the Steinheim Basin (Preuss in AG Ries 1963: 18; Preuss 1969b: 397).

Preuss felt reluctant to interpret the Steinheim Basin as an impact crater because fields of two or more craters were thus far known only as small structures, up to a hundred metres in diameter. Larger craters had so far all been only single specimens. For this reason, Preuss saw himself forced to assume two

---

[53] 'Na ja, ich haben natürlich Herrn Dehm in Erinnerung. Er war immer sehr vorsichtig mit seinen Äußerungen. Er hat alle möglichen Bedenken immer angemeldet. Er hat selbst 20, 30 Jahre im Ries kartiert und hat auch jede Menge Leute gehabt, die im Ries kartiert haben. Und er hat also auch immer viele Bedenken geäußert.' (Pohl interview).

[54] '... ein echtes magmatisches Gestein fehlt, aber ein pseudovulkanisches Gestein, der Suevit, bestehend aus zertrümmerten und angeschmolzenen Komponenten kristalliner Gesteine, enthält gedrehte Bomben und Lapilli und führt ungewöhnliche Mineralien, u.a. Hochdruckformen des Quarzes; er ist jünger als die Trümmermassen' (Dehm in AG Ries 1963: 6).

separate impact events, an assumption that he immediately rejected as simply too improbable (Preuss in AG Ries 1963: 10).

The argument for an impact at the Ries Basin that impressed Preuss most was the tremendous amount of energy such an event was able to provide, and which explained the widespread brecciation so much better than any volcanic scenario (Preuss in AG Ries 1963: 10). Like Gifford and Quiring, he tried to visualise the effect through a comparison with chemical explosives:

> One meteorite of 16,000 t ([corresponding to] an iron sphere of 16 m or a stony sphere of 20 m in diameter), and an assumed medium velocity of 22 km/s, also brings with it one million tons of TNT energy (that is 60 times that of the same amount of chemical explosive). This energy is sufficient to melt 70 times the amount of rock or to blast 2,000 times the amount of rock with a velocity of 1000 m/s. These values are basically dependent on the assumed velocity of the meteorite. The rapid triggering of such high energy is an argument for the impact theory; the size of the Ries, therefore, rather supports than rejects it. (Preuss in AG Ries 1963: 10–11)[55]

In February 1962, a second high-pressure modification of SiO_2 was found in rock samples from Meteor Crater and from the Ries Basin. This new mineral was called stishovite[56] and had an even higher density (requiring correspondingly higher formation pressure) than coesite, further strengthening the impact theory (Chao et al. 1962; Chao and Littler 1963; see also Preuss in AG Ries 1963: 10).

It was now realised that the pressure reached by such an event would have been enormous. First estimates suggested that the projectile responsible for the Ries Crater might have only been some 500 to 1,000 m in diameter, less than one twentieth of the crater size, which nevertheless liberated energy equivalent to some 150,000 megatons of TNT (Preuss in AG Ries 1963: 14). 'The effects

---

[55] '1 Meteorit von 16 000 t (Eisenkugel von 16 m Ø oder Steinkugel von 20 m Ø) und einer geschätzten mittleren Meteoritengeschwindigkeit von 22 km/sec bringt ebenfalls 1 Mt TNT Energie mit (also das 60-fache der gleichen Sprengstoffmenge). Diese Energie reicht aus, um etwa das 70-fache an Gestein einzuschmelzen oder das 2 000-fache an Gestein mit 1000 m/sec Geschwindigkeit auszuwerfen. Diese Werte hängen weitgehend von der angenommenen Meteoritengeschwindigkeit ab. Die schnelle Auslösung einer so hohen Energie ist ein Argument für die Einschlagstheorie, die Größe des Rieses spricht deshalb eher dafür als dagegen'. (Preuss in AG Ries 1963: 10–11).

[56] The mineral stishovite had been predicted theoretically, and was synthesised in 1961 by two Russian physicists at the Institute of High-Pressure Physics at the USSR Academy of Sciences, Sergey M. Stishov (born 1937) and Svetlana V. Popova, before it was discovered in impact craters. Its specific gravity is 4.287 g/cm^3 (<http://en.wikipedia.org/wiki/Stishovite>, <http://de.wikipedia.org/wiki/Stishovit> both accessed 27 December 2011; Mark 1995: 122; Stishov 1995).

must have achieved global significance; the masses of dust should be distributed over widest distances. Characteristics might be spherules of silicates or magnetite or finest fragments of crystals. At any rate, one should be prepared to notice such conspicuous layers in drill cores of [Miocene age], which might serve as a marker horizon' (Preuss in AG Ries 1963: 14).[57]

A second, and for Preuss, equally impressive argument was that the interpretation of the Ries Basin as an impact crater was about to solve another geological riddle in Europe: The moldavites (tektites that were to be found some 300 km away in Bohemia, and that Preuss had worked on for his doctoral degree), were now brought 'into a surprising context with the Ries' (Preuss in AG Ries 1963: 16).

In 1961, immediately after Shoemaker and Chao's publication on coesite in the Ries Basin, Alvin J. Cohen (1918–91), an American geochemist from the University of Pittsburgh (Cassidy 1993), had suggested that the moldavites were generated by the Ries impact (Cohen 1961). This idea was now supported by Wolfgang Gentner's working group at the Max Planck Institute for Nuclear Physics in Heidelberg, who had performed chemical analyses of moldavites and potential source material and had executed radiometric age determinations, which pointed to an identical age of moldavites and Ries suevite (Zähringer in AG Ries 1963: 35–6).

As to the relationship between the physical impact mechanism and the crater morphology, Preuss's understanding was still hazy, as he (like probably most of his colleagues) still took the crater morphology as it appeared at present, without taking into consideration the effects of elastic rebound or even simple gravitative reorganisation and settling of the heavily disturbed and brecciated initial crater rim:

> The impacting meteorite has hammered out of the Mesozoic beds of the Keuper [Late Triassic] and the Jurassic a crater of some 20 km diameter, corresponding to its energy. The ejected masses rest dispersed atop the [Swabian] Alb as polymictic breccia and as [fine-grained, monomictic limestone breccia]. From the crystalline basement, which is considerably more rigid, the meteorite was only able to excavate a correspondingly smaller crater of some 8 to 10 km diameter. The debris masses of the crystalline heap themselves around the 'crystalline crater' and form

---

[57] 'Die Auswirkungen müssen weltweiten Umfang erreicht haben, die Staubmassen dürften über weiteste Strecken verteilt sein. Kennzeichen könnten Silikatkügelchen oder Magnetitkügelchen sein oder feinste Kristallinbruchstücke. Jedenfalls sollte man auch in Bohrkernen des [Miocene] auf derartige, auffällige Lagen achten, die einen Leithorizont abgeben könnten' (Preuss in AG Ries 1963: 14).

the 'crystalline dyke' of Reich and Horrix,[58] still within the 'Jurassic crater' (Preuss in AG Ries 1963: 17–18)[59]

The Munich Professor of Petrography Georg Fischer[60] attended the 1963 Ries Working Group meeting, but did not show much interest. Neglecting the work of Richard Löffler, he pointed out that the dissertation of his PhD student, Wilhelm Ackermann (1958; see Chapter 3), had been the only one concerned with the crystalline rocks of the Ries Basin (Fischer in AG Ries 1963: 32). With this, Fischer obviously considered his 'homework' done. Ackermann, according to Fischer, had shown that the various glasses found at the Ries Basin did not represent a uniform magma, but were inhomogeneous and had developed within the rock itself through rising temperatures. Their refraction was distinct from normal magmatic glasses, as Ackermann had shown; possibly due to the fine-crystalline formation of coesite and stishovite (Fischer in AG Ries: 33). Fischer's tone conveyed the message: 'I told you so'.

In contrast, Fischer's colleague from Tübingen, Wolf von Engelhardt, recognised the potential of new petrographic research at the Ries Basin. He already had two PhD students working on the crater: Friedrich Hörz (see above and Chapter 12), who was to investigate the *Flädle*, dark melt fragments in suevite and their minerals, while the other was to research the sedimentary inclusions in suevite (Engelhardt in AG Ries 1963: 28). 'I confidently believe that a thorough mineralogical analysis of suevite will allow the conditions of its formation to be determined, better than was ever possible through the

---

[58]   Wilhelm Horrix (born 1920) obtained a PhD from Munich in 1952, and subsequently worked as a geophysical prospector. In 1958, he started his own engineering office as a geophysicist in Krefeld (Freyberg 1974: 73).

[59]   'Der einschlagende Meteorit hat aus dem mesozoischen Schichten des Keuper und Jura entsprechend seiner Energie einen Krater von ca. 20 km Durchmesser herausgeschlagen. Die ausgeworfenen Massen liegen über die Alb verteilt, als Bunte Bresche und als Grieße. Aus dem kristallinen Grundgebirge, das erheblich fester ist, konnte der Meteorit nur einen entsprechend kleineren Krater von 8 bis 10 km Durchmesser herausschlagen. Die Trümmermassen des Kristallins häufen sich um den 'Kristallinkrater' und bilden noch innerhalb des 'Jurakraters' den 'Kristallinen Wall' von Reich und Horrix' (Preuss in AG Ries 1963: 17–18).

[60]   Georg Fischer (born 1899) served as a soldier from 1917 to 1918 and then studied in Munich, where he also obtained his PhD. He *habilitated* in 1927. From 1928 to 1945 Fischer worked at the Prussian Geological Survey in Berlin. In 1946, he switched to the Office of Water Resources in Munich, and in 1948 he became Professor of Petrography without chair at the University of Munich. He retired in 1967 (Freyberg 1974: 42).

speculations that were the custom so far. In this tuff-like rock all the secrets of the Ries formation lie hidden' (Engelhardt in AG Ries 1963: 30).[61]

Engelhardt at that early time supported the push for a scientific drilling campaign mainly in order to gain access to fresh, unweathered material, but had not yet considered the possibility of investigating the impact-based metamorphic zonation of the crater, which later brought him and several of his PhD students international respect and acclamation (Engelhardt in AG Ries 1963: 30; see Chapter 12).

Drilling the Ries Crater

Among the foremost tasks that the Ries Working Group undertook was a drilling campaign near Wörnitzostheim, some 10 km west of Nördlingen, where in the mid-1950s a strongly negative magnetic anomaly had been detected. Based on the old paradigm, it had been assumed that a reversely magnetised basaltic intrusion at shallow depth was the culprit. Now, however, it was fairly admitted that there were also other possibilities, such as a deficit in normally magnetised matter or reversely magnetised suevite. A borehole of some 120 m depth was deemed sufficient to test the various alternatives (Angenheister in AG Ries 1963: 41).

Following the 1963 colloquium, the Ries Working Group received grant money from the *Deutsche Forschungsgemeinschaft* for a drilling campaign at Wörnitzostheim, which eventually reached some 180 m in depth (Vidal 1969: 9). Although this was far from reaching the depth of a well that had been drilled close to ground zero by the German petroleum company *Deutsche Erdöl AG* (DEA) in winter 1953–54 (see Chapter 8), it proved to be more valuable for research, because it was situated within the crystalline ring and thus encountered only 25 m of lake sediment, which were then followed by 75 m of suevite and 80 m of brecciated basement rock (BGL 1999).

The older, 350 m deep, drilling of the DEA (330 m of lake sediment and 20 m of suevite; BGL 1999) had been encouraged by the petroleum geologist and director of the Federal Geological Survey in Hannover Alfred Bentz (Vidal 1969: 9; Vidal 1974: 7). Scientific interest in the drill cores was, however, negligible. They were taken to Gießen by Rudolf Mosebach, director of the institute for mineralogy and petrology of the University of Gießen, who published on them only briefly, a full ten years later, in an obscure journal. In his paper, the drill

---

[61] 'Ich glaube zuversichtlich, daß es möglich sein wird, auf der Grundlage einer genauen mineralogischen Analyse des Suevit seine Bildungsbedingungen besser festzulegen, als dies im Rahmen der bisher üblichen Spekulationen möglich war. In diesem tuffähnlichen Gestein liegen alle Geheimnisse der Riesentstehung verborgen' (Engelhardt in AG Ries 1963: 30).

cores seemed to be merely a pretext for publishing a review of the contemporary Ries debate, in which he kept careful neutrality by following an 'equal space' policy (Mosebach 1964).

It was Wolf von Engelhardt who in 1965 eventually remembered the existence of these cores, and retrieved them with the help of the DEA. They subsequently became the subject of a PhD dissertation, while Angenheister arranged for geomagnetic measurements. Engelhardt's PhD student, Ulrich Förstner,[62] nicely demonstrated that the suevite blanket underneath the lake sediments represented a single cooling unit, and thus was unlikely to have been formed by several independent volcanic events. Textural peculiarities were now interpreted as reworking and slumping within the upper part of the suevite deposit (Förstner 1967).

The core from the new borehole near Wörnitzostheim was most ardently awaited, because this was the locality where the old volcanists expected to find a plugged up volcanic vent or a shallow intrusion. Still the research agenda was dictated by the basic question of whether the crater was due to volcanic or cosmic forces.

It turned out that there was no trace of the basalt that had been inferred from geomagnetic data. Instead, it was found that the lake sediments were underlain by some 80 m of strongly magnetised suevite, which readily explained the exceptional magnetic anomaly at this locality, followed more deeply by chaotic breccia. '[T]here is now no longer any reason to look for rock bodies other than the existing suevite bed in order to interpret the anomalies' (Hahn 1969: 347).[63]

Georg Wagner noted: 'Near Wörnitzostheim within the Ries, 80 m of suevite have now been drilled, and below it polymictic breccia? And near Steinheim steep Jurassic beds in some 80 m depth. My physical-chemical-astronomical views are no longer sufficient ... I do not yet see a possibility for a reasonable

---

[62]   Ulrich Förstner studied geology and mineralogy at the universities of Tübingen and Vienna from 1960 onwards. He received his doctoral degree in 1967 from the University of Tübingen, where he had been a PhD student under Wolf von Engelhardt. In 1967 he started working as a scientific assistant at the institute for sedimentology at the University of Heidelberg, with a short interruption as a lecturer for mineralogy at the University of Kabul, Afghanistan. In 1971, Förstner obtained his qualification for a professorship in Heidelberg. From 1974 to 1982 he was a lecturer and professor without chair in Heidelberg. From 1982 until his retirement in 2005, he was a professor and head of the department of environmental techniques at the Technical University of Hamburg-Harburg (<http://www.tu-harburg.de/iue/doc/dt/mitarbeiter/foerstner_ulrich/foerstner_lebenslauf_dt.html> accessed 6 October 2012; <http://www.tuhh.de/iue/mitarbeiterinnen/ulrich-foerstner.html> accessed 8 March 2013).

[63]   ' ... es besteht nun kein Grund mehr, zur Deutung der Anomalien nach einem anderen Gesteinskörper als der vorhandenen Suevitlage zu suchen' (Hahn 1969: 347).

explanation' (letter from Georg Wagner to Helmut Hölder, 12 November 1965, GA 3/222).[64]

While the old guard was disappointed, the impactists felt stimulated and confident that the Ries crater was 'just one of the numerous known meteorite craters' (Preuss 1969a: 12).[65] The results of the drilling campaign were published in the 'thick blue book' with which Wagner was so frustrated (BGL 1969).

It had finally been realised that the suevite outcrops, only present in isolated batches (hitherto interpreted as representing individual volcanic centres) were erosional remnants of a once-continuous ejecta blanket (Hüttner 1969: 190), and that this unusual rock also filled the whole Ries Basin to a thickness of some 50 to 500 m! This basin infilling showed inverse remnant magnetism, since it had been formed at a time when the Earth's magnetic field was inverse to its present orientation. Thus, it was the source of the negative magnetic anomalies at the Ries Basin (Pohl and Angenheister 1969):

> Since the suevites derived from the crystalline rocks, new ferromagnetic ore phases must have been formed by the Ries event. It has been shown that magnetite[66] is the main carrier of the magnetization. This magnetite partly can only have been formed in an oxidising atmosphere at temperatures above 1,400°C. (Pohl and Angenheister 1969: 335)[67]

Such an enormous temperature, considerably higher than ordinary volcanic temperatures, precluded the interpretation of suevite as a volcanic rock.

Gravimetric measurements conducted by scientists at the Institute of Geophysics in Kiel during the years 1962, 1965 and 1966 confirmed that the Ries Basin was not a tectonic horst as the old Branca and Fraas theory stated (or the only slightly modified model the Munich school had proclaimed), but that an enormous amount of rock had been expelled by the Ries explosion, leading to a mass deficit of some 70,000 million tons (Jung et al. 1969: 341).

---

[64] 'Im Ries hat man jetzt bei Wörnitzostheim 80m Suevit erbohrt, darunter Bunte Breche [sic]? Und bei Steinheim steil gestellten Jura in etwa 80m. Meine physikalisch-chemisch-astronomischen Vorstellungen reichen nicht mehr aus. ... ich sehe noch keine Möglichkeit einer vernünftigen Klärung' (letter from Georg Wagner to Helmut Hölder, 12 November 1965, GA 3/222).

[65] 'Heute ist das Ries nur einer der zahlreichen bekannten Meteoritenkrater' (Preuss 1969a: 12).

[66] The mineral magnetite is an iron oxide Fe_3O_4.

[67] 'Da die Suevite aus den kristallinen Gesteinen entstanden sind, müssen beim Riesereignis neue ferrimagnetische Erzphasen gebildet worden sein. Es zeigt sich, daß Magnetit der Hauptträger der Magnetisierung ist. Dieser Magnetit kann z. T. in oxydierender Atmosphäre nur bei Temperaturen über 1400°C gebildet worden sein' (Pohl and Angenheister 1969: 335).

The crater underneath the lake sediment filling, however, was not a simple bowl shaped depression. Seismic investigations[68] revealed a considerable topography beneath the base of the suevite (Angenheister and Pohl 1969: 324) that would later be interpreted as a ring structure developed during the elastic rebound.

Engelhardt and his students reported a new world of deformation effects on minerals, as well as newly formed minerals produced by shock wave metamorphosis (Engelhardt et al. 1969; see Chapter 12).

Scientists from the Max Planck Institute for Nuclear Physics in Heidelberg performed age determinations (K/Ar as well as fission track dates) on fresh suevite samples from the drill core and confirmed an identical age for the moldavites (the Bohemian tektites) and the impact glasses of the Ries Crater. They also showed that a bentonite horizon within the beds of the Upper Freshwater Molasse was also of the same age. They consequently interpreted this thin, glass-bearing clay layer no longer as volcanic ash, but as distal Ries fallout in the form of a rain of micro-tektites formed by condensation from the plume of vaporised rock (Gentner and Wagner 1969).

Rudolf Hüttner summarised the geological evidence from his own PhD mapping project, from his late friend Gerold Wagner's field notebooks and from their joint field campaign at *Altenbürg* quarry and elsewhere in the Ries. He thoroughly described and classified the rock types generated during the Ries event according to macroscopic criteria. He explained how Gerold Wagner's efforts to obtain statistical tectonic data within the polymictic breccia had been thwarted by the completely chaotic rock fractures and joints. Hüttner then nicely illustrated this chaos by drawing representative fractures in rock cubes (Figure 11.3). This intense brecciation and (in places) steep contact between the breccia and suevite suggested to Hüttner that suevite emplacement must have immediately followed the breccia. Closer inspection of dyke-shaped occurrences of breccia led to the conclusion that they were not dyke-shaped at all, but rather irregular shapes, somehow squeezed in and kneaded like dough, as Hüttner put it. Based on a revision of the suevite at the quarry of Aumühle (which according to Ahrens and Bentz had been deposited not in a single event but in several stages: see Chapter 3), it was now interpreted as having acquired its bedding through secondary processes, such as redistribution by water action, by slumping and by compaction of still-unconsolidated suevite. Rudolf Hüttner also published the first evidence of shatter cones in the Ries Basin (Hüttner 1969).

---

[68]   Refraction seismic profiles commissioned by Hermann Reich and Alfred Bentz in 1948, 1949 and 1952 (six radial from the Ries centre and one tangential): two refraction seismic profiles crossing the Ries by Gustav Angenheister and Jean Pohl from the Institute of Applied Geophysics in Munich in 1967; and an additional reflexion and refraction seismic survey in 1968, not yet fully evaluated when the paper was published (Angenheister and Pohl 1969).

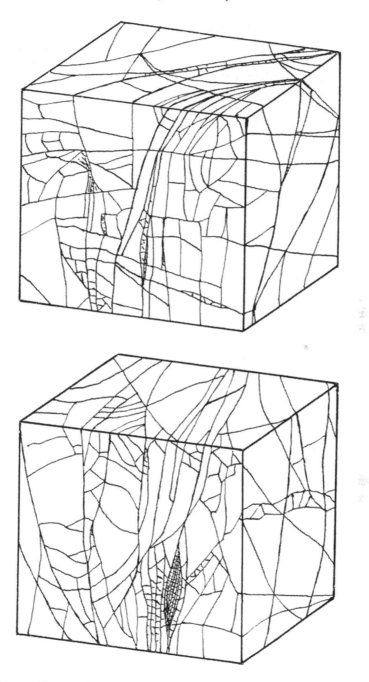

Figure 11.3   Chaotic fractures in Ries rock drawn by Rudolf Hüttner (side length of the cubes: 8 cm; from Hüttner 1969: 147, Figure 8).

Erwin David, a physicist (see Chapter 12), explained what was known about the physical parameters of an impact and thus provided educated guesses about processes during the Ries event. He impressed on his readers the enormous, sudden energy release, leading to a shock wave moving with hypersonic velocity. The energy injected at point zero, David explained, was several times the amount necessary to vaporise the meteoritic projectile as well as the target rock. Consequently, no meteorites were to be expected – only minerals incorporating atoms from the rock vapour bore witness to the impact explosion. Because the amplitude of the shock wave penetrating the ground diminished rapidly with distance, various zones of differing shock wave effects were to be expected. Any lack of initial symmetry due to inclined impact, irregular shape of the projectile or target morphology, was already lost within the zone of vaporisation. Thus, concentric zones of impact metamorphism would be generated. When the shock wave had passed, David explained, the compressed country rock would rebound to a certain extent, reorganising the crater and forming, for example, the central mountain of the Steinheim Basin. Suevite was the product of an incandescent explosion plume being hurled into the sky some 20 km high, which meant some two to three minutes of time before it fell back onto the polymictic breccia that had been emplaced along the ground. Moldavites had been generated by shock-melting of the surface right at the centre of the impact, in an area some 100 m in diameter. The beds in which moldavites were found also contained abundant charcoal, a testimony to the extensive wildfires that were ignited by the rain of hot melt fragments (David 1969).

While for many researchers the new paradigm proved to be stimulating, and many of the bits and oddities of the great Ries puzzle finally clicked into place (revealing not only the solution to an old question, but also an intriguing picture of new and fruitful lines of investigations), there remained some who conservatively clung to what they thought they knew.

A common and perhaps understandable reaction among the more conservative researchers had been to downplay the geological significance of the few disturbing seconds of the Ries event in comparison to the calm millions of years of evolution of the Tertiary Molasse Basin, and to continue to discuss its stratigraphy without paying attention to the double impact of the Ries Crater and Steinheim Basin (see above).

However, the geologist Herold (from the Bavarian Geological Survey), in the 'thick blue book', continued to plough his conservative furrow in the face of the evidence, preferring to evoke hypothetical agents and processes, being content with them, if only they were of terrestrial origin. At Niedertrennbach, some 30 km east of the town of Landshut, blocks and slivers of fragmented Upper Jurassic limestone were found ballistically embedded in fluviatile marly sands of the Upper Freshwater Molasse, at the same stratigraphic level as the bentonite layers[69] of the Molasse Basin (Herold 1969: 413).

According to Herold, the limestone blocks either came – like Reuter's Blocks – from the Ries Basin some 150 km away, or 'some hundred metres up to a few kilometres from the locality, where they were found, an explosively active volcano penetrated the Mesozoic and Tertiary sedimentary cover and produced this Jurassic limestone from some 500 m depth' (Herold 1969: 424).[70]

For unspecified petrographic reasons and because of the contemporaneity of the bentonites, which Herold interpreted as volcanic, he considered it far less probable that the limestone blocks were derived from the Ries, and supported their origin from local volcanism (Herold 1969: 424–5):

> The point of eruption probably lies far from the point of impact [of the Jurassic limestone blocks] at Niedertrennbach, because on the map sheet of Frontenhausen, despite good outcrop conditions, no other xenolith ejecta and no volcanic tuffs were detected in the blanket of marly sands. Neither investigations in the area surrounding the place of discovery nor geological mapping and cursory inspection has offered any additional clues as to the location of the Upper Miocene volcano. ... It must now be left to chance whether one day, perhaps in the course of larger earthworks in the Tertiary area of Lower Bavaria, diatremes, calderas or a crater wall will be excavated or cut and recognized as such. (Herold 1969: 425)[71]

Research Drilling at Nördlingen

Papers like the one by Herold had a political value, if not a geological one. They allowed Helmut Vidal, President of the Bavarian Geological Survey, to argue that, although the last decade had provided a wealth of information from

---

[69] In the same volume, one of these bentonite layers was interpreted as distal fallout from the Ries event (Gentner and Wagner 1969; see above).

[70] 'oder ein explosiv tätiger Vulkan hat einige hundert Meter bis einige Kilometer von der heutigen Fundstelle entfernt das mesozoische und tertiäre Deckgebirge durchbrochen und diesen Jurakalk aus etwa 500m Tiefe zutagegefördert' (Herold 1969: 424).

[71] 'Die Eruptionsstelle liegt sehr wahrscheinlich weit vom Einschlagspunkt bei Niedertrennbach entfernt, denn auf Blatt Frontenhausen ließen sich in der Sandmergeldecke trotz guter Aufschlußverhältnisse keine weiteren Nebengesteinsauswürflinge und auch keine vulkanischen Tuffe nachweisen. Weder aus Untersuchungen in der Umgebung der Fundstelle noch bei geologischen Kartierungen und Übersichtsbegehungen ergaben sich irgendwelche zusätzlichen Anhaltspunkte über die Lage des obermiozänen Vulkans. ... Es muß nun dem Zufall überlassen bleiben, ob eines Tages vielleicht bei größeren Erdbewegungsarbeiten im Tertiärgebiet Niederbayerns Durchschlagsröhren, Calderen oder Kraterwälle freigelegt oder angeschnitten werden und als solche erkannt werden' (Herold 1969: 425).

various sub-disciplines of the geosciences concerning the Ries Crater, it was still not enough to solve the 'Ries problem' once and for all, even though the impact theory was now considered the most probable. Further investigations were urgent, and documenting the deeper structures of the crater was mandatory. Deep-drilling reaching two to three thousand metres depth was what Vidal considered desirable (Vidal 1969: 10–11).

Helmut Vidal (1919–2002) had studied geology at the University of Munich, where he also obtained his doctoral degree. He then began work at the agricultural *Bayerische Landesanstalt für Moorwirtschaft und Landkultur* and became its director in 1963. In 1966 he became president of the Bavarian Geological Survey, a position that he held until his retirement in 1984 (Schmid 2003).

By 1973 Vidal had raised the money required for another drilling campaign near Nördlingen. Half a million German Marks were provided by the Bavarian State, another 250,000 DM came courtesy of the joint effort of all West German Geological Surveys, and the DFG had provided 100,000 DM. Drilling began on 29 June 1973 and lasted until 15 January the following year. The hole reached 1,206 m in depth, which the newspapers for some reason hailed as 'deeper than planned' (*Neue Züricher Zeitung*, undated; Steinert 1973, 1974: all newspaper clippings from JME: Off-print collection).

The drilling project recovered a near-complete drill core, consisting of some 325 m of lake sediments – sandstones and clays – deposited after the Ries event; then some 281 m of suevite, which represented the fall-back of the explosion plume; followed by another 600 m of strongly fragmented and shocked crystalline rocks, mingled with fine-grained suevite in fissures and clefts. 'A decrease in the fragmentation with depth, or even a hint of a continuum toward 'undisturbed' crystalline basement up to the final depth of the borehole, is not to be detected' (Gudden 1974: 11).[72]

Results were published again in a special volume of the *Geologica Bavarica* (BGL 1977), which this time had a considerably more international touch compared to the 'thick blue book' following the Wörnitzostheim drilling campaign (BGL 1969). The proceedings of the research drilling at Nördlingen contained thirty-four papers, seven of which were in English (mostly summaries of the individual sections, but also a few original papers by English-speaking authors). There was a brief introductory chapter, followed by twelve articles about the post-impact lake sediments, nine on petrography and shock metamorphism, seven geophysical papers, two not easily assignable to one of the other categories and three papers dealing with impact processes.

---

[72] 'Eine Abnahme der Zertrümmerung nach unten oder gar die Andeutung eines Überganges in 'unbeanspruchtes' kristallines Grundgebirge ist bis zur Bohrloch-Endteufe nicht zu erkennen' (Gudden 1974: 11).

One of the most significant results was that the complexity of the crater had previously been underrated. The drilling led to the acknowledgment that an enormous reorganisation had accompanied and/or followed the excavating stage of the impact explosion. There was no bowl-shaped cavity that was filled in by fallback breccia, suevite and sediments. Such a cavity, if it ever truly existed, was only a brief and transient phenomenon that was rapidly reorganised by elastic rebounding of the compressed basement and by gravitative slumping of the heavily fragmented and disrupted country rock. The most important tool for assessing such movements was an understanding of shock metamorphism, which had been acquired by various researchers including Wolf von Engelhardt and his students (Chapter 12). This knowledge (about which type of deformation, and which new minerals, required which pressure range in order to form) allowed shock metamorphism to be used as an impact barometer (Engelhardt and Graup 1977: 267):

> The observed shock effects indicate that the rocks in the upper range of the profile were affected by dynamic pressures of at least 100 kbar, but less than 150 kbar. These values are not in accordance with the pressure of 40 kbar at the bore hole site (3.8 km from the center) derived from model calculations by David (this vol.). On the other hand, the rocks at the bottom of the hole (1,206 m) sustained shock pressures that were not much greater than approximately 10 kbar. The resulting pressure gradient for the whole profile is much greater than expected for the pressure decay in a propagating shock wave.
>
> We therefore assume that the crystalline rock series, as exposed in the drill hole, is not autochthonous but composed of several disconnected slices. We suppose that the rocks that now comprise the lower range of the profile (below about 670 m) slid down from a region close to the original crater rim and underran more greatly shocked rocks coming from an original position closer to the center. (Engelhardt and Graup 1977: 257)

This mechanism of crater reorganisation, of course, also explained what had struck Alfred Bentz as incompatible with a central explosion (Bentz 1929; see Chapter 3), namely the downfaulted and slumped blocks and slivers along the periphery of the Ries Basin.

As the physicist Erwin David explained (David 1977b), the pressure wave penetrated the ground in the shape of a half-sphere:

> From the shock pressure experienced by the rocks in the drill cores, established mineralogically, it can be determined how far from the centre the respective rock originally resulted. [Rocks at] 600 m depth [were] displaced about 1 km towards the periphery of the Ries; [rocks at] 700 m depth [were] displaced 0 km; [rocks

at] 1,200 m depth [were] displaced towards the centre by about 1 to 2 km. This fits nicely into the general picture of the impact process: in the upper layers, ejecting movement occurred; below, centripetal movement occurred due to shock wave rebound, refilling the centre.

If we instead assume that, following one's first impression, the rock of the drill core was nearly undisplaced, then the reconstruction of the shock wave leads inescapably to absurd consequences. (David 1977b: 459)[73]

David stressed the contemporaneity of ejection and rebound. In his opinion, there had never been a bowl-shaped transient cavity comparable to Meteor Crater, and thus it was false to name the area within the crystalline ring an 'inner crater'. The modern rim of the Ries on the other hand was 'the limit up to which the basement rock had been displaced along fracture planes', and thus the limit of the centripetal movement of the surrounding rocks due to rebound (David1977b: 469).[74]

Stöffler (1977) based his cratering model of the Ries impact structure on field evidence as well as on the physical constraints delineated by David. According to him, melt was produced from the meteorite as well as from the target rock during the penetration phase, while at the same time highly shocked material was expelled by lateral escape. The polymictic breccia was produced by more or less horizontal transport along the Earth's surface, while the suevite was emplaced ballistically. Stöffler spoke of a 'transient crater', probably not to contradict David, but for clarity's sake (more neatly separating the various stages of the impact process in the mind of the reader), as he too stressed the 'collapsing of the transient cavity by immediate rebounding and by slumping of the rim zone' (Stöffler 1977: 456).

Hüttner (1977) demonstrated that various phases of the impact process were indeed macroscopically discernible from the deformation of the rock and thus also accessible to the traditional field geologist. He distinguished four phases

---

[73] 'Aus den mineralogisch ermittelten Stoßdrucken des Gesteins der Bohrkerne ist deshalb zu entnehmen, aus welcher Ursprungsentfernung vom Zentrum das betreffende Gestein stammt. Teufe 600m: riesauswärts um etwa 1km verschoben; 700m: Verschiebung 0; 1200m: Verschiebung rieseinwärts um etwa 1 bis 2km. Das fügt sich gut in das allgemeine Bild des Einschlagsvorgangs ein: Obere Schichten Auswurfbewegung, darunter von der Stoßwellenrückfederung ausgehende Einwärtsströmung, die das Zentrum wieder auffüllt. // Nimmt man dagegen entsprechend dem ersten Eindruck an, das Gestein der Bohrung wäre nahezu unverschoben, so führt die Rekonstruktion der Stoßwelle unvermeidlich auf absurde Konsequenzen' (David 1977b: 459).

[74] ' ... die Grenze, bis zu der das Gestein des Untergrundes entlang Bruchflächen verschoben wurde' (David1977b: 469).

of the impact: first, the formation of shatter cones[75] marked the passing of the shock wave. Next, dyke-shaped fillings of impact breccia were caused by the reactivation of old fissures and fault zones during rebound. Then, small-scale, irregular joints (virtually shattering the rocks) were witness to the fall-back after the rebound had raised the country rock higher than its present height. Finally, fine, more-or-less horizontal scratch-marks were attributable to centripetal mass movements.

Crater reorganisation by rebound was, however, not commonly accepted. Edward Chao, of all people, saw no necessity for large-scale secondary modifications of an initially bowl-shaped transient crater. He specifically argued against general rebound as well as down-faulting and consequent subsidence of the rim, stressing instead the importance of horizontal mass movements. For Chao, rebounding was a superfluous crutch to save an untenable model of a bowl-shaped initial crater (Chao 1977b).

Although Stöffler's and Chao's models seem to contradict each other markedly in this and several other respects, it may well be that they both represent valid visualisations of the same physical modelling as expressed in David's essay (David 1977b): Stöffler stressed the analytical approach of mentally dividing the impact process up into a series of distinct model stages, whereas Chao holistically saw the impact as a continuous process, forming the present Ries Basin 'in one go', so to speak.

Measurements of thermal conductivity of the drill cores were used to calculate cooling time of the hotly emplaced suevite ejecta blanket, thus also providing a time frame for the reclamation of the area by plants and animals. It was found that cooling for a 200 m thick suevite bed from 600°C to 100°C took about two thousand years (Pohl et al. 1977: 327).

Ahmed El Goresy and Edward Chao presented clear evidence for the presence of meteoritic material. Ahmed El Goresy studied mineralogy and petrology at the University of Heliopolis in Cairo and at the University of Heidelberg, where he obtained his PhD degree in 1961 with a thesis on petrology of ore minerals in igneous rocks. From 1961 to 1963 he worked as a research mineralogist for the Egyptian Geological Survey. In 1963 he returned to Germany as a postdoctoral researcher at the Max Planck Institute for Nuclear Physics in Heidelberg. In 1967 he began a postdoctoral position in the Geophysical Laboratory of the Carnegie Institution in Washington D.C. In 1969 El Goresy returned as senior scientist to the Heidelberg Max Planck Institute. After he qualified for a professorship in 1970, El Goresy also became Professor of Petrology at the University of Heidelberg. He retired in 1998, but kept working scientifically as honorary

---

[75] See David 1977a. There was now also ample evidence for shatter cones in the Ries, because they were found on fine-grained amphibolite fragments from the drill core (Gudden 1974: 19).

professor at the Max Planck Institute for Chemistry in Mainz until 2005, and afterwards as a guest scientist at the Bavarian Geoinstitute in Bayreuth.[76]

El Goresy and Chao had found ultra-fine veinlets in crushed and heavily brecciated rocks, which represented the highest surviving country rock just below the deepest level of penetration. The veinlets were infilled with iron, chromium and nickel containing high contents of osmium and iridium. Clearly, these opaque mineral phases were all that remained from the meteoritic projectile. They represented metallic melt from the meteorite itself, which was forcefully injected into the finest joints and cracks in minerals when the projectile penetrated the target rock (El Goresy and Chao 1977; Chao and El Goresy 1977):

> Not surprising, but satisfying as the last keystone in the chain of proofs for the non-volcanic origin of these crater types …, is the total absence of all volcanic phenomena. Any remaining doubters who might have hoped to find basaltic dykes during deep drilling might now be converted. The traces of the unequivocally mechanical brecciation are too evident – even if we did not have the mineral-optical and physical evidence for highest shock wave pressure up to the deepest level [of the drill hole]. (Steinert 1974)[77]

The Steinheim Basin

> Two craters, adjacent to each other and showing no noticeable age difference, might be caused by a double meteorite. No clues contradict it, although among the known 1,500 small planets so far no double planetoid has been observed … therefore it is of great interest if the Steinheim Basin is also meteoritic, because we then must expect the existence of double meteorites, and these might also be observable elsewhere.

---

[76] <http://www.uni-protokolle.de/nachrichten/id/22774/> accessed 12 June 2012; <http://www.bgi.uni-bayreuth.de/?page=4&mode=s&id=128&lng=en> accessed 12 June 2012.

[77] 'Nicht überraschend, doch befriedigend als letzten Schlußstein in der Beweiskette für die nichtvulkanische Entstehung dieser Kratertypen … ist das völlige Fehlen aller vulkanischen Phänomene. Letzte Zweifler, die vielleicht bei der Tiefbohrung Basaltgänge zu finden hofften, dürften bekehrt sein. Die Spuren der eindeutig mechanischen Zertrümmerung sind zu klar – auch ohne daß bis in die Tiefe hinein die mineraloptischen und -physikalischen Nachweise für Stoßwellenhochdrücke vorliegen' (Steinert 1974).

However, we may not conclude, based on this reflection alone, that the Steinheim Basin must be a meteorite crater like the Ries. The presentation of evidence must be made separately. (Preuss 1969b: 392)[78]

While from 1963 onwards the Ries Crater became an object of 'Big Science', its smaller companion, the Steinheim Basin, remained somewhat neglected. 'The Steinheim Basin ... was not part of the investigations, because a rock comparable to suevite is missing' (Groschopf and Reiff 1971: 521).[79]

The Steinheim Basin was still mainly seen as an unusual volcano, although it lacked even the smallest trace of volcanic rocks. Preuss, an unconditional supporter of the new paradigm regarding the Ries Basin, initially assumed the origin of the Steinheim Basin to be volcanic 'not only because of its conspicuous relation to a widespread, positive magnetic anomaly, but also because of its inner, central rise, which according to all experience has so far not been observed at meteorite craters' (Preuss 1964: 289).[80] Despite the observance of central rises in explosion craters (correctly interpreted by Ives in 1919 as a form of rebound (see Chapter 2) and published in German by Trusheim in 1940; see Chapter 4), Preuss consequently interpreted the Steinheim Basin and the diatremes of the Swabian Volcano as secondary volcanism, somehow caused by the impact at the Ries Basin (Preuss 1964).

Its location in Baden-Württemberg still seemed to force a separate treatment of the Steinheim Basin, different from the mainly Bavarian Ries Crater. When the geologists Paul Groschopf[81] (1909–2000: from the Baden-Württemberg

---

[78] 'Zwei Krater, die benachbart sind und keinen erkennbaren Altersunterschied zeigen, könnten von einem Doppelmeteoriten herrühren. Es spricht nichts dagegen, trotzdem unter den bekannten 1500 Kleinen Planeten bisher kein Doppelplanetoid beobachtet ist ... Es ist deshalb von großem Interesse wenn auch das Steinheimer Becken meteoritisch ist, da dann mit der Existenz von Doppelmeteoriten gerechnet werden muß und diese auch anderswo zu beobachten sein könnten. // Aus dieser Überlegung kann aber nicht geschlossen werden, daß das Steinheimer Becken ein Meteoritenkrater wie das Ries sein muß. Die Beweisführung muß getrennt erfolgen' (Preuss 1969b: 392).

[79] 'Das Steinheimer Becken ... wurde nicht in diese Untersuchungen einbezogen, weil dort ein dem Suevit vergleichbares Gestein fehlt' (Groschopf and Reiff 1971: 521).

[80] ' ... nicht nur wegen seiner auffälligen Lage zu einer ausgedehnten positiven magnetischen Anomalie, sondern auch wegen seiner inneren zentralen Erhebung, die nach allen Erfahrungen bisher an Meteoritenkratern nicht beobachtet wurden' (Preuss 1964: 289).

[81] Paul Groschopf obtained his PhD in 1935 from the University of Kiel, and afterwards worked successively for the German Society of Rock Mechanics in Berlin, as an assistant at the Institute for Oceanography of the University of Kiel, and at the German Geological Survey. During World War II, he was a soldier. In 1946 he became a geologist at the Baden-Württemberg Geological Survey, first in Stuttgart and later in Freiburg (Freyberg 1974: 55;

Geological Survey in Freiburg) and Winfried Reiff[82] (born 1930: the former director of the Baden-Württemberg Geological Survey in Stuttgart) published some of their results in the 'thick blue book', as Georg Wagner used to call it, they felt it necessary to apologise for presenting a paper on the Steinheim Basin in a volume dedicated to Ries research (Groschopf and Reiff 1969).

At the Steinheim Basin, low budget research prevailed. Hans Berckhemer (born 1926), geophysicist at the University of Stuttgart, undertook gravimetric investigations and detected a negative gravity anomaly within the Steinheim Basin, due to the mass deficit and the reduced rock density caused by brecciation of the underground (Berckhemer in AG Ries 1963: 38). Parallel to the activities of the Ries Working Group, Groschopf and Reiff began independent investigations at the Steinheim Basin, whenever their time permitted (Reif 1976; Adam 1980: 20).

It took them more than a decade to document all existing outcrops and study them for a new geological map of the Steinheim Basin. From 1964 onwards the Baden-Württemberg Geological Survey provided money for a drilling campaign. Twenty-two mostly shallow bore-holes were drilled and the tailings investigated. Cores were not recovered. For financial reasons, a depth of up to 150 m was reached only rarely. Consequently, research was mainly restricted to the crater filling. The drillings were supplemented by a geophysical survey in collaboration with the Niedersachsen Geological Survey, and directed by the physicist G. Hildebrand (Groschopf and Reiff 1969: 405).

Below the sediments of the crater lake, a hitherto-unknown rock was found. It was a breccia of smaller and larger rock fragments, set in a matrix of grey, finely crushed marl. The polymictic breccia contained various limestones, marls, clays and sandstones from the Upper- and Mid-Jurassic beds disturbed by the crater. The sandstones were investigated by Engelhardt, who found no coesite but ample planar deformation features in the quartz grains (Engelhardt et al. 1967b). 'They have only been described from confirmed and assumed meteorite craters, but may also be generated artificially by shock waves. To the present stage of our knowledge, they are therefore regarded as proof for the action of a meteorite' (Groschopf and Reiff 1969: 407).[83]

<http://www.steinheimer-becken.de/steinheimer_becken_vips.html> accessed 16 January 2012).

[82]   <http://www.steinheimer-becken.de/steinheimer_becken_vips.html> accessed 16 January 2012.

[83]   'Sie sind nur von gesicherten und vermuteten Meteoritenkratern beschrieben, können jedoch auch experimentell durch Stoßwellen erzeugt werden. Sie werden deshalb nach dem heutigen Stand unserer Kenntnis als Beweis für die Einwirkung eines Meteoriten angesehen' (Groschopf and Reiff 1969: 407).

Groschopf and Reiff also reported that the aragonite-bearing limestone, which had formerly been interpreted as a precipitate of thermal springs, had meanwhile been recognised as the product of algae:

> To summarize, it may be said that thanks to recent investigations and research work, a number of arguments that were formerly set up for a volcanic origin may now be interpreted differently. ... Proofs or concrete clues for a volcanic origin of the Steinheim Basin have not been obtained. No final proof for the meteorite theory has so far been obtained either; however, regarding our present state of knowledge, all clues speak for it. (Groschopf and Reiff 1969: 409, 411)[84]

Despite seismic investigations, the extent of the damage inflicted by the impacting projectile was still underestimated in the late 1960s, and so 'final proof' was expected to come from a new borehole some 500 m deep:

> If we summarize and weigh the various arguments against each other, the meteorite theory is to be preferred, although we too lack final proofs.[85] We were able to provide some new facts towards the solution of the Steinheim riddle, but the final solution is still missing. The path to be undertaken, however, can be discerned clearly. The main problem might be solved by a drilling up to 500 m depth, as has already been demanded by earlier geologists. At this depth the *Lias* [Lower Jurassic] must already be drilled through. Should this really be present undisturbed, then a volcanic origin is precluded. On the other hand a meteorite impact would be questionable, if at this depth rocks from farther down are encountered, such as displaced *Keuper* [Upper Triassic] or even [volcanically] altered rocks. (Groschopf and Reiff 1966: 167)[86]

---

[84] 'Zusammenfassend darf man sagen, daß durch neuere Untersuchungen und Arbeiten eine Reihe der früher für vulkanische Entstehung angeführten Argumente jetzt anders gedeutet werden können'. (Groschopf and Reiff 1969: 409). 'Beweise oder konkrete Anhaltspunkte für eine vulkanische Entstehung des Steinheimer Becken haben sich nicht ergeben. Zwar läßt sich auch für die Meteoriten-Theorie noch kein schlüssiger Beweis erbringen, doch sprechen – nach unserem heutigen Kenntnisstand – alle Anzeichen dafür'. (Groschopf and Reiff 1969: 411).

[85] Throughout the Ries literature the term *Beweis* (proof) is often used. Of course, this does not mean a mathematical proof, but the term is rather used in a colloquial way, meaning a clue or an argument that an author, for more or less well-defined reasons, finds convincing.

[86] 'Wägen wir die verschiedenen Argumente zusammenfassend gegeneinander ab, so ist der Meteoritentheorie der Vorzug zu geben, obwohl letzte Beweise auch uns bislang noch fehlen. Zur Lösung der Rätsel um Steinheim konnten wir zwar einige neue Erkenntnisse beisteuern, doch steht die endgültige Lösung noch aus. Der einzuschlagende Weg zeichnet sich aber klar ab. Durch eine Bohrung bis in rund 500m Tiefe, wie sie auch schon von früheren Geologen gefordert wurde, könnten die Hauptprobleme gelöst werden. In dieser

However, since the shock wave penetrates the country rock in roughly the shape of a half-sphere, at least the radius of the crater is needed before anything remotely resembling undisturbed basement rock may reasonably be expected. In the case of the Steinheim Basin, this would be at least 1.7 km in depth.

In 1970, two boreholes were drilled in the centre of the Steinheim Basin, one reaching 603 m and the other 353 m (Groschopf and Reiff 1971: 524). And even the deepest rocks were far from undisturbed because – speaking with hindsight – the drilling was still much too shallow. Nevertheless, the authors disposed of their previous prediction without further ado, and instead still supported the impactist view, because the drilling result proved to be much more complex than initially assumed:

> According to our initial opinion, we should have reached at about 500 m depth the lower limit of the impact-disturbed beds of the *Lias*. In fact, *Lias* mixed with *Dogger* [Mid-Jurassic] already turned up close to the surface (some 40 m below ground level) and between 135 to 335 m in large, nearly vertical slivers. Below followed (to a depth of 406 m) marl and especially sandstone of the Mid-*Keuper* beds, which – like the beds of *Dogger* and *Lias* above – rest more than 200 m higher than expected according to normal stratigraphy. Surprisingly, below the *Keuper* the drilling encountered a dark-grey, uniform, vertical series of beds, the stratigraphic setting of which provided considerable difficulties. According to the sparse occurrence of microfossils, it is now certain that underneath the *Keuper Dogger* and *Lias* again occur, but this time some 70 m deeper, as would be consistent with the undisturbed stratigraphy. Regrettably, the maximum depth was prompted by the performance of the drilling equipment. From geophysical measurements, we know however that the rocks resting below are still strongly disturbed until 1000 to 1100 m depth. (Groschopf and Reiff 1971: 524)[87]

Tiefe muss der Lias bereits durchteuft sein. Ist dieser tatsächlich ungestört vorhanden, dann wird eine Entstehung durch Vulkanismus ausgeschlossen. Ein Meteoriteneinschlag ist andererseits unwahrscheinlich, wenn in dieser Tiefe Gesteine aus tieferen Schichten, wie verlagerter Keuper oder gar veränderte Gesteine angetroffen werden' (Groschopf and Reiff 1966: 167).

[87] 'Nach unseren ursprünglichen Ansichten hätten wir in rund 500 m Tiefe die Untergrenze der durch den Einschlag gestörten Schichten im Lias erreichen müssen. Tatsächlich trat der Lias mit Dogger vermengt bereits nahe der Oberfläche (rund 40m u[nter] G[elände]) und von 135 bis 335 m in großen, nahezu senkrecht stehenden Schollen auf. Darunter folgten bis 406 m Teufe Mergel und vor allem Sandsteine des Mittleren Keuper, der ebenso wie die darüber liegenden Schichten des Dogger und Lias mehr als 200 m höher liegt als nach der normalen Schichtenfolge anzunehmen war. Überraschenderweise traf die Bohrung unter dem Keuper eine dunkelgraue, eintönige, steilstehende Schichtserie, deren stratigraphische Einstufung erhebliche Schwierigkeiten bereitete. Nach dem spärlichen Vorkommen von Mikrofossilien steht nun fest, daß unter dem Keuper wieder Dogger und

Even though the drilling did not oblige with the expected results, it provided additional evidence for an impact at the Steinheim Basin:

> While the quartz grains in the primary basin breccia and in the *Dogger* sandstone still showed planar deformation features, within the *Keuper*, they were only broken. The decrease of the pressure effect from above to below proves, however, that the pressure originated by a cosmic and not by a volcanic event, because in the latter case then it must be the other way round. (Groschopf and Reiff 1971: 524)[88]

Groschopf and Reiff had started their investigations with the conviction that the Steinheim Basin, like the Ries Crater, owed its origin to a meteorite impact, and in the end, they were able to proudly state that 'within ten years of research piece for piece, we were able to confirm it' (Groschopf and Reiff 1971: 521).[89]

While the various drilling programmes at the Ries and Steinheim Basins continued, the research there had obtained a general significance that earlier generations of researchers would never have dreamt of:

> While this volume [BGL 1969] was in the making, the first humans have landed with their spacecrafts in lunar craters and have returned Moon rocks to Earth. As the lunar surface is mainly covered by smallest to largest craters – the same has been confirmed for Mars by radio pictures – comparative investigations of the Moon rocks with those from a deep drilling at the Ries may perhaps allow for certain conclusions concerning the origin of the craters of the Moon. (Vidal 1969: 11)[90]

Lias liegen, diesmal jedoch rund 70 m tiefer als der ungestörten Schichtfolge entsprechen würde. Leider war die Endteufe durch die Leistungsfähigkeit des Bohrgerätes bedingt. Aus geophysikalischen Messungen wissen wir jedoch, daß die darunter liegenden Gesteine noch bis in 1000 bis 1100 m Tiefe stark gestört sind' (Groschopf and Reiff 1971: 524).

[88] 'Zeigten die Quarzkörner in der primären Beckenbrekzie und im Dogger-Sandstein noch planare Elemente, so waren die Quarzkörner des Keuper-Sandsteins nur zerbrochen. Die Abnahme der Druckeinwirkung von oben nach unten beweist aber, daß der Druck durch kosmisches und nicht durch vulkanisches Geschehen entstanden ist, da es sonst umgekehrt sein müsste' (Groschopf and Reiff 1971: 524).

[89] 'In zehnjähriger Forschungsarbeit konnten wir Stück für Stück den Nachweis dafür bringen' (Groschopf and Reiff 1971: 521).

[90] 'Während des Entstehens dieses Bandes sind die ersten Menschen mit ihren Raumschiffen in Mondkratern gelandet und haben Mondgestein zur Erde zurückgebracht. Da die Mondoberfläche überwiegend von kleinsten bis größten Kratern bedeckt ist, gleiches haben Funkbilder vom Mars bestätigt, könnten vielleicht vergleichende Untersuchungen der Mondgesteine mit solchen aus einer Tiefbohrung im Ries gewisse Rückschlüsse auf die Entstehung der Mondkrater ermöglichen' (Vidal 1969: 11).

Chapter 12
Ries Crater – A Terrestrial Proxy for the Moon

Parallel to the new and intensive Ries research with its interdisciplinary and international scope, impact research rekindled the discussion about the origin of lunar craters. In 1961, President John F. Kennedy had promised to land an American on the Moon before the decade was over, and suddenly lunar geology became more than just a pleasant pastime for hobby astronomers. It had previously been something to speculate about; nothing more. It was never imagined that one could hope to lay hands on actual lunar rocks:

> You probably know that I have been intensely occupied with, and fascinated by, the geology of the Moon for more than ten years. ... And as to the Moon, for me it becomes ever more exciting. I basically do nothing else but amass arguments against the exclusive meteoritic hypothesis; that is I am interpreting the lunar landscape actualistically; that is geologically. By the way, the dispute about 'meteorites versus volcanism' is a highly interesting phenomenon both of the history of science and psychology and as such somewhat – depressing. "Nothing is created so idiotic that it won't be started again!" (Letter from Kurd von Bülow [see Chapter 2] to Walter Carlé, 18 October 1964, GA 11/36)[1]

In the 1960s, 'the interpretation of the various crater formations on the Moon still equivocated between volcanic formations and meteorite impacts [as their underlying cause]' (Preuss in AG Ries 1963: 20).[2] This was so despite the fact that the concept of impacts on the Moon had originated in the nineteenth century (Proctor 1873; Gilbert 1893), with remarkable insights into the key issues and

---

[1] 'Sie wissen ja wohl, daß mich die Geologie des Mondes seit mehr als 10 Jahren intensive beschäftigt und fasziniert. ... Und was den Mond angeht, so wird er für mich immer aufregender. Ich tue eigentlich nichts anderes, als Argumente gegen die exklusive Meteoritenhypothese zu häufen, d.h. die Mondlandschaft aktualistisch zu deuten, d.h. geologisch. Übrigens ist der Streit Meteoriten c/a Vulkanismus ein hochinteressantes, wissenschaftsgeschichtliches bzw. psychologisches Phänomen und als solches etwas – deprimierend. "Es ist nichts so blöd gesponnen, es wird doch von vorn begonnen"!' (Letter from Kurd von Bülow to Walter Carlé, 18 October 1964, GA 11/36).

[2] 'Die Deutung der vielgestaltigen Kraterbildungen auf dem Mond schwankt noch zwischen vulkanischen Bildungen und Meteoriteneinschlägen'. (Preuss in AG Ries 1963: 20).

processes. Ives (1919) argued (based on the experience of World War I) for an explosive origin of lunar craters, and gave a correct interpretation of the central mountains. Gifford (1924; see Chapter 6) understood the rapid transformation of kinetic energy into heat, thus recognising that the required projectile would be much smaller than the crater. From the mid-1920s onwards, discoveries of terrestrial impact craters (in addition to Meteor Crater) highlighted the fact that a potential impact history was not unique to the Moon (Schultz 1998).

In Germany, there had also been early advocates for an impact origin of lunar craters. These included Franz Meinecke (born 1885)[3] in 1909, who made this argument based on an analogy with Meteor Crater (Hoyd 1987: 155); and Alfred Wegener, who, due to a cardiac defect, had spent World War I in the weather service of the German army. Thanks to several balloon ascents, 'he [Wegener] was able to supplement his knowledge of something that every infantry soldier knew a good deal about – the morphology of shell craters'. Based on this experience, as well as simple experiments, he concluded that the lunar craters had an impact origin (Greene 1998: 113).

Most astronomers, however, remained sceptical:

> The majority of astronomers explain the craters of the moon by volcanic eruption – that is by an essentially geological process – while a considerable number of geologists are inclined to explain them by the impact of bodies falling upon the moon – that is by an essentially astronomical process. This suggests that each group of scientists find the craters so difficult to explain ... that they have recourse to ... another field than their own. (Grove K. Gilbert according to Davis 1922; quoted after Torrens 1998: 184)

In 1949, Ralph B. Baldwin published his book *The Face of the Moon*. Among its most notable arguments is the treatment of the depth to width relationship of the lunar craters. As others before him had also noted, a mathematical law unaffected by local conditions governed the depth-to-width ratio of the craters. Larger craters were comparatively shallower than smaller ones. According to Baldwin, this mathematical correlation argued for the application of the same genetic force in crater formation. Remarkably, when Baldwin added measurements of man-made bomb craters to his data on terrestrial impact craters, he found that they both plotted along the same curve, and therefore he concluded that all of these craters were produced by explosions. Because '[o]n the earth, naturally occurring explosion craters were caused by the impact of meteorites. On the lunar surface,

---

3   Franz Meinecke obtained his doctorate in geology from the University of Halle in 1910. In 1911, he took a teaching examination and afterwards worked as a schoolteacher. From 1915 to 1918, he was a soldier in World War I. He returned to his teaching position after the end of the war (Freyberg 1974: 102).

too, they must be meteorite craters'. Baldwin considered this argument for the explosive origin of lunar craters 'unassailable' (Mark 1995: 106). Nevertheless, up to the early twentieth century most scientists opted for a volcanic origin of lunar craters (Marvin 1986: 151; Chapter 8).

This was the state of affairs when Kennedy opened the race to the Moon in 1961. Eugene Shoemaker succeeded in talking his superiors at the US Geological Survey into forming a department for 'astrogeological' studies, which he headed between 1961 and 1968, and which was mostly financed by NASA (Marvin 2002). In 1963 his department was transferred to Flagstaff, Arizona, and in 1966 it was renamed Center for Astrogeology. From 1961 to 1970, Shoemaker was also a co-investigator in the Ranger Television Experiment, and a principal investigator for the Surveyor Television Experiment and the geological parts of the Apollo 11, 12 and 13 flights (Barsig and Kavasch 1983: 30).

In August 1964, the space probe Ranger 7 reached the Moon, sending pictures until it impacted on the lunar surface. These pictures up to the last one showed smaller and smaller circular craters, scaling down all the way to the limits of resolution. For the first time, these images made it convincingly clear that these craters could not be of volcanic origin. At least, that is how Shoemaker felt. But the debate went on.[4]

Therefore, the recovery of actual rocks from the surface of the Moon by the Apollo expeditions, showing all the impact features that had been described in the Ries suevite and other similar rocks elsewhere became crucial for settling the long-standing argument about the possible impact origin of lunar craters.

Shock Metamorphism

The term shock metamorphism has been created to describe all changes in rocks and minerals resulting from the passage of transient, high-pressure shock waves. ... Such a process is quite distinct from normal geological metamorphism, and it becomes necessary to think in terms of microseconds instead of megayears ... peak pressures reach several megabars, and cubic miles of rock may be affected by pressures in the 10–100 kb range; (3) temperatures may locally reach 10,000°C, and reactions observed in preserved fragments of melted rock indicate that temperatures in these exceeded 2,000°C; (4) reactions are so rapid and quench times so brief that chemical and physical disequilibrium effects are dominant and kinetic factors become extremely important. (French 1968a: 2)

---

[4] Interview of Dr Eugene Shoemaker by Ronald Doel on 17 June 1987, Niels Bohr Library and Archives, American Institute of Physics, College Park, MD USA, <www.aip.org/history/ohilist/5082_3.html> accessed 25 February 2011; see also Chapter 8.

The concept of shock metamorphism was a communal effort of quite a number of international researchers, physicists, mineralogists and geologists. This new field of investigation took shape in the early 1960s when, through the discovery of high-pressure minerals, coesite and stishovite, the impact origin of hitherto-debated geological structures could be confirmed. Now, strange mineralogical features that had previously been described as mere curiosities could be investigated in the light of their unusual origin. Geological research was supported by the contemporary interest in shock wave effects during nuclear test explosions (providing an experimental setting in which the explosive forces reached the same order of magnitude as the smaller impact events); and by the efforts of NASA and the USGS Astrogeology Group to prepare for the future investigation of Moon rocks.

In April 1966, a conference on 'shock metamorphism of natural materials' took place at the Goddard Space Flight Center with some ninety participants. Forty papers were presented (French and Short 1968: vi). Bevan French had invited Wolf von Engelhardt to attend the conference and give a lecture on his work at the Ries crater. The conference volume, which was published two years later, contained two papers with Engelhardt as first author (Engelhardt and Stöffler 1968; Engelhardt et al. 1968) and also a paper concerning experimental work performed by Friedrich Hörz (Hörz 1968), Engelhardt's PhD student.

The international scope of the research by Engelhardt and his working group was much admired by those staying back in Tübingen. That he was part of a large group of international researchers, however, remained obscure to many of his Tübingen colleagues. They did not perceive Engelhardt as one notable scientist among others equally distinguished. Instead, he was judged in traditional, hierarchical terms. The fact that he was invited to foreign conferences led to the 'natural' assumption that he was *the* Expert to be called in for consultations. Consequently, Engelhardt's fame became larger than life and a source of local pride:

> Research by [Engelhardt] and his collaborators has developed and given scientific birth to a completely new geo-scientific discipline – 'impact metamorphism' – including the description, investigation and interpretation of all those mineralogical phenomena that result from the influence of shock waves on minerals and rocks. To the foundations of this discipline, which has now become indispensable with the outreach of man into space, also belong experience-based, time-consuming investigations with the polarization microscope. In this field too, Wolf von Engelhardt has accomplished pioneering work and amassed an enormous amount of knowledge. (Pichler 2000)[5]

---

5   'Seine und seiner Mitarbeiter Forschungen haben eine ganz neue geowissenschaftliche Fachdisziplin, nämlich die 'Impakt-Metamorphose', ins wissenschaftliche Leben gerufen

Engelhardt's working group established a pace in Ries research that before was unheard of. Engelhardt was well organised and assigned specific tasks to his students, thus gaining immensely in efficiency. A completely new style of research evolved. Division of work flanked by close cooperation within a working group was its key strategy. Engelhardt also helped his students to succeed, writing letters of recommendation for them, and some eventually gained positions that in turn helped Engelhardt and opened international doors for him. As was common practice, he often took first authorship on the more important, international publications, although the actual lab work was probably performed by his PhD students or research associates:

> Mr Stöffler was really the most important of his [Engelhardt's] students, [Dieter] Stöffler and [Friedrich] Hörz. Mr Hörz now works for NASA, but at these working group meetings in Tübingen Mr Stöffler has already played an important role, and alongside Engelhardt he really had the deciding role. Of course, in this phase of mineralogical investigation, when the mineralogical investigations that had been neglected for decades had been basically made up for, all this was discovered in a rather condensed way, in one go. (Hüttner interview)[6]

Dieter Stöffler (born 1939) studied geology and mineralogy at the University of Tübingen. He obtained his PhD in mineralogy in 1963. He remained as staff scientist at the Institute of Mineralogy and Petrography in Tübingen, where he '*habilitated*' in 1970 and became associate professor. From 1971 to 1972 he was Senior Postdoctoral Research Associate at the NASA Ames Research Center in Moffett Field, California. In 1974 he became Professor of Petrography and Economic Geology at the Institute of Mineralogy in Münster. In 1987 he became Professor of Cosmic Mineralogy and Director of the Institute of Planetology in Münster. In 1993 Stöffler was appointed Director of the Natural

---

und erschlossen, umfassend die Beschreibung, Untersuchung und Deutung all jener mineralogischen Erscheinungen, die durch die Einwirkung von Stoßwellen auf Mineralien und Gesteine hervorgerufen werden. Zu den Grundlagen dieser mit dem Griff des Menschen ins Weltall unentbehrlich gewordenen Fachdisziplin gehören auch erfahrungsgeprägte, zeitaufwendige Arbeiten mit dem Polarisationsmikroskop. Wolf von Engelhardt hat auch auf diesem Felde Pionierarbeit geleistet und ein enormes Wissen gesammelt' (Pichler 2000).

6   'Der Herr Stöffler war ja eigentlich so der wichtigste Schüler von ihm, der Stöffler und der Hörz. Der Herr Hörz der ist ja jetzt bei der NASA, aber der Herr Stöffler hat dann ja bei diesen Zusammenkünften, bei diesen Arbeitstagungen in Tübingen auch schon mit eine wichtige Rolle gespielt und der hat dann doch nebst Engelhardt die entscheidende Rolle da gehabt. Man hat natürlich in dieser Phase der mineralogischen Untersuchungen, wo man gewissermaßen die mineralogischen Untersuchungen nachgeholt hat, die durch Jahrzehnte nicht erfolgt sind, da kam ja das doch alles sehr komprimiert auf einmal in dieser Phase'. (Hüttner interview).

History Museum in Berlin and Professor of Mineralogy and Petrography at the Humboldt University in Berlin. He retired from the museum directorship in 1999 and from the professorship in 2005. Between 1980 and 1995 he was in residence once as Visiting Senior Scientist at the Lunar and Planetary Institute and the NASA Johnson Space Center in Houston, Texas, and several times at the Planetary Geosciences Division at the University of Hawaii at Manoa, Honolulu:[7]

> Globalization ... decades before it was discovered by the world economy, was then already practised by Wolf von Engelhardt in the way that he sent his students to Canada and other non-European countries to work on impact structures there. ... [C]onsequently, the scientific quotations and exchange rates in the stock exchange of fame in the geosciences rose for Wolf von Engelhardt and his Tübingen Institute [of mineralogy] to a high, stable international level. (Pichler 2000)[8]

Thus, at the local level, Engelhardt was much admired and celebrated, although at the global level his achievements – though still remarkable – are on a somewhat more human scale. However, the fact remains that (mineralogical) Ries research, as practised by Engelhardt and his working group, was now an international affair fuelled by the American effort to reach the Moon:

> NASA's preparations for the Apollo missions to land on the Moon since about 1960 have been the initiating factor, not only for the establishment of the impact theory of the Ries ..., but also for the internationalization of Ries research. ... Today the Ries researcher is no longer focused exclusively on the local geological history of the Nördlinger Ries; but is focused foremost on the universal elucidation of the impact process that formed the Ries. (Stöffler 1987: 276)[9]

---

7   <http://www.naturkundemuseum-berlin.de/en/institution/mitarbeiter/stoeffler-dieter/stoeffler-curriculum-vitae/> accessed 24 July 2012; Engelhardt and Hölder 1977: 55.

8   'Die Globalisierung ... wurde, Jahrzehnte bevor die Weltwirtschaft sie entdeckte, schon damals von Wolf von Engelhardt praktiziert, dergestalt, daß er Schüler nach Kanada und in andere außer-europäische Länder schickte, um an dortigen Impaktstrukturen zu arbeiten. ... dafür stiegen die wissenschaftlichen Notierungen und Kurse an der Bekanntheitsbörse der Geowissenschaften für Wolf von Engelhardt und sein Tübinger Institut auf ein hohes, krisenfestes internationales Niveau' (Pichler 2000).

9   'Die Vorbereitungen zu den Apollo-Mondlandemissionen der NASA seit etwa 1960 waren nicht nur das auslösende Element für die Begründung der Impakttheorie des Ries ... sondern auch für die Internationalisierung der Riesforschung. ... Der Blick des Riesforschers ist daher heute nicht mehr ausschließlich auf die lokale geologische Geschichte des Nördlinger Ries gerichtet, sondern in allererster Linie auf die allgemeingültige Erklärung des für das Ries ursächlichen Impaktprozesses' (Stöffler 1987: 276).

A number of different criteria for shock metamorphism, and thus diagnostic features for impact craters, were developed through investigation of well-preserved and well-established impact craters, including the Ries Crater. These criteria were used successfully to assess more strongly eroded structures. They included the presence of high-pressure minerals such as coesite, stishovite and diamond; high strain-rate effects such as planar deformation features (PDFs) (Figure 12.1) and diaplectic glasses; high-temperature effects with remnants of melting processes requiring temperatures well above 1,500°C (too high for normal geological processes) (French 1968a: 2); and shatter cones, which were already usable macroscopically in the field (Dietz 1968).

The pioneering work on the application of high-pressure minerals such as coesite and stishovite had been performed by Chao et al. (1960), Shoemaker and Chao (1961a), Chao et al. (1962) and Chao and Littler (1963). Planar deformation features:

> [e]nigmatic planar structures, often occurring as multiple sets in single quartz grains, were first noted by McIntyre (1962) in his study of specimens from Clearwater Lakes, Quebec. Short (1966a) suggested that the multiplicity of these features is unique and might constitute a criterion for recognition of shocked quartz, von Engelhardt and Stöffler (1965) reported similar features in specimens from the Ries Basin in Germany.[10] Carter (1965) described them in detail. (Carter 1968: 457)

Diaplectic[11] glasses, glassy pseudomorphs of quartz and feldspar, had already been described from the Ries by Oberdorfer (1904) and Ackermann (1958). However, they had been interpreted at eutectic, volcanically generated melts. Chao (1964)[12] and Stöffler (1966) then interpreted the isotropisation of the minerals as caused by the impact shock wave (Bunch and Cohen 1968: 509); that is due to destruction of the crystal lattices by the intense shock. Engelhardt and Hörz (1964) and Hörz (1965; Chapter 11) pointed out the high density of these glasses, which indicates that they were formed at high pressure (Robertson et al. 1968: 433). In contrast, true impact melts had been demonstrated to be extremely heterogeneous (Shoemaker and Chao 1961). In 1961 Chao reported nickel-bearing metallic spherules from the Ries Crater in a talk at the DMG

---

[10]   After Bunch and Cohen (1964) described such features in quartz from Meteorite Crater and Wabar.

[11]   A term coined by Engelhardt and Stöffler (1968: 163): 'It is therefore proposed to apply the term diaplectic to all products produced by shock waves, in such a way that morphological characteristics of the liquid state are lacking'. In this original definition, PDFs were included as well, but eventually, it was retained for diaplectic glasses only, distinguishing them from glass that had been produced by melting.

[12]   See also Chao and Littler (1963).

Figure 12.1   PDFs in quartz from the Ries Crater (Germany): Thin-section photomicrograph (in crossed polars [CP]) of a shocked quartz grain with two sets of planar deformation features (PDFs) (Sample NHMW-L4665; Suevite from the Otting quarry, Ries, Germany). (Credit: Ludovic Ferrière, NHM Wien).

meeting in Tübingen. These studies were confirmed by Schüller and Ottemann (1963; Chapter 11) and El Goresy (1964, 1965), who documented '[h]igh-temperature fusion and decomposition reactions among opaque minerals, indicating temperatures probably in excess of 1,500°C' (El Goresy 1968). Reviews and summaries of impact metamorphism were published by Chao (1966, 1967a, b) and by Short (1966b). However, these features were not described (and criteria developed) only at the Ries Crater. Research at a number of Canadian impact craters also contributed prominently to the increasing body of knowledge about shock metamorphism (Bunch et al. 1967; Dence 1964, 1965; French 1968b).

The newly developed criteria were particularly reliable because they were buffered by observations on the craters produced by experimental nuclear and large chemical explosions (French 1968a: 7):

> The recognition and use of petrographic shock criteria have, in the last ten years, changed the status of meteorite impact from that of a curious process forming

rare structures less than a mile across into a mechanism seriously considered as the agent of formation of nearly a hundred structures on the earth, some of which are tens of miles in diameter and have been the focus of geological controversies and economic development for more than fifty years. (French 1968a: 7)

Although there was now agreement within the community of impact crater researchers about the indicators of extremely high shock pressures, there was still much work to be done regarding 'quantitative information about the conditions of pressure and temperature under which they form' (French 1968a: 12). Nevertheless, several shock metamorphic stages were distinguished, with increasing intensity of deformation from the periphery of the crater towards the centre.

At the Ries Crater, Stöffler (1965, 1966; Engelhardt and Stöffler 1968) distinguished four hemispherical shock zones, depending mainly on the metamorphic alteration of quartz and feldspar, whereas similar work was done at Canadian craters concluding 'that as many as four subdivisions of the deformation of quartz in Stöffler's facies could be recognized in many of the Canadian craters' (Dence 1968: 170; Robertson et al. 1968). Chao, too, recognised various degrees of shock metamorphism (summarised in Chao 1967a, b):

> For any mineral, the shock effects can range from none to intense. For example, quartz in shocked rocks can be classified as follows: 1, unshocked; 2, fractured; 3, has planar features; 4, has planar features of diverse index of refraction (shock lamellae) and is partly transformed to glass; 5, completely transformed to silica glass in the solid state; 6, silica glass with recrystallized quartz along fractures; 7, silica glass with cristobalite; 8, silica glass containing a small amount of coesite; 9, silica glass containing coesite and stishovite; and 10, flowed lechatelierite. (Chao 1968: 137)

Chao provided similar classifications for other minerals (most of them less detailed than for quartz) and tried to correlate various effects between the minerals. He developed a table of minerals (quartz, feldspars, biotite, amphibole, magnetite, ilmenite, titanite, rutile, zircon), effects (from unaffected to fully molten or disintegrated) and seven degrees of metamorphism to be enlarged by further investigations (Chao 1968: 138, Figure 1). While Engelhardt and Stöffler (1968) developed their concept of shock-metamorphic facies mainly with regard to the Ries crater, Chao (1968) envisioned a wider use of his concept suitable for all sorts of target rock mineralogies. While Engelhardt and Stöffler provided a strongly descriptive argument, Chao argued with more theoretical background, providing mineralogical and crystallographical reasons for the behaviour of certain minerals under shock deformation. This allowed predictions about the behaviour of additional minerals to be investigated in the future.

Perhaps unavoidably, there developed during the 1960s a certain rivalry between the Engelhardt working group on the one hand and Edward Chao on the other. This became evident to contemporary observers in 'that [Chao] followed his [own] thoughts, and there was little correspondence with others. ... Mr El Goresy and Mr Chao have worked together, but [with] the Tübingen institute [of mineralogy] there was always rivalry – less at the beginning, but then it escalated increasingly' (Hüttner interview).[13]

It seems that colleagues were aware of the simmering conflict, because French (1968a) in his introduction to shock metamorphism remained carefully neutral as to priorities: 'The Ries structure, which has become probably the best-established impact crater after Meteor Crater itself, has recently been the object of numerous petrographic and stratigraphic studies (Chao, 1967a, 1967b; Engelhardt, 1967; Engelhardt et al., 1967[a]; Förstner, 1967; Stöffler, 1967)' (French 1968: 11). In contrast, Engelhardt and Stöffler (1968) clearly stress Stöffler's priority: 'By means of microscopic studies of the basement rock fragments, a series of stages of increasing alteration can be established (Stöffler, 1965, 1966; Chao 1967[b])'. Dence (1968: 169), on the other hand, states that 'Examination of Cenozoic craters attributed to hypervelocity meteorite impact has produced comprehensive studies of distinctive petrographic features identified as the results of shock deformation or transformation (Chao 1964, 1967a, 1967b)', while Bunch and Cohen (1968: 509) credited Chao (1964, 1966, 1967[a, b]) with the discovery of diaplectic glasses[14] at the Ries crater, adding that 'Stöffler (1966) described similar glassy transformations of feldspar, which he also attributed to shock, from a large number of Ries localities' (Bunch and Cohen 1968: 509).

Chao (1968) ignored the Engelhardt working group completely, while El Goresy (1968) at least mentioned Hörz (1965) in the list of references, although not in the text itself. Otherwise, he too ignored any work by Engelhardt and his co-workers.

Thus there is some evidence that a certain rivalry existed, although as to the underlying reasons for it we can only speculate. Chao's Ries engagement was part of his larger involvement with the USGS Astrogeology Group and NASA, so the Ries Crater was only one of several impact craters that he researched. Research in Germany involved considerable travel, and so there was never much time for him to establish himself locally as an authority, nor would that have been reasonable in the light of his other workload. Most of his results were then

---

13   ' ... dass [Chao] seine Gedanken verfolgt hat und wenig Korrespondenz mit anderen war. ... Herr von El Goresy und der Herr Chao, die haben dann gemeinsam zusammengearbeitet und das Tübinger Institut, da war eben immer eine Konkurrenz – am Anfang ja weniger aber immer mehr hat sich das dann herausgewütet' (Hüttner interview).

14   For which Chao then proposed the term *thetomorphic* (Chao 1967a: 212), which, however, did not stick.

presented at conferences (Chao and Littler 1963) or went straight into internal reports (Chao 1964), rather than being published right away. As a result, Chao was outpaced by the sheer manpower of the Engelhardt working group, although – and this must have been the cause of some bitterness for him – he had clearly pioneered mineralogical impact research at the Ries Crater in 1960–61 and provided Engelhardt with the necessary foothold with his talk in Tübingen (in 1961) and his correspondence in those early years (see Engelhardt papers at GA; Chao and Littler 1963).

Suevite Samples From the Moon

When NASA called for proposals to scientifically investigate the lunar rock samples they expected to collect through the Apollo programme, Wolf von Engelhardt was among those who applied for access to samples. When Apollo 11 successfully returned from the Moon in July 1969, it carried with it 21.7 kg of rock and soil samples from *Mare Tranquilitatis*. After a quarantine for safety reasons and preliminary investigations by NASA, samples were distributed to some 140 scientists including Engelhardt; one of seven Principal Investigators in West Germany. In the end, Apollo 11 to 17 recovered a total of 381.69 kg of lunar rock, divided into 2196 samples (Taylor 1975: 4–8).

Engelhardt's intention was to undertake petrographic studies 'to determine shock effects'. As principal investigator, he was responsible for proper and safe storage of the samples and for returning them after use accompanied by detailed documentation as to their treatment, possible contamination, exposure to radiation and so on. His co-investigators were his three postdoctoral students Dieter Stöffler, Wolfgang Müller[15] and Jörg Arndt.[16] Engelhardt received a sample of regolith, 'unconsolidated impact breccia produced by multiple

---

[15]   Wolfgang Friedrich Müller (born 1939) studied physics and geology in Munich, and then mineralogy and physical and inorganic chemistry in Tübingen. He obtained his PhD in Tübingen at the Institute for Mineralogy and Petrology in 1965. Afterwards, he became a research associate to Wolf von Engelhardt, pioneering transmission electron microscopy of shock metamorphic features. In 1971 he obtained a scholarship to conduct research in the Department of Materials Science and Engineering at the University of California at Berkeley for eighteen months. After his return, he worked at the Institute of Crystallography at the University of Frankfurt a. M., where he qualified for professorship in 1975. In 1977 he was appointed Professor of Mineralogy at the Technical University in Darmstadt (Aken 2001).

[16]   Jörg Arndt (born 1938) obtained his mineralogical doctoral degree in 1964 from the University of Tübingen. From 1964 to 1966 he was a research engineer at Siemens & Halske AG in Munich. From 1966 to 1975 he became research associate, research assistant, and lecturer at the University of Tübingen, obtaining his qualification for professorship. In 1978 he became Professor of Mineralogy and Crystallography at the University of Tübingen

hypervelocity impact of meteoritic bodies', and his working group found that 'Shock effects are lamellar deformation structures in plagioclase, pyroxene, olivine and ilmenite, isotropisation of plagioclase and various glasses (fragments and regular bodies) formed by shock-induced melting' (ZERIN: Engelhardt-papers, Folder Apollo 11).

Engelhardt applied for and received samples from all of the Apollo missions, in order to perform the same investigations on shock metamorphic features. Although later samples came unceremoniously by airmail, Engelhardt flew in person to Houston to receive his very first Apollo 11 samples (ZERIN: Engelhardt-papers, Folder Apollo 11; Pichler interview):

> When in July 1969 Neil Armstrong and Edwin Aldrin brought the first Moon samples back to Earth, Wolf von Engelhardt (as one of the few chosen non-US researchers) was among the recipients and investigators of these precious extra-terrestrial goods. The delivery of these samples from the airport at [Stuttgart-]Echterdingen to the [mineralogical] institute, under special police protection with blue light and siren, was like a mineralogical triumphal procession. Wolf von Engelhardt had specially flown to Houston to transfer these 'holy' relics of the Moon to Tübingen. (Pichler 2000)[17]

> It was a major event for Tübingen. ... Escorted by the police, von Engelhardt was driven to Tübingen with the container of lunar samples. It was only a very small amount, and he himself showed them, the samples, the container – he was quite moved – in front of the institute. (Pichler interview)[18]

> I still have his trembling voice in my ears, when he exclaimed in front of the institute: *"Grains from the Moon! How precious!"* (Pichler 2000)[19]

(<http://onlinelibrary.wiley.com/doi/10.1002/piuz.19870180302/abstract> accessed 23 July 2012).

[17] 'Als im Juli 1969 Neil Armstrong und Edwin Aldrin die ersten Mondproben zur Erde brachten, gehörte auch Wolf von Engelhardt als einer der wenigen auserwählten Nicht-US-Forscher zu den Empfängern und Bearbeitern dieses kostbaren außerirdischen Gutes. Die Lieferung dieser Proben vom Flughafen Echterdingen zum Institut unter besonderem Polizeischutz mit Blaulicht und Sirenen glich einem mineralogischen Triumphzug. Wolf von Engelhardt war eigens nach Houston geflogen, um diese Reliquien des Mondes nach Tübingen zu überführen' (Pichler 2000).

[18] 'Das war ein Großereignis für Tübingen. ... Mit Polizeibegleitung kam von Engelhardt nach Tübingen gefahren mit dem Behälter der Mondproben. Das war ja nur eine ganz kleine Menge und er selbst hat – er war ja richtig bewegt – vor dem Institut hat er sie gezeigt, die Proben, die Behälter' (Pichler interview).

[19] 'Ich habe noch seine bebende Stimme in den Ohren, als er vor dem Institut ausrief: *"Körnchen vom Mond! Welche Kostbarkeit!"* (Pichler 2000).

Among those from Tübingen who during the following days knocked at von Engelhardt's door to see the crystals was also Ernst Bloch:[20] He touched them "with great reverence" and according to Carola Bloch refrained from washing his hands for some days. (JME: newspaper clipping from *Schwäbisches Tagblatt*, 18 November 1993)[21]

Engelhardt subsequently published some thirty papers, often together with Jörg Arndt and others, on the lunar samples (Pichler 2000). It rapidly turned out that, as regards to impact phenomena, lunar rock looked exactly as the impactists had predicted. It was full of impact metamorphic features, mostly derived from multiple impact events. Planar deformation features in and diaplectic glasses of the various minerals in lunar basalts and anorthosites were described by Engelhardt and other researchers. There was now no way to retain the idea that the lunar craters were all volcanoes, and so the Apollo Space Program became crucial for convincing the remaining sceptics.

In a somewhat circular argument, newspaper reporters concluded that now that the nature of lunar craters was known, it was quite clear that the Ries Crater had originated from the same sort of event: 'The riddle of the Nördlinger Ries is solved ... Since American Moon travellers have returned from the Earth's satellite with rock samples, it is now certain, after thorough comparison, that the meteorite crater has originated by the same method as the surface of the Moon'.[22] Consequently, from then on, the Ries Crater served science as a 'proxy to the Moon'[23] (JME: Off-print collection, newspaper clipping from *Eichstätter Kurier*, 3 April 1974).

This was also what intrigued the public during the research drilling at Nördlingen in 1973–74. Research on the Ries Crater was no longer a local problem interesting only a handful of specialists, but had gained planetary significance and now interested people on a global scale:

---

[20] Ernst Bloch (1885–1977), a German philosopher and author, who at that time was working at the University of Tübingen.

[21] 'Unter den Tübingern, die in den darauffolgenden Tagen bei von Engelhardt anklopften, um die Kristalle sehen zu dürfen, war auch Ernst Bloch: Er befühlte sie 'mit großer Ehrfurcht' und soll sich, Carola Bloch zufolge, danach einige Tage lang die Hände nicht gewaschen haben'. (JME: newspaper clipping from *Schwäbisches Tagblatt*, 18 November 1993).

[22] 'Das Rätsel um das Nördlinger Ries wurde gelöst ... Nachdem amerikanische Mondfahrer vom Erdtrabanten mit Gesteinsproben zurückkehrten, steht nun nach eingehenden Vergleichen fest, daß der Meteoritenkrater genauso entstanden ist, wie die Oberfläche des Mondes'. (JME: newspaper clipping from *Eichstätter Kurier*, 3 April 1974).

[23] 'Das Nördlinger Ries dient Wissenschaft als 'Mondersatz'' (JME: newspaper clipping from *Eichstätter Kurier*, 3 April 1974).

The Ries Crater can serve as a model for 'impact crater formations', the origin of craters through the impact of large meteorites on the Moon and planets, on which the non-volcanic craters are among the most important surface features. The Moon, Mars, presumably Venus and (as we learned a few days ago), Mercury too, are densely covered by craters. Drilling into the interior of such a crater on Earth should first of all provide deep insights into the phenomena surrounding the origin of such craters. (Steinert 1974)[24]

The suevite was now likened to the lunar regolith: 'the Moon soil, [consisting of] rock dust, molten rock congealed as glass and coarse rubble with all degrees of shock wave deformation of the mineral structure, with pressures up to millions of atmospheres' (Steinert 1974).[25]

In the summer of 1976, NASA undertook its own research drilling programme at the Ries crater, putting down a series of shallow boreholes within the polymictic breccia under the supervision of Engelhardt's former PhD student, Fred [Friedrich] Hörz from Houston (Hörz et al. 1977). 'The Ries of Nördlingen is no longer a mere locally interesting special case of a volcano that exploded instead of erupting. Instead, it is now, far beyond this scenario, the ... emblematic case of the – not only terrestrial but especially cosmic – phenomenon of a meteorite impact' (Dehm 1977: 10).[26]

Astronauts at the Ries Crater

The Apollo programme increased the prestige of Ries Crater research in a very direct and public way. It was now not just something done by NASA experts far away in the USA. It happened right there in Tübingen and finally in Nördlingen and the quarries of the Ries Basin. Astronauts, celebrated like football stars (or

---

[24]   'Der Nördlinger-Ries-Kater kann als Modell für die 'Impaktkraterbildung', die Entstehung von Kratern durch den Einschlag von großen Meteoriten auf Mond und Planeten dienen, bei der die nichtvulkanischen Krater zu den wichtigsten Oberflächenformen gehören. Mond, Mars, vermutlich die Venus und, wie man seit einigen Tagen weiß, auch der Merkur sind mit Kratern dicht bedeckt. Eine Bohrung in das Innere eines solchen Kraters auf der Erde sollte vor allem tiefgehende Einblicke in die Phänomene bei der Entstehung derartiger Krater liefern' (Steinert 1974).

[25]   '... dem Mond-'Boden', der aus Gesteinsstaub, geschmolzenem und zu Glas erstarrtem Gestein und groben Trümmern mit allen Graden der Stoßwellendeformation von Drücken bis zu Millionen von Atmosphären der Mineralstruktur' (Steinert 1974).

[26]   'Das Nördlinger Ries ist nun nicht mehr nur ein lokal interessierender Sonderfall eines zur Explosion statt zur Eruption gelangten Vulkans, sondern jetzt weit darüber hinaus der ... Musterfall des nicht nur irdischen, sondern vor allem kosmischen Phänomens eines Meteoreinschlags' (Dehm 1977: 10).

Figure 12.2   Astronauts and instructors at field training in the Ries Crater. From left to right: M. Abadian, Wolf von Engelhardt, unknown, Edgar Mitchell, Joe Engle, Friedrich Hörz, Eugene Cernan, Dieter Stöffler, Alan Shepard.(Photo courtesy of NASA).

even more so, as they appealed to all levels of the populace from more plebeian levels to intellectuals), actually came to the Ries crater for field training (Figure 12.2):

> The Moon and its matter rushing through the valley of the river Neckar [towards Tübingen] have really marked and distinguished Wolf von Engelhardt and his contemporary [mineralogical] institute – the university [of Tübingen] also took notice of this, with astonishment. This especially found its expression ... (and media-wide) in the 1971 working visit of the three [sic] astronauts of the future Apollo 14 mission, including their instructor Dr Friedrich Hörz – he was a student of Engelhardt as well – to Tübingen and the Ries. Those were great days, in which one was elevated to worldwide attention and esteem – Tübingen and the Ries, for some hours, was the centre of the world. (Pichler 2000)[27]

---

[27]   'Der durch das Neckartal brausende Mond und seine Materie haben Wolf von Engelhardt und sein damaliges Institut – auch die Universität, die dies erstaunt zur Kenntnis nahm – wahrlich gezeichnet und ausgezeichnet. Das fand seinen Ausdruck ... vor allem aber, mediendurchschlagend, im Arbeits-Besuch der drei Astronauten der künftigen Apollo-14-Mission samt ihrem Instrukteur Dr. Friedrich Hörz – auch er ein Engelhardt-Schüler – 1971 in Tübingen und im Ries. Es waren große Tage, in denen man hochgestiegen war in

In August 1970, Hörz accompanied the future Apollo 14 astronauts who were to land on the Moon in February 1971 (Alan B. Shepard (1923–98) and Edgar Dean Mitchell (born 1930)), as well as the members of their back-up crew (Eugene A. Cernan (born 1934) and Joe H. Engle (born 1932)) on a one week trip to Germany. Cernan was later Commander of the Apollo 17 mission, while Engle was replaced as Lunar Module Pilot by the geologist Harrison Schmitt (born 1935).[28] The astronauts attended private geological briefing lectures by Wolf von Engelhardt, and from August 10 to 14 they went to see the Ries rocks in the field. Apart from Engelhardt himself, they were accompanied by Dieter Stöffler, Friedrich Hörz and the NASA geologist Michael [Mike] McEwen, a member of the Manned Spacecraft Center's Geology Branch, who was responsible for the geological training of the Apollo 14 crew. The purpose of this training at the Ries crater was 'later in the Apollo 14 Mission, during sample-taking on the surface of the Moon, to recognise those rocks that had been exposed to meteorite impacts or that were generated by them' (ZERIN: Engelhardt papers, Folder Apollo 11).[29] The astronauts themselves viewed this task similarly. When Shepard was interviewed, he said: 'Field-training here will certainly be very interesting, since here we are going to learn which rocks we have to bring back from the Moon' (Pflugmacher 1970a).[30]

NASA and the USGS astrogeology group viewed sample-taking on the Moon in a less relaxed way, and were well aware of the difficulties, stressing the importance of training for the astronauts: 'In this, as in every other aspect of the sampling process, the astronaut must use judgment based on a full understanding of the problem' (Shoemaker et al. 1965: 33):

---

weltweiter Aufmerksamkeit und Ansehen – Tübingen und das Ries für Stunden der Nabel der Welt' (Pichler 2000).

[28]   Harrison Schmitt obtained a BSc degree from the California Institute of Technology in 1957, studied one year at the University of Oslo, Norway, and then went to Harvard University, where he received a doctoral degree in geology in 1964. He worked briefly at the USGS Astrogeology Center at Flagstaff and became a NASA science astronaut in 1965. Schmitt provided geological expertise to Apollo crews. He was eventually lunar module pilot on Apollo 17, the mission with the longest lunar surface extravehicular time (some 22 hours) and the largest amount of lunar samples returned to Earth (some 115 kg) (<http://www.jsc.nasa.gov/Bios/htmlbios/schmitt-hh.html> accessed 26 June 2012).

[29]   '... später bei der Apollo 14-Mission bei der Probenentnahme auf der Mondoberfläche die Gesteine erkennen zu können, die Meteoriteneinschlägen ausgesetzt waren oder durch sie erzeugt worden sind' (ZERIN: Engelhardt papers, Folder Apollo 11).

[30]   '... das Feldtraining hier wird sicherlich sehr interessant. Hier werden wir nämlich erfahren, welche Steine wir vom Mond mitbringen sollen'. (Pflugmacher 1970a). The English quote above is retranslated from the newspaper article by Pflugmacher, thus it is almost certainly not Shepard's original wording but rather reflects what Pflugmacher understood and how he related his understanding to his readers. This also holds true for the other quotations from astronauts in the newspaper reports below.

One of the principal problems presented by a surface covered by a fragmental layer of the kind described above will be to select appropriate samples for return to Earth. With a wide variety of rock fragments to choose from, the astronaut will be faced with the difficult task of deciding what specimens best represent the material at hand. He will need to estimate the relative abundance of different rock types in order to select representative samples, and, in addition, he will want to collect as many of the infrequently occurring rock types as possible, as these will provide information on the more distant, and possibly deeper, parts of the moon.

In certain respects the model described for the fine structure of the plains surfaces is similar to terrain on Earth which has been covered by deposits left by melting of a continental glacier. Like glacial drift, the debris layer on the moon obscures the 'bedrock' in most places, is heterogeneous in character and irregular in thickness, and contains rock fragments of widely diverse individual histories. Thorough study of such a layer in a local area either on the moon or on the Earth can provide a great amount of information about the geology and history of a broad segment of the planetary crust, provided that the origin of the fragmental layer is understood. Failure to understand the nature and origin of the fine structure of the moon, on the other hand, can lead to serious scientific misinterpretation and confusion. For this reason we regard the field geological investigation and the role of the astronaut as a trained field observer to be fundamental to most of the lunar scientific investigations to be performed in Project Apollo, and of paramount importance to the success of later lunar missions. (Shoemaker et al. 1965: 27–8)

In order to provide the necessary knowledge of geology, time for geological training somehow had to be added to the already full schedules of the astronauts:

Pre-flight requirements on the time of the astronauts include time for 1) training in geological field methods and 2) practice of the geological field investigation in simulated Early Apollo scientific missions. To date, individual astronauts have spent from 9 to 43 days training in geological field methods. About half of the astronauts have completed 34 days of training. Approximately 12 days additional training in geological field methods is planned. An average of 12 days of practice of the geological field investigation in simulated Early Apollo scientific missions is planned. (Shoemaker et al. 1965: 37)

'In part, the field exercises will be carried out at localities where the geologic features correspond to features that may be encountered on the moon' (Shoemaker et al. 1965: 39). One of these terrestrial localities with lunar similarity was the Ries Crater. Other training sites had been in Mexico, Alaska and Iceland (Pflugmacher 1970b).

The arrival of the astronauts created immense publicity for Ries Crater research. The Swiss newspaper *Neue Züricher Zeitung* (incorrectly) hailed the Ries Basin as the largest known impact crater on Earth, which 'at present is the most interesting geological-astrogeological research object in Germany and has become a key to solving cosmic and terrestrial problems' (JME: newspaper-clipping from *Neue Züricher Zeitung*, undated):[31]

> Where, exactly, in the Ries Basin rocks similar to the Moon will be dug for, is so far reckoned to be a secret. The scientific companions of the astronauts, headed by the well-known German geologist Wolf von Engelhardt, won't disclose the localities, as they fear veritable migrations of the inquisitive, all wanting to marvel at the astronauts from close-up. (Pflugmacher 1970a)[32]

Selected newspaper reporters, however, had the opportunity to see a few glimpses of the fieldwork:

> The journey to the Moon begins within the Ries! ... Since Tuesday around noon, the four space travellers have landed in seven quarries and gravel pits around Nördlingen, without noticeably losing their zeal. Never tiring, the Moon travellers hit at gravel walls, crush blocks of stone to closely view the broken edges and interestedly listen again and again to the explanations of their scientific coaches; as a rock scientist, Professor Wolf von Engelhardt from Tübingen is involved as Principal Investigator in the analyses of Moon rock commissioned by the American space agency NASA. ...
>
> Because it will be impossible at present to take scientists as passengers in the space ships to the Moon, it is thereby attempted to provide to the astronauts sufficient geological base knowledge, so that they can bring from their landings on the Earth's satellite more rock samples than just a few sacks full of fleetingly shovelled surface material as had been the case in the previous two lunar landings. ... Especially from the next voyage to the Moon early in 1971, space experts expect as many

---

[31] 'Damit ist das Ries das zurzeit wohl interessanteste geologisch-astrogeologische Forschungsobjekt Deutschlands und ein Schlüssel zur Lösung kosmischer und irdischer Probleme geworden' (JME: newspaper-clipping from *Neue Züricher Zeitung*, undated).

[32] 'Wo nun genau im Rieskessel überall nach mondähnlichem Gestein geschürft wird, gilt bisher noch als geheim. Die wissenschaftlichen Begleiter der Astronauten, an ihrer Spitze der bekannte deutsche Geologe Wolf von Engelhardt, geben die Orte nicht preis, weil sie wahre Völkerwanderungen von Neugierigen befürchten, die alle die Astronauten aus nächster Nähe bestaunen möchten' (Pflugmacher 1970a).

and varied rock samples as possible. These are expected to offer scientific insight into the composition of the Moon, its origin and its age. (Pflugmacher 1970b)[33]

Although the main purpose of the trip, the geological training of the astronauts, proceeded nicely, the cultural aspect of the trip seems to have been more disappointing, especially for the town of Nördlingen, because the organisers of the trip completely neglected the human interest factor:

> On their own the four astronauts explore the pretty town with its cosy corners and dreamy alleys. Of course they are most impressed by the completely preserved town wall with the still walkable patrol path. ...

> The town would have loved to provide the astronauts with guides well versed in history for their walks through town. However, the German and American coaches of the Moon men declined this strictly, on the basis that a guided tour through town wouldn't interest the astronauts. The coaches, allegedly experts of the NASA department for science and information, also resisted a reception by the town. The principal mayor Dr [Hermann] Keßler, who specifically interrupted his vacation on the [Italian] Riviera because of the American astronauts, without hesitation joined the Moon travellers on their bus and personally arranged a reception for them in the town hall. There, the Apollo Commander Alan B. Shepard, on behalf of his comrades, heartily expressed thanks for the friendly and considerate way of the Ries populace with the words: "We have never been bothered or disturbed in our work". ...

---

[33] 'Die Fahrt zum Mond beginnt im Ries! ... Seit Dienstag mittag [sic] sind die vier Weltraumfahrer in sieben Steinbrüchen und Kiesgruben rund um Nördlingen gelandet, ohne daß bisher ihr Eifer sichtbar nachgelassen hat. Nimmermüde klopfen die Mondfahrer Kieswände ab, klettern auf steilen Felswänden herum, zertrümmern Steinbrocken, um die Bruchstellen in Augenschein zu nehmen und lauschen immer wieder interessiert den Erläuterungen ihrer wissenschaftlichen Begleiter; Professor Wolf von Engelhardt aus Tübingen ist als Gesteinswissenschaftler im Auftrag der amerikanischen Weltraumbehörde Nasa als 'Principal Investigator' in die Analysen von Mondgestein eingeschaltet. ... // Nachdem es vorläufig noch nicht möglich sein wird, Wissenschaftler als Passagiere in den Raumschiffen zum Mond mitzunehmen, wird damit versucht, den Astronauten so viele geologische Grundkenntnisse zu vermitteln, daß sie von ihren Landungen auf dem Erdtrabanten mehr Gesteinsproben mitbringen als nur einigen Säcke voll flüchtig zusammengeschaufelten Oberflächenmaterials wie dies bei den beiden bisherigen Mondlandungen geschehen ist. ... Gerade von der kommenden Mondfahrt anfangs 1971 erwarten sich Weltraumexperten möglichst viele und unterschiedliche Gesteinsproben. Diese sollen der Wissenschaft dann Aufschluß geben über die Zusammensetzung des Mondes, über seine Herkunft und über sein Alter' (Pflugmacher 1970b).

Following this compliment for the consideration of the Ries populace, the behaviour of the German scientists, who try to turn every one of the thirteen quarry visits into a state secret, seems ever more bizarre. Allegedly, for fear that hordes of spectators might disturb the scientific fieldwork of the astronauts, neither the head of the district, Dr Eberhardt Schmidt, is informed, nor are the most well-known scientists of Bavaria invited. "The poor pretext cannot hide the fact that the group of geologists from Tübingen simply did not want anybody to steal their show. They have turned the astronauts' visit into a matter of prestige" says a high-ranking personality in Nördlingen. ...

However, on their tour the rock expert Wolf von Engelhardt and his team from Tübingen also annoy the owners of quarries and gravel pits, whose premises they invade without greeting, dig around in, and then disappear again, likewise without any formality. When quarry owner Egon Bschor in Ronheim tried to film the astronauts, he was hindered by the NASA 'busybodies'; but only until he resolutely declared: "I am the owner here, and if you dislike something, you may leave at once!" Director Dr Merker, too, learns about the arrival of the astronauts only when the scientists are already in the quarry. ...

The astronauts have no idea of the wrangling of overly keen NASA officials, among which the Tübingen geologists also count themselves. ... When they discover on a picture the boy drummers of the Boys Music Band of Nördlingen, and learn that even a short presentation of the young musicians has been declined by NASA officials, they shrug regrettably and grin to the words: "Well, there are people who lamentably think only in rocks".

"We have enjoyed the additional things too, although, of course, we have come first of all to the Ries Basin to study this unique meteorite crater and its interesting rock formations. On the Moon, everything is different, but in some ways we nevertheless hope to discover similar things. Our landing site in the Mare Imbrium will also be a meteorite crater. Not such a beautiful one, which is so nice to live in like here in the Ries. The rock, however, might be of a material similar to what we see here in the ejecta hills of the crater rim" says Allan B. Shepard, who in all probability will stand on the Moon in February 1971. (Pflugmacher 1970c)[34]

---

[34] 'Auf eigene Faust ziehen die vier Astronauten in der reizvollen Stadt mit ihren stimmungsvollen Winkeln und verträumten Gassen auf Entdeckung aus. Natürlich imponiert auch Ihnen am meisten die völlig erhaltene Stadtmauer mit dem immer noch begehbaren Wehrgang. ... // Liebend gern hätte die Stadt den Astronauten geschichtskundige Führer mit auf den Stadtbummel gegeben, doch die deutschen und amerikanischen Begleiter der Mondmänner lehnen dies strikt mit der Begründung ab, eine Stadtführung interessiere die Astronauten nicht. Die Betreuer, angebliche Könner der Nasa-Abteilungen für Wissenschaft und Information, sträuben sich auch gegen einen Empfang durch die Stadt. Kurzentschlossen

Despite all of the squabbles and misunderstandings, the actual field training proceeded without unsolvable problems, and geology for once stood in the limelight of (borrowed) fame, providing Ries research with a brief spell of prestige like it had never seen before. As a tourist destination, the Nördlinger Ries still owes a debt of gratitude to the beginning of space exploration, highlighting space geology in its *Rieskratermuseum*, which also has a sample of lunar rock on display.

What the astronauts' training at the Ries crater did for the geological sampling on the Moon is more difficult to assess. Obviously, a proper geological university education cannot be replaced by a few days of special training. No matter how hard they tried, there is no way on Earth to really prepare for what it must be like

---

steigt darauf hin Oberbürgermeister Dr. Keßler, der eigens seinen Urlaub an der Riviera wegen der amerikanischen Astronauten unterbrochen hat, zu den Mondfahrern in den Bus und verabredet mit ihnen persönlich den Empfang im Rathaus. Dort ist es dann Apollo-Kommandant Alan B. Shepard, der sich auch im Namen seiner Kameraden für die nette und zurückhaltende Art der Ries-Bevölkerung herzlich mit den Worten bedankt: "Wir sind bei unserer Arbeit nie belästigt oder gestört worden". ... // Nach diesem Kompliment an die Besonnenheit der Rieser Bevölkerung wirkt das Benehmen der deutschen Wissenschaftler, die jeden der insgesamt 13 Steinbruchbesuche zu einem Staatsgeheimnis zu machen versuchen, um so merkwürdiger. Angeblich aus Angst, Zuschauermassen könnten die wissenschaftliche Feldarbeit der Astronauten stören, werden weder Landrat Dr. Eberhardt Schmidt verständigt noch die bekanntesten geologischen Wissenschaftler Bayerns eingeladen. "Der fadenscheinige Vorwand kann die Tatsache nicht verbergen, daß die Tübinger Geologengruppe sich einfach die Schau nicht stehlen lassen wollte. Sie haben aus dem Astronautenbesuch eine Prestigesache gemacht", sagt eine hochgestellte Persönlichkeit Nördlingens. ... // Der Gesteinsexperte Wolf von Engelhardt und sein Tübinger Team verärgern auf ihrer Tour durchs Ries aber auch die Steinbruch- und Kieswerksbesitzer, in deren Anlagen sie grußlos einfallen, herumbuddeln und dann ebenso formlos wieder verschwinden. Als Steinbruchbesitzer Egon Bschor in Ronheim versucht, die Astronauten zu filmen, wird er von den Nasa-'Gschaftelhubern' daran gehindert, freilich nur, bis er resolut erklärt: "Ich bin hier der Besitzer, und wenn Ihnen etwas nicht paßt, können Sie sofort verschwinden!" Auch Direktor Dr. Merker erfährt von der Astronautenankunft erst, als die Wissenschaftler längst im Steinbruch sind. ... // Von einem Gerangeln übereifriger Nasa-Beauftragter, zu denen sich auch die Tübinger Geologen zählen, haben die Astronauten keine Ahnung. ... Als sie auf einem Bild die Bubentrommler der Nördlinger Knabenkapelle entdecken und erfahren, daß selbst ein Ständchen der jungen Musiker von Nasa-Offiziellen abgelehnt worden ist, zucken sie bedauernd mit den Schultern und grinsen zu dem Ausspruch: "Es gibt halt Leute, die denken leider nur in Steinen". // "Uns hat das Drumherum schon auch gefallen, obwohl wir natürlich in erster Linie hier im Rieskessel sind, um diesem einmaligen Meteoritenkrater mit seinen interessanten Gesteinsformationen zu studieren. Auf dem Mond ist zwar alles anders, aber in einigen Dingen hoffen wir doch, ähnliches zu entdecken. Auch unsere Landestelle im Regenmeer wird ein Meteoritenkrater sein. Kein so schöner, in dem man so angenehm leben kann wie hier im Ries. Das Gestein aber könnte von ähnlichem Material sein, wie wir es hier in den Auswurfhügeln des Kraterrandes finden", sagt Allan B. Shepard, der voraussichtlich im Februar 1971 auf dem Mond stehen wird' (Pflugmacher 1970c).

to totter around in a clumsy space-suit under unfamiliar gravitational conditions or do geology on a different celestial body under strange lighting conditions and with practically no time to think about what one is doing. In the end, rock sampling on the Moon necessarily had to be a more or less random process, and the training had been just enough to confuse and to give the astronauts a frustrating sense of what they were missing:[35]

> It wasn't really a matter of being able to describe what we saw in this particular case; because, at that point, we were so rushed that all we were trying to do was see different things and grab it without really noting how it necessarily differed. The only thing that I recall about these craters – or these boulders was that there were inclusions or variations within the rocks; and I assume that they were crystals within the rock, or some crystalline forming in the rocks. I don't know that that's true; they might have been, for example, a breccia with just a conglomerate in them, and I don't know whether that's true or not either. There simply wasn't time to look at them in that detail; so, we just grabbed, photographed, and ran; and I would be kind of at loss to give you an articulate description of really what those rocks are like. (Pre-lift-off phase on the return leg of Apollo 14; Mitchell answering questions of scientists; from Bailey and Ulrich 1975: 05/18/15/07/LMP)

While the reality of geology on the Moon is somewhat sobering for a trained geologist, those who at that time saw only the news on television felt elated and dreamt of a thrilling future, carrying Ries geology out to space: 'to boldly go, where no geologist had gone before':

> While this volume was in the making, the first humans have landed with their spacecrafts on the Moon and have returned Moon rocks to Earth. … Certainly it won't be long until, alongside the astronauts, geoscientists too will walk on the Moon, and then as lunologists they will begin the investigation of the Moon with the aid of geological, petrographic-mineralogical and geophysical methods that were tested on Earth.

> Then Ries research will build a bridge to the geo-scientific exploration of extraterrestrial space, and the work of German geoscientists in the Ries might

---

[35]   From Apollo 15 onwards, the geology training of astronauts was much intensified and also involved the capcoms and science backroom geologists to achieve a common language and overcome problems of communication, while sampling and observation routines were practised so that they would not hamper actual geological observations. The improved training added greatly to the success of the later missions (Lofgren 2008).

[then] indirectly be a non-negligible German contribution to research in space, and provided with relatively modest public money. (Vidal 1969: 11)[36]

Georg Wagner Finally Convinced

From 31 March to 4 April 1970, the *Oberrheinischer Geologischer Verein* once again met in Nördlingen; it had done so before in 1903 and 1924, and the Ries Basin had additionally been a field trip destination in 1938 and 1948 (Metz and Treibs 1970: 8):

> The Ries of Nördlingen has [again] been chosen as the meeting place, because during the time of the first landings on the Moon it seemed desirable to get to know, under expert guidance, a circular structure on Earth that is similar to lunar craters. Earlier conferences of the *Oberrheinischer Geologischer Verein* have visited the Ries as field trips, on which occasions this structure was always regarded with the view of other genetic causes. (Metz and Treibs 1970: 5)[37]

It was pointed out to the delegates that during recent years, various experts of different fields (geologists, mineralogists, petrographers, geophysicists, geochemists, geochronologists, astronomers and physicists) had reviewed the phenomena of the Ries Basin in light of the new evidence of the high-pressure minerals coesite and stishovite (Metz and Treibs 1970: 5).

The weather was beastly, with late snow, but nevertheless more than 250 conference members swarmed the fields and quarries (Vidal 1974: 6). Among them was Georg Wagner, frail and old, amiably and politely assisted by Helmut

---

[36] 'Während des Entstehens dieses Bandes sind die ersten Menschen mit ihren Raumschiffen in Mondkratern gelandet und haben Mondgestein zur Erde zurückgebracht. ... Es wird sicher nicht mehr lange dauern, bis neben den Astronauten auch Geowissenschaftler den Mond betreten und dann als Lunologen mit den auf der Erde erprobten geologischen, petrographisch-mineralogischen und geophysikalischen Methoden mit der Erforschung des Mondes beginnen werden. // Damit schlägt die Ries-Forschung eine Brücke zur geowissenschaftlichen Erforschung des extraterrestrischen Raumes, und die Arbeiten deutscher Geowissenschaftler im Ries könnten damit indirekt ein nicht unwesentlicher und mit relativ bescheidenen öffentlichen Mitteln geleisteter deutscher Beitrag zur Erforschung des Weltraumes werden' (Vidal 1969: 11).

[37] 'Das Nördlinger Ries war als Tagungsraum gewählt worden, weil es wünschenswert erschien, in der Zeit der ersten Mondlandungen eine Mondkratern ähnliche Kreisstruktur auf der Erde unter sachkundiger Führung kennen zu lernen. Schon frühere Tagungen des Oberrheinischen Geologischen Vereins hatten das Ries mit Exkursionen besucht, dabei war diese Struktur stets unter dem Blickwinkel anderer Entstehungsursachen betrachtet worden' (Metz and Treibs 1970: 5).

Vidal (letter from Georg Wagner to Helmut Vidal, 26 May 1970, UAT 605/343). All the German paragons of new Ries research gave talks in Nördlingen, among them the physicist Erwin David.

Erwin Johann Hermann David (1911–94) had studied physics in Hamburg, Munich and Tübingen. He had studied mainly under Arnold Sommerfeld (1868–1951), who from 1906 to 1935 was Professor of Theoretical Physics at the University of Munich; and Wilhelm Lenz (1888–1957), a physics professor at the University of Hamburg and a former student of Sommerfeld (who recommended David to Sommerfeld as an exceptionally gifted student). After his PhD (possibly in 1933–34), David worked at Peenemünde under Wernher von Braun (1912–77; see Neufeld 1995) as one of a large number of engineers and physicists in the development of rocket technology for the German military. When in August 1943 the Royal Air Force (UK) bombarded Peenemünde, David was injured. It seems likely that he quit his job as a consequence. In the late 1960s, he gave his work address as the *Deutsch-Französisches Forschungsinstitut Saint-Louis* (ISL) [German-French Research Institute in Saint-Louis, France]. Unfortunately, however, there are no papers concerning David preserved at the archive there. The ISL in Saint-Louis is a bi-national institute for military research, founded in 1959 but developed from a French institution that in May 1945 had invited a group of German physicists, who had been active in military research, to work for France. In the late 1960s at ISL, the behaviour of high-velocity bodies upon impact was investigated:[38]

> Anyhow, Mr David was the decisive man who often came to the working group meetings ... and from Mr David we first learned something about such an impact event, such a shock wave: what actually happens in such an event. And therefore, he was also very important, at least for us. To understand how something like that works. ... Mr David was such a small, lively man – it is a pleasure for me to remember him. (Hüttner interview)[39]

---

[38] Hüttner interview; pers. comm. Einwohnermeldeamt Steinen 4 May 2011; pers. comm. Magdalena Kaufmann, Deutsch-Französisches Forschungsinstitut Saint-Louis, by e-mail 18 March 2010; letters from Wilhelm Lenz to Arnold Sommerfeld 29 April 1932 and 5 May 1932 at the archive of the Deutsches Museum Munich HS 1977–28/A,199; letter from Werner Romberg (Oslo) to Arnold Sommerfeld 28 December 1943 at the archive of the Deutsches Museum Munich HS 1977–28/A,290; <http://www.isl.eu/Content/geschichte-isl.aspx?mode=print> accessed 10 March 2010.

[39] 'Jedenfalls war dann der Herr David der entscheidende Mann, der immer wieder zu so Arbeitstagungen kam ... und vom Herrn David haben wir dann erst mal richtig etwas gelernt über so ein Impaktereignis – über so eine Stoßwelle – was da eigentlich so passiert und deswegen ist er auch sehr wichtig – für uns zumindest gewesen. Für das Verständnis wie so etwas funktioniert. ... Es war so ein kleiner lebhafter Mann der Herr David – ich erinnere mich gerne an ihn' (Hüttner interview).

As a physicist, David was able to recognise and assess the force of a hypervelocity impact, and as a person gifted with pedagogic qualities, he was also able to explain and visualise the event for others:

> It is really something, which a physicist can see – what a gargantuan event this is, especially if one imagines that there is something moving in, impacting with a velocity exceeding twice, three, four times the velocity of sound, that is the velocity of [seismic P] waves. This must achieve an enormous transformation of energy. Matter – the solid matter! – will be compressed to a small part of its normal volume. ... [The energy] cannot leave. And then there will be a condition equivalent to a most highly explosive blasting agent, and this then will go bang most horribly (Signer interview)[40]

The conference, and especially David's talk, finally turned the tide for Georg Wagner. David developed the impact scenario in such a way that Wagner for the first time found it sufficiently illustrative. This was no sudden conversion, though. During the months before, Wagner had already shown occasional signs of rethinking, after he had received a favourable report about another talk by David from Karl Schädel, geologist at the Baden-Württemberg Geological Survey in Freiburg (letter from Georg Wagner to Karl Schädel, 11 September 1969, UAT 605/343).

> Already in 1904 E. Werner had ... thought of comets as the cause of the Ries. In 1960 the Chinese Chao found the heavy quartz coesite in the Ries, which also occurs in the Arizona crater, and [its formation] required pressures and temperatures that we can only assume for the impact of a comet. However, when he calculated how many tons of coesite are present in the Long Daniel of Nördlingen, he found not much belief in those that were troubled before by political propagandists. But our mineralogists found in the Ries deposits so much coesite that cannot be explained otherwise. David, a geophysicist, has now calculated the whole thing and depicted the meteorite impact in drawings, which

---

[40] 'Es ist ja schon etwas, das man als Physiker dann schon sieht, was das für ein Wahnsinnsereignis ist. Vor allem wenn man sich vorstellt, dass da irgendetwas kommt, aufschlägt mit einer Geschwindigkeit die zwei-, drei-, viermal so groß ist wie die Ausbreitungsgeschwindigkeit, also die Wellengeschwindigkeit. Das muss ja eine wahnsinnige Energieumwandlung geben. Die Materie, die feste Materie wird zusammengedrückt auf einen kleinen Bruchteil ihres Normalvolumens. ... [Die Energie] kann nicht weg. Und dann gibt es einen Zustand, der eben einem höchst explosiven Sprengstoff entspricht, und das knallt dann ganz greulich' (Signer interview).

also allow an understanding of the [geological] features, so that we have no other possibility of interpretation. (Undated note by Georg Wagner, UAT 605/343)[41]

A mere five days after the conference, Wagner wrote to a former colleague: 'The incandescent hot comet, which here has hit the Earth, is hardly to be doubted anymore' (letter from Georg Wagner to his colleague Schneider, 9 April 1970, UAT 605/343).[42]

Wagner's immediate concern as a convert to the new paradigm was the responsibility of educating the general populace, especially the teachers, who would need a good understanding of the impact event in order to teach their schoolchildren. After the conference, Wagner thus wrote a whole series of near-desperate letters to colleagues pleading with them to do something about it, but was disappointed:

> Without a comet there is no longer a way. Fifteen million years ago, it was a gigantic catastrophe, after which no life existed for a thousand kilometres around Nördlingen! Years will pass, until this is translated into the comprehensible and is evaluated. Ordinary mortals can hardly cope with raising velocity, mass, temperature, pressure to the power of whatever. And until we can translate and evaluate, we do not need journalists, but arduous pioneer work in the area of the catastrophe. Until now, none of the researchers have dared to give a lively, comprehensible representation. Raising physical entities to the power of something does not here help us further. (Letter from Georg Wagner to Georg Fahlbach 7 April 1970, UAT 605/343)[43]

---

[41] 'E. Werner hatte schon 1904 ... an Kometen als Erzeuger des Rieses gedacht. 1960 hatte der Chinese Chao im Ries den Schwerequarz Coesit gefunden, der auch im Arizonakrater vorkommt und Drucke und Temperaturen voraussetzt, wie wir sie nur bei einem Kometeneinschlag annehmen können. Als er aber vorrechnete, wie viel Tonnen Coesit im Langen Daniel von Nördlingen stecken würden, fand er bei den von politischen Propagandisten Leidgeprüften wenig Glauben. Aber unsere Mineralogen fanden in Riesablagerungen so viel Coesit, der sonst nicht zu erklären ist. Ein Geophysiker David hat nun die Sache durchgerechnet und in Zeichnungen den Meteoriteneinschlag dargestellt, der auch die Formen verstehen läßt, so saß wir keine andere Deutungsmöglichkeit haben' (undated note by Georg Wagner, UAT 605/343).

[42] 'An dem glühend heißen Kometen, der hier die Erde traf, ist kaum noch zu zweifeln'. (Letter from Georg Wagner to his colleague Schneider, 9 April 1970, UAT 605/343).

[43] 'Ohne Kometen geht es nicht mehr. Aber vor 15 Mio Jahren war es eine Riesenkatastrophe, nach der es 1000 km um Nördlingen kein Leben mehr gab! Bis das nun ins Verständliche übersetzt und ausgewertet ist, vergehen noch Jahre. Denn mit den Zehnerpotenzen von Geschwindigkeit, Masse, Temperatur, Druck können gewöhnliche Sterbliche kaum etwas anfangen. Und bis wir übersetzen und auswerten können, braucht man keine Journalisten sondern mühselige Pionierarbeit im Katastrophengebiet. Bis jetzt hat noch keiner der Forscher es gewagt, eine lebendige, verständliche Darstellung zu geben.

The journalists are quite plucky, but do not understand nearly enough. Most of the survey geologists, on the other hand, are too lazy at writing or insufficiently well-expressed with their pen. The teachers, however, lack the knowledge. Therefore, I must mediate. (Letter from Georg Wagner to a younger colleague, 9 April 1970, UAT 605/343)[44]

Eventually, Wagner became fairly desperate: 'Who else can I address? By all means, I cannot disgrace myself so much that I ask Munich for help?' (Letter from Georg Wagner to a younger colleague, 9 April 1970, UAT 605/343).[45] In the end, however, he saw no other choice and actually wrote to Vidal in Munich:

You will think ill of me, because I have not contacted you after so much of your friendliness. In the eighty-fifth year, however, everything goes slower, and I was too agitated to get myself orientated. But now, I am convinced of the meteorite. Only I wrestle for an understandable depiction. And this, despite your beautiful blue book, is very difficult. I asked various friends for such a summary that a non-expert can also understand; but received only negative responses. ... Why has nobody developed my four diagrams of the Ries further? (Letter from Georg Wagner to Helmut Vidal, 26 May 1970, UAT 605/343)[46]

Munich too would not offer immediate help, and thus Wagner remained discontented. The pace of Ries research was still so fast that scientific publications tended to become rapidly outdated, while the newest results were communicated within the small community of Ries researchers by letter, telephone,[47] or at

---

Denn die physikalischen Zehnerpotenzen helfen uns hier eben nicht weiter'. (Letter from Georg Wagner to Georg Fahlbach, 7 April 1970, UAT 605/343).

[44] 'Die Journalisten haben wohl Schneid, verstehen aber viel zu wenig. Die meisten Landesgeologen sind dagegen zu schreibfaul oder zu wenig federgewandt. Den Schulmeistern aber fehlt es an Wissen. Drum muß ich eben vermitteln'. (Letter from Georg Wagner to a younger colleague, 9 April 1970, UAT 605/343).

[45] 'An wen könnte ich mich denn sonst wenden? Ich kann mich doch nicht so blamieren, daß ich München um Hilfe bitte'. (Letter from Georg Wagner to a younger colleague, 9 April 1970, UAT 605/343).

[46] 'Sie werden schlecht von mir denken, weil ich nichts hören ließ auf all Ihre erwiesene Freundlichkeit. Aber im 85. Jahr geht alles langsamer und ich war zu erregt, um mich zurecht zu finden. Aber ich bin jetzt vom Meteoriten überzeugt. Nur ringe ich um eine verständliche Darstellung. Und die ist trotz Ihres schönen blauen Buchs sehr schwierig. Ich bat verschiedene Freunde um eine solche Zusammenfassung, die uach [sic] ein Nichtfachmann verstehen kann; erhielt aber nur Körbe. ... Warum hat niemand meine 4 Blockbilder vom Ries weiter entwickelt?' (Letter from Georg Wagner to Helmut Vidal, 26 May1970, UAT 605/343).

[47] ' ... this account illustrates the great importance of the X-ray and telephone to modern mineralogists' (Pecora 1960: 18).

conferences and working group meetings. For the few and most active members of the new impact crater research, there was no time to devote to popular depictions, while outsiders struggled with a lack of proper information. It took quite some time for popular treatments to become available (Gall 1976; Chao et al. 1978; Grau and Höfling 1978), and Wagner fretted about it:

> The Ries won't leave me. We reckoned only with terrestrial forces, with Pluto. And those were insufficient. But the few meteorite falls were inadequate. It was necessary to fall back on cosmic forces of a completely different order of magnitude. I should still write a concluding chapter on the Ries, because the Ries is of a completely different magnitude, Zeus versus Pluto! The reports of meteorite impacts [meteorite falls?] were too meagre for us to allow building a [mental] bridge to the Ries. The journeys to the Moon have opened our eyes. And we have also learned to know different magnitudes of forces! (Letter from Georg Wagner to a colleague, late June 1970, UAT 605/343)[48]

A New Routine

When the research drilling at Nördlingen was completed in 1974, and the Apollo programme was prematurely truncated due to a lack of funding, and a lack of political necessity, after the USA had made their point, Ries research rapidly faded from the headlines and a new routine set in.

Engelhardt and his students continued to work on impact metamorphism, and Engelhardt lectured on extraterrestrial petrology, but the general interest diminished and other equally short-lived sensations replaced the hype about thrilling drilling, lunar proxies and astronauts hammering Ries rocks.

The slowing pace on the other hand finally allowed the fulfilment of Georg Wagner's dream. Finally, there was time to give thought to popular depictions of the Ries and Steinheim event. In 1976, the *Jura-Museum Eichstätt* opened, and – despite its name 'Museum of the Jurassic' – dedicated some exhibition space to the Ries Crater and the impact event. In 1978 the *Meteorkratermuseum* in Sontheim (Steinheim Basin) opened its doors, dedicated to the geology of the

---

[48] 'Das Ries läßt mich nicht los. Wir rechneten eben nur mit irdischen Kräften, mit Pluto. Und die reichten nicht aus. Aber die wenigen Meteoritenfälle genügten nicht. Man mußte kosmische Kräfte, ganz andere Größenordnungen hearziehen. [sic] Eigentlich müßte ich noch ein Schlußwort zum Ries schreiben. Denn das Ries ist eben von ganz anderer Größenordnung Zeus gegenüber Pluto! Die berichte [sic] von Meteoriteneinschlägen waren zu kümmerlich, als da9 [sic] wir die Brücke zum Ries hätten schlagen können. Die Fahrten zum Mond haben unsere Augen geweitet. Und haben auch andere Zehnerpotenzen von Kräften kennen gelernt!' (Letter from Georg Wagner to a colleague, late June 1970, UAT 605/343).

Steinheim Basin from the pre-impact geology, through the impact event and its rocks, to the post-impact lake sediments and their fossils. Additionally, a footpath of 9 km length, with explanations in the field, now allowed the interested visitor to explore the crater and its rocks directly and in a hands-on manner.

In 1990 the *Rieskratermuseum* in Nördlingen was opened, displaying Ries geology as well as planetary science, meteorites and a sample of Moon rock. The exhibition also placed the crater in the context of other impact craters worldwide. The section on the history of Ries geology has recently been replaced by an exhibition on the K/T-boundary and mass extinctions via impacts. In the Ries Basin too, numerous quarries and other outcrops are accessible to those who wish to see the impact crater and its rocky phenomena with their own eyes. Guide booklets provide easy access to individual excursions by car or bicycle.

In 1998 the ZERIN (Centre for Ries Crater and Impact Research in Nördlingen) was opened as a collaboration between the Humboldt University in Berlin and the town of Nördlingen. Here researchers find support, access to drill cores of the various Ries drillings, a collection of impact rocks, pictures, maps and outcrop descriptions. Summer-schools are also held here to educate geoscience students about the geology of impact craters.[49]

While museums and popular publications indicate that Ries research has finally reached the calmer waters of 'normal science', much of the excitement and stimulus of the 1960s and early 1970s is gone, although some unsolved geological questions do remain:

> Very true, this has not been worked off by palaeontology, because this [type of event] must be detectable. There must be possibilities ... – Look at the pumice ash from the [volcano] Laacher See, which can be found as far as Geneva in the peat profiles, and it was not great masses of it that were drifted by the wind, but only centimetres or less. This is still [work] to be done. That must be acknowledged. You know, one only addresses such issues if one is convinced that it has been so. There was an impact and ... there must have been catastrophic consequences for the environment. If one is convinced of this, ... then one has also the interest to say that now we must have a look at it. What did these consequences really look like? For the animals, for example? (Pichler interview)[50]

---

[49]   <http://www.naturkundemuseum-berlin.de/sammlungen/mineralogie-petrographie/zentrum-fuer-rieskrater-und-impaktforschung-noerdlingen-zerin.html> accessed 26 September 2012.

[50]   'Ganz richtig, das ist nicht aufgearbeitet von paläontologischer Sicht, denn das muss nachweisbar sein. Da muss es Möglichkeiten geben ... – Schauen Sie mal die Sache an mit den Laacher Bims-Aschen, die man ja weit in Genf in den Moorprofilen findet und das war ja auch keine große Masse, die da verweht worden ist, sondern das waren ja nur Zentimeter oder weniger. Das wäre noch zu tun, aber das muss man anerkennen. Wissen Sie, da geht man nur wirklich ran, wenn man überzeugt ist, das war so; da hat es reingekracht und ... da muss es

The effect of the Ries and Steinheim impacts on the environment have been investigated by palaeontologists of the Bavarian State Collection of Palaeontology and Geology and the University of Munich (Heißig 1986; Böhme et al. 2001) with negative results: 'So on the base of our present knowledge, the fact must be accepted, that there is no evidence of any changes caused by the Ries impact event. The faunal changes, reflecting environmental and climatic changes, seem to be gradual and continuous until the stratigraphic gap above the 'sand-marl-cover'' (Heißig 1986: 178):

> The Ries impact event had no long-time effect on the composition of the fauna and the flora in Southern Germany. The radius of destruction was less than the area of distribution of all of the documented species. The ecosystem was quickly reconstituted after the catastrophic impact. The probably fast reorganization of the fluviatile ecosystems of the Upper Freshwater Molasse is explained by their dynamic properties. The immigration of the faunal and floral taxa was accelerated by fluviatile drift from refugia. (Böhme et al. 2001: 228)

The methods used, however, were strangely inadequate in addressing the question at hand, namely: What actually was the effect of the impacts on the environment? Both papers operate with basically a single outcrop, the quarry of Unterneul, some 60 km SSE of the Ries Crater. This proximal outcrop is the only known site that spans the catastrophe itself, as well as some time before and after it:

> Unterneul ... shows some traces of the catastrophe preserved within a channel that was later covered by a small lake and its sediments. This channel was incised through 5 m of sand into a greyish marl bed. In this situation the impact event occurred. The upper 0.2 m of the marl are heavily disturbed by the falling blocks and the compression wave. The sediment, even if already consolidated, was partly reworked and formed a matrix filling the space between undissolved debris of marl. There are horizontal shearing planes with a faint glimmer of oriented mica grains on the surfaces, cutting through fossil snails. Medium-sized to small blocks intruded with sharp edges into the surface of the marl.

> The whole channel was later covered by fine clay, or marginally, by the gravels of a small delta. These contain lots of flattened trunks, remains of the broken logs that escaped burning by falling into the water. On the other hand, well rounded wooden blocks were found that have a higher degree of carbonization, gagate

---

katastrophale Umweltfolgen gehabt haben. Wenn man davon überzeugt ist, ... hat man auch Interesse, dass man sagt, jetzt müssen wir uns das mal ansehen. Wie sind denn die Folgen wirklich gewesen? Tierwelt zum Beispiel' (Pichler interview).

near the surface, lignite in the centre, with folded wood structure. These pieces of wood have been compressed before being imbedded in the sediments and were, therefore, no more compressible. Probably they were thrown out from the crater or its immediate neighbourhood, compressed by the immense pressure of the explosion. During their flight into the foreland all edges and splitters burned away until they fell into the water of the channel. Where the immediate cover of the Ries debris is clay, a few centimetres above we find the first shells of the mussel *Margaritifera*, sometimes broken over the edge of a bigger block by the compression of the clay. Half a meter above that layer follows a well-preserved and rich leaf flora. (Böhme et al. 2001: 218–19)

Otherwise, the authors only describe outcrops and their fauna and flora below or above the event horizon without adequate control on the exact stratigraphic age. 'The stratigraphy within the Upper Freshwater Molasse is based mainly on large mammals. Dehm (1951) distinguished three series by the means of Proboscidea' (Heißig 1986: 171). No great palaeontological expertise is needed to realise that the use of elephants as guide fossils is somewhat limited in terms of resolution. Up to the present (2014), there has been no systematic review of available drill cores and their microfossil contents that may allow for a time resolution high enough to assess such a (geologically speaking) short-term event.

None of the impact researchers in southern Germany ever claimed that the Ries and Steinheim impacts produced a mass-extinction, and the truth of this has been demonstrated by Heißig (1986) and Böhme et al. (2001). For anything else, the methods used were unfortunately inadequate.

While mainstream palaeontology in the area still seems curiously reluctant to take the impact events into account, others clearly overshot the mark. Any number of freeloaders took up the fashion and made themselves a nuisance to geologists:[51]

> Well, then a lot of true, probable and alleged meteorite craters were found, and a genuine fashion was created. Everybody wanted to have his own meteorite crater. There have been really bizarre ideas, for example, people have interpreted round lakes in the Alps – they are lakes in sink holes, which have formed over gypsum – as meteorite craters; and so on. However, that did not preclude that there really exist true ... impact structures. (Trümpy interview)[52]

---

[51]   This continues well into the present; see the alleged 'Chiemgau impact' (<http://web.archive.org/web/20070927230744/>, <http://download.naturkundemuseum-berlin.de/presse/ Chiemgau.pdf> or <http://de.scribd.com/doc/55715772/Offener-Brief-zum-Chiemgau-Impakt-und-zu-Aktivitaten-des-Chiemgau-Impact-Research-Teams-in-der-Offentlichkeit> all accessed 14 November 2012).

[52]   'Nun, es wurde dann eine Menge von echten, vermutlichen und vermeintlichen Meteorkratern gefunden und es bildete sich eine eigentliche Zeitmode. Jeder wollte

On the other hand, sometimes it proved difficult to assess at first glance the scientific fertility of regarding unsolved geological problems in the light of the impact theory:

> During the past fifty years that it has been under discussion, meteorite impact has been called on to explain practically every mystery in the geological column including continental drift, faunal extinctions, and climate changes. Such speculation runs well ahead of available data, but it must not blind geologists to the fact that meteorite impacts may be the best answer for some of the unexplained problems of the earth. (French 1968a: 13)

Researchers such as French and Dietz frowned upon such speculations, because they feared that the enthusiasts would do more harm than good to the impact theory, 'gross overstatement [for impact] will give more comfort to the critics of impact geology than its advocates' (Robert Dietz criticising René Gallant, quoted by Torrens 1998: 184), but as French had pointed out, for some of the unexplained problems, meteorite impact may indeed provide an answer:

> Impacts then have played a big role in palaeontology. This began with the discovery by Walter Alvarez[53] in Gubbio in central Italy that there the boundary layer between the Cretaceous and Tertiary, which can be established very well by pelagic foraminifera, has an abnormally high content of iridium. And in collaboration with his father, a famous physicist,[54] he then formulated the hypothesis that at the Cretaceous/Tertiary-boundary a huge celestial body fell to Earth and that this celestial body was responsible for the mass extinction at the end of the Cretaceous. (Trümpy interview)[55]

---

seinen Meteorkrater haben. Es gab darunter ganz bizzare Ideen, z.B. hat man runde Seen in den Alpen, das sind Dolinenseen, die sich über Gips gebildet haben, als Meteorkrater und so weiter gedeutet. Das konnte aber nicht davon abhalten, dass es eben doch echte ... Impaktstrukturen gab' (Trümpy interview).

[53]   Walter Alvarez (born 1940) received his PhD in geology at Princeton. He then worked for the petroleum industry. In 1977 he joined the faculty at the University of California at Berkeley (<http://eps.berkeley.edu/development/view_person.php?uid=90> accessed 14 November 2012).

[54]   Luis W. Alvarez (1911–88), an American experimental physicist and inventor, spent most of his professional career on the faculty of the University of California, Berkeley. In 1968 he was awarded the Nobel Prize in Physics (<http://www.nobelprize.org/nobel_prizes/physics/laureates/1968/alvarez-bio.html> accessed 14 November 2012).

[55]   'Nun die Impakte, die haben dann natürlich auch eine große Rolle gespielt in der Paläontologie. Das fing an mit der Entdeckung von Walter Alvarez in Gubbio in Mittelitalien, dass dort die Grenzschicht zwischen Kreide und Tertiär, die sich aufgrund der pelagischen Foraminiferen sehr gut festlegen ließ, einen abnorm hohen Gehalt an Iridium

Although a large end-Cretaceous impact is now well-established, and even the crater itself has been found, there is still some debate about the actual link between the impact and the mass extinction event at the K/T-boundary. Geologists, mineralogists and micro-palaeontologists have in general been more inclined to accept a causal link, while reluctance seems more common among the community of vertebrate palaeontologists (Glen 1994: 52–4; see also Keller and Smit et al. 2004; Marvin 1999).

Little Green Prussians

As early as 1963, Josef Zähringer (1929–70)[56] drew attention to inconsistencies within the stratigraphy of the Molasse Basin north of the Alps. Although the Ries impact and the moldavites were of the same radiometric age, local stratigraphies placed them in different formations of allegedly different biostratigraphic ages (Zähringer in AG Ries 1963: 35). Palaeontologists and stratigraphers, however, saw no reason for doing anything about it.

The problem of stratigraphy again became an issue when in 1996 shocked quartz and suevite pebbles where found at the base of the *Graupensand*, another of the classic 'problems' of the geology of southern Germany (Buchner et al. 1996). This coarse-grained quartz sand deposit was supposed to be some three million years older than the Ries and Steinheim impact event.

The ensuing debate between apologetic molasse basin stratigraphers and the challengers was intense (Lemcke 1997; Buchner et al. 1998; Reichenbacher et al. 1998a; Reichenbacher et al. 1998b; Buchner et al. 2003), but did not lead to a satisfactory conclusion. For an outsider, two schools seem to be pitted against each other. Although Reichenbacher et al. (1998a), sixteen authors arguing by authority, confirm the molasse stratigraphy to be well-established, radiometric

---

aufweist. Und in Zusammenarbeit mit seinem Vater, einem berühmten Physiker, hat er dann die Hypothese aufgestellt, dass dieses Iridium ein Hinweis sei, dass da an der Kreide-Tertiär-Grenze ein riesiger Himmelskörper auf die Erde gefallen sei und dass dieser Himmelskörper für das Massensterben am Ende der Kreide verantwortlich sei' (Trümpy interview). For an account of this discovery see Alvarez (1997).

[56] Josef Zähringer had studied physics, mathematics, chemistry and mineralogy at the universities of Freiburg and Göttingen. From 1955 to 1956, he was an assistant at the Institute of Physics in Freiburg, where he also obtained his doctoral degree in 1956. He then joined the faculty of the Brookhaven National Laboratory in the USA, and in 1958 became an assistant at the Max Planck Institute for Nuclear Physics in Heidelberg, later obtaining his qualification for professorship in 1963 at the University of Heidelberg. In 1964 he became a full staff member of the Heidelberg Max Planck Institute, and was appointed its director one year later (<http://edoc.mpg.de/62408>, <http://www.mpg.de/1163440/S004_Rueckblende_096–097.pdf> both accessed 14 November 2012).

dating (^{40}Ar/^{39}Ar laser probe age determinations) of two glass particles from *Graupensand* components yielded identical ages to the Ries suevite (Buchner et al. 2003).

It is clearly beyond the scope of this book to express any verdict in what is an ongoing debate, especially since it currently seems to be locked in some sort of intellectual siege. It is, however, interesting to see how two outsiders, a former palaeontologist and a former volcanologist, reacted to the issue before the radiometric age dating was published:

> In 1967 I was ... in Australia, and there I have ... seen small meteorite craters, where iron was still preserved; the Henbury Craters and especially the Gosses Bluff ... with all the trimmings, with polymictic breccia, with shatter cones, with everything the heart desires, ... and there for me the impact nature was completely unequivocal and afterwards I also no longer uttered doubts about the Ries Basin ... .

> The whole story then became again topical about three years ago [1996] via the 'famous' *Graupensand* affair. Within the *Graupensand* ... minerals were found that show traces of a shock metamorphism as it occurred in the Ries Basin. Proposals to derive these from pegmatites in the Bohemian Massif were not convincing, and Mr von Engelhardt, who really can be expected to know this, has assured me that these components in the *Graupensand* show effectively the same features as the Ries impactites. And there of course the question of the age of the *Graupensand* is raised: Was the *Graupensand* ... of approximately Lower Miocene – that is approximately Ottnang age? Then it should be about three million years older, that is some 18 million years old, than the Ries event 15 million years ago. And then we should have two impacts shortly after each other, three million years apart in more or less the same area, and that seemed to me relatively improbable. I have therefore written to Kurt Lemcke[57] in Munich that I can understand the affair only if, on board of such an asteroid, there happened to be some sort of ... malicious creatures, for instance little green Prussians, who with constant malice every three million years hurled pieces of said asteroid on Bavaria. ... Then a great controversy ensued in which other geologists voiced the opinion that the *Graupensand*s were much younger and of Mid-Miocene age, that

---

[57] Kurt Lemcke (1914–2003) studied geology in Heidelberg and Freiburg. He obtained his PhD in 1937 from the University of Jena. Until 1939, he was an assistant at the Mecklenburg Geological Survey, which was transformed into the Geological Institute of the University of Rostock. From 1940 to 1945, he worked for the German Geological Survey and partook in World War II. After the war, he mapped as a freelance geologist in Württemberg, and in 1948 he became a petroleum geologist at the Elwerath Company in Hannover. From 1969 onwards, he also lectured at the Technical University of Munich, where he was appointed honorary professor in 1974 (Freyberg 1974: 94; <http://www.dgg.de/cms/front_content.php?idcat=161> accessed 14 November 2012).

is of the same age or somewhat younger than the Ries event, and that they simply contained reworked impactites of the Ries as well as polymictic breccia. Well, of course I cannot decide this question. If the *Graupensand*s are of equal age or somewhat younger than the Ries event, then the world is completely OK, then we do not need to revise our opinion. Should, however, the *Graupensand* be older, then the story becomes embarrassing again, because to me it still seems extremely improbable that three million years apart, two times, such a celestial body should burst into the same area. It is really funny that this highly interesting question is now again reduced to quite simple, conventional stratigraphy, and the *Graupensand* problem is really a very important problem. For me, as a former palaeontologist, guide fossils – in the Tertiary this means mouse teeth – are of greatest relevance. ... I didn't work in this area, but this conventional placement of the age of the *Graupensand* deposit is extraordinarily important to assess the nature of the Ries event. Explosive volcanism, of course, can happen repeatedly without problems. Repeated impacts in the same area are not impossible; however, they are extremely improbable.

As I said, after I had seen the Ries myself and after I had seen Gosses Bluff, I have put my suspicion that it might be some sort of unknown volcanism after all nicely back into the drawer, and now three years ago came this *Graupensand* problem, which has made the whole story again topical.

... Mr Lemcke by the way has never answered my letter with the little green Prussians. ... And now, as I have said, very much depends on the decisive dating of the pebbles; that is physical [radiometric] dating of the pebbles in the *Graupensand*. ... if it is possible to date these components of Ries rock within the *Graupensand*, then this would be a direct clue. (Trümpy interview)[58]

---

[58] '1967 war ich ... in Australien und habe dort ... kleine Meteoritenkrater gesehen, wo noch Eisen vorhanden war, die Henbury-Craters und dann vor allem den Gosses Bluff ... mit allem Drum und Dran, mit bunter Brekzie, mit Shatter Cones, mit allem was man sich nur wünschen kann, ... und dort war also die Impaktnatur für mich völlig eindeutig und ich habe dann auch keine Zweifel mehr am Ries geäußert ... . // Die ganze Geschichte wurde dann etwa vor drei Jahren wieder aktuell, durch die berühmte Graupensandgeschichte. Also im Graupensand wurden ... Mineralien gefunden die Spuren einer Schockmetamorphose, wie sie im Ries vorkommt, zeigen. Erklärungsversuche, das aus Pegmatiten im Böhmischen Massiv herzuleiten, waren nicht überzeugend und Herr von Engelhardt, der das schließlich wissen muss, hat mir versichert, dass diese Komponente in den Graupensanden effektiv dieselben Erscheinungen zeigen wie die Riesimpaktite. Und da stellte sich natürlich die Frage nach dem Alter des Graupensandes: War der Graupensand ... untermiozänen, ungefähr Ottnang-Alters? Dann müsste er um etwa drei Millionen Jahre älter sein, also etwa 18 Millionen Jahre, gegenüber dem 15 Millionen Jahre alten Riesereignisses. Und dann müssten wir zwei Impakte kurz, das heißt drei Millionen Jahre, nacheinander im mehr oder

Although as a former palaeontologist Rudolf Trümpy voiced much confidence in classical stratigraphic methods, and thus fell back into doubting the impact origin of the Ries Crater rather than concluding that there must be something wrong with the local stratigraphy, the ex-volcanologist Hans Pichler readily accepted the mineralogical evidence, while being quite relaxed in the light of the possible overthrow of decades of stratigraphic work:

> At the base of the *Graupensand* channel pebbles have been discovered, and some curious geologist from Stuttgart [Elmar Buchner] ... thinks it might be worthwhile to look more closely at these fragments; he has thin-sections made; he looks into them and is indeed capable of recognising that the quartz therein presents impact structures, and sends the whole stuff to von Engelhardt. "What do you think, great master?" – since he is really an expert in these matters – and

weniger gleichen Gebiet gehabt haben, und das erschien mir relativ unwahrscheinlich. Ich habe deshalb an Kurt Lemcke in München geschrieben, dass ich mir die Sache nur erklären könne, falls an Bord eines solchen Asteroiden irgendwelche ... böswillige Lebewesen, zum Beispiel kleine grüne Preußen säßen, die mit konstanter Bosheit alle drei Millionen Jahre Stücke dieses Asteroiden auf Bayern heruntergeschmissen hätten. ... Nun entspannt sich daraus eine große Kontroverse in dem andere Geologen nun die Ansicht vertraten, die Graupensande seien viel jünger und seien mittelmiozänen Alters, also gleich alt oder etwas jünger als das Riesereignis und sie hätten einfach Impaktite des Ries und bunte Brekzie wieder aufgearbeitet. Nun selbstverständlich kann ich diese Frage nicht entscheiden. Wenn die Graupensande gleich alt oder etwas jünger als das Riesereignis sind, dann ist die Welt völlig in Ordnung, dann müssen wir unsere Ansichten nicht revidieren. Sollten die Graupensande aber älter sein, dann wird die Geschichte wiederum peinlich, denn es erscheint mir nach wie vor extrem unwahrscheinlich, dass im Abstand von drei Millionen Jahren zweimal ein solcher Himmelskörper ungefähr in dieselbe Gegend geplatzt hätte. Es ist also lustig, dass diese im Grunde genommen hochinteressante Frage sich nun wieder auf ganz simple konventionelle Stratigraphie reduziert und das Graupensandproblem ist wirklich ein sehr wichtiges Problem. Als ehemaliger Paläontologe würde ich das größte Gewicht auf Leitfossilien, das heißt im Tertiär Mausezähnchen, legen. ... Ich habe in dieser Gegend nicht gearbeitet, aber diese konventionelle Lage des Alters der Graupensandrinne, die ist außerordentlich wichtig, um die Natur des Riesereignisses abzuklären. Wiederholter Explosionsvulkanismus, kann natürlich ohne weiteres auftreten. Wiederholte Impakte im selben Gebiet sind nicht unmöglich, aber extrem unwahrscheinlich. // Also wie gesagt nachdem ich das Ries selbst angesehen hatte und nachdem ich den Gosses Bluff angesehen hatte, habe ich meinen früheren Verdacht, dass das doch eine Art unbekannter Vulkanismus sein könne, wieder schön in die Schublade gelegt und jetzt kam vor drei Jahren dieses Graupensandproblem, dass die ganze Geschichte irgendwie aktuell machte. // ... Der Herr Lemcke hat mir nie geantwortet auf meinen Brief mit den kleinen grünen Preußen. ... Und jetzt, wie gesagt, hängt sehr viel ab von der entscheidenden Datierung dieser Graupensande und von der wenn immer möglichen Datierung von Geröllen, also physikalischen Datierungen von Geröllen im Graupensand. ... wenn man diese Komponenten von Riesgestein im Graupensand datieren kann, dann wäre das ein direkter Hinweis' (Trümpy interview).

Engelhardt confirms that they are Ries pebbles; and thereby, the whole, or let's say, the upper part of the stratigraphy of the molasse basin is overturned, and a thick paper is published [Reichenbacher et al. 1998a] ... where all the opinions run together and ... good Lemcke became really angry with Engelhardt, although they really had been colleagues since the Elwerath time ... and it is practically a relaunch of the Ries debate of the 1960s.

... All in all the two pieces of pebble have practically mixed up a complete stratigraphy.

... Yesterday I talked to Engelhardt and he told me: "You know what happened? Now from another location a number of pebbles have been discovered from the *Graupensand* channel and I have investigated this too, and it is the same stuff as in the Ries". There are now no more doubts ...

That it has somehow been swindled into [the locality] or gotten into it by accident, by anthropogenic nature or such, seems not to be the case. This can be excluded. Consequently, a stratigraphic fact is 'overturned again' – in inverted commas. ... I find that splendid; something new again ... so that is highly interesting, and Engelhardt is somewhat frustrated that it causes so much hubbub. ...

[Engelhardt] quite coolly describes the features in the thin sections of these pebbles. They are unambiguously shocked things that must be connected to the Ries event; and [he] does not at all discuss that these should be stratigraphically different; instead, he says, that's it: full stop. So, now, take it up, contest this! Similar to what the Americans have said: This is such and such, therefore the Ries is a structure that was formed by an impact. Very interesting story; I only laughed at it.

... Thus the molasse stratigraphy is not final, but must be ordered anew. Why not? Now it's their turn to come up with something! (Pichler interview)[59]

---

[59] 'Es werden also am Boden der Graupensandrinne Gerölle entdeckt, und irgend ein neugieriger Stuttgarter Geologe ... meint es wäre wert, dass man diese Bruchstücke mal genauer ansieht, lässt also Dünnschliffe machen, schaut rein und ist tatsächlich fähig zu erkennen, dass die Quarze darin Impaktstrukturen aufweisen und schickt das Ganze zu von Engelhardt. "Was meinen Sie denn, großer Meister?" – denn er ist wirklich Fachmann auf diesem Gebiet – und Engelhardt bestätigt, es sind Riesgerölle und damit wird die gesamte oder sagen wir mal der obere Teil der Stratigraphie des Molassebeckens über den Haufen geworfen und es gibt eine dicke Schrift, ... da laufen die ganzen Meinungen zusammen und ... der gute Lemcke, der ist dem Engelhardt richtig böse geworden, obwohl die eigentlich Kollegen seit der Elwerath-Zeit sind. ... und es ist praktisch das eine neue Auflage des Riesstreites von 1960. // ... Im Endeffekt haben diese beiden Geröllstücke praktisch

As noted above, radiometric dating confirmed that the age of the Ries suevite and suevite pebbles from the *Graupensand* are identical within analytical uncertainties (14.3–14.4 Ma). Thus, although 'little green Prussians' are no longer to be reckoned with, and therefore the impact origin was again acceptable for Rudolf Trümpy, the pressure on the stratigraphy of the molasse basin has severely increased; to what effect is no longer part of history, but belongs to the future.

---

eine ganze Stratigraphie durcheinander gebracht. // ... Ich habe mit Engelhardt gestern gesprochen und er hat mir gesagt: "Wissen Sie was passiert ist? Jetzt sind also von einem anderen Fundort soundso viel Gerölle aufgetaucht von der Graupensandrinne und das habe ich auch untersucht und das ist das gleiche Zeug wie das Ries". Es gibt also nun keinen Zweifel mehr ... // Dass das da irgendwie reingeschwindelt worden ist, oder durch Zufall da reingekommen ist, anthropogener Art oder so was, das ist wohl nicht der Fall. Das kann ausgeschlossen werden. Damit 'stürzt wieder' – in Anführungszeichen – eine stratigraphische Tatsache zusammen. ... Ich finde das prima, mal was Neues ... das ist also hochinteressant und Engelhardt ist also etwas frustriert darüber, dass es so viel Wirbel macht. ... // [Engelhardt] beschreibt da ganz cool die Erscheinungen im Dünnschliffbild dieser Gerölle. Das sind eindeutig geschockte Sachen, die mit dem Ries-Event im Zusammenhang stehen; und lässt sich auch gar nicht weiter aus, dass das stratigraphisch anders sein müsste, sondern er sagt, das ist es: Punkt. So macht mal was daraus, versaut das mal. So ähnlich wie die Amerikaner gesagt haben, das ist so und so, das Ries ist damit eine Struktur, die eben von einem Impakt gebildet wird. Sehr interessant die Geschichte; ich habe nur gelacht. // ... Auf die Weise ist die Molassestratigraphie nichts Endgültiges, sondern muss neu geordnet werden. Warum denn nicht? Sollen sie sich mal was einfallen lassen!' (Pichler interview).

Chapter 13

From Local Patriotism to a Planetary Perspective

Having taken our narrative close to the present, it is now time to sum up. As we have seen, for much of the twentieth century geologists in Germany struggled to understand the Ries Basin and Steinheim Basin. The 'Ries Problem' has been a staple diet of several generations of geologists in southern Germany, and what at first glance looked like the product of some enigmatic volcanic activity turned out to be exceedingly complex and of far-reaching consequence for both scientific thinking and methodology. Traditional efforts to understand the Ries and Steinheim Basins from the local geological history and framework eventually failed, when the discovery of high-pressure phases of SiO_2 and the ensuing thorough testing of the new theory alongside the old paradigm, forced German geologists to accept the reality of catastrophic impact events.

Dismissal of the impact theory, which was suggested as early as 1904 and discussed to some extent outside the mainstream during the early 1930s, was only partly articulated with factual arguments, such as seeming clues to the existence of thermal springs after the formation of the Ries Basin that appeared to point to an extended period of volcanic activity. The enduring activity of the 'Ries volcano' was also supported by an alleged poly-phase deposition of suevite 'tuff', whereas the negative gravity anomaly and absence of actual meteoritic material would not fit the proposed impact scenario (Chapter 3). Thus, in those early days, the impact hypothesis had to face rather strong contradictory arguments, whereas volcanology was not advanced enough to create sufficient doubt as to the notion of obtaining the necessary explosive energy from within the Earth by 'ordinary' volcanic forces.

However, the most prevalent of factual arguments was the special, historically predisposed location of the Ries Basin, which was believed to make a direct meteorite hit here of all places most unlikely. This last argument – essentially alone among factual arguments – remained in discussion and appeared convincing until the very last resistance died away in the 1970s (Chapter 10).

Otherwise, the motivation for the blatant dismissal of the impact idea and refusal to even discuss it boiled down to structural problems rather than scientific arguments. Scientific isolation of Germany in the 1930s and 1940s prevented exchange with the developing impact ideas in the USA and Canada, and a rigid actualism/uniformitarianism (although the alternative volcanic models were themselves non-actualistic) rejected catastrophic theories with an extraterrestrial cause out of philosophical principles rather than because of scientific data. A habit of thinking in terms of 'geological myths' further hampered systematic research, invoking crude analogies to volcanoes like Krakatoa or Bandai-San, without actually having worked there, thus using the concept of 'theory' for presenting a 'narrative' without the intent to test it.

Rather trivial problems such as the location of the Ries Basin at the historical border between Bavaria and Württemberg, which were and are served by two separate Geological Surveys, and rivalry between the universities of Tübingen and Munich, did not improve matters. At the same time, a lack of mineralogical work on the two structures, possibly curbed by the financially insufficient state of the Tübingen Mineralogical Institute as well as the traditional interest of the Munich palaeontologists rather than geologists and mineralogists in the Ries Basin, did not allow for the early recognition of the unusual mineralogy and geochemistry of the Ries Basin rocks. The concentration of palaeontologists on the Ries Basin also imposed a stratigraphic rather than a structural approach to Ries research. Especially during the earlier debate in the 1930s, acceptance of an impact theory as an interesting alternative to the predominate volcanic theories was additionally hampered by deficits in the development of volcanology and a lack of physical understanding, which still hampered the understanding and visualisation of the impact process later on (Chapters 3, 6 and 10).

Hölder (1962) stressed as a key feature of the paradigm shift that within the new impact theory the Earth was no longer a closed system, but was vulnerable to cosmic events (Chapter 9). Engelhardt and Zimmermann (1982) agreed to a certain extent, when they pointed out that the impact theory violated an unspoken principle of geology, the 'principle of endogenic causes' (Engelhardt and Zimmermann 1982: 357–8). These authors additionally explained the strong resistance against the impact theory by a severe violation of the principle of uniformitarianism and of actualism, which the impact theory implied (Engelhardt and Zimmermann 1982: 357). Although there is some truth in this perception (as, for example, Seemann strongly objected to both the impact and volcanic Ries theories because they were irreconcilable with his gradualist worldview: Chapter 3), the Kranz–Löffler theory of a central volcanic explosion was just as non-actualistic and non-uniformitarian as any impact theory might be.

Remarkably, Engelhardt and Zimmermann in their philosophical treatise violate their own definitions of these principles and their consequences (Engelhardt and Zimmermann 1982: 354–5). In notable consensus with Beurlen (1935) and Becksmann (1939), these authors regard actualism and uniformitarianism as 'ahistoric' concepts, in contrast to what they call 'historical non-actualism and non-uniformitarianism':

For actualistic and uniformitarian theories the Earth's past is 'everlasting present'. Their intent concerning objects recorded from the past – rocks, minerals, fossils, structures and so on – is to strip them at first of their seeming strangeness, to recognize the present in them and to reconstruct 'past forms of the present' from them. The unique past events and conditions as such are irrelevant for the actualist and uniformitarianist, because for them everything geological also occurs and exists in the present. What is relevant – from the actualistic or uniformitarian point of view – are the laws of nature, which manifest themselves in present and past things alike.

For non-actualistic and non-uniformitarian theories, however, the Earth's past is not everlasting present, but the history of 'other times' and events 'alien to the present'. Alongside the laws of nature, which concern things present, past and future alike, the uniqueness of singular events, conditions and époques and their succession in time are the true topic of research, as long as it is oriented in a non-actualistic and non-uniformitarianistic way. In theories with this emphasis, geoscience departs furthest from physics and chemistry, constituting itself as historical science and becoming related to human history, which is more interested in single, specific events and in new things appearing in time than in general laws. (Engelhardt and Zimmermann 1982: 355)[1]

Thus, according to these authors, both views (actualism and non-actualism) have different 'concepts of geological time'. Historical geoscientific research

---

[1] 'Für aktualistische und uniformitaristische Theorien ist Erdvergangenheit 'immerwährende Gegenwart'. Ihr Vorhaben hinsichtlich der aus der Vergangenheit überlieferten Objekte – der Gesteine, Mineralien, Fossilien, Strukturen usw. – besteht darin, dieselben ihrer zunächst erscheinenden Fremdheit zu entkleiden, an ihnen die Gegenwart wiederzuerkennen und aus ihnen 'vergangene Gegenwart' zu rekonstruieren. Die singulären vergangenen Ereignisse und Zustände sind für den Aktualisten und Uniformitarier nicht als solche relevant, da sich für ihn alles Geologische auch in der Gegenwart abspielt und vorhanden ist. Worauf es – aktualistisch oder uniformitaristisch gesehen – ankommt, sind die sich im Gegenwärtigem und Vergangenem gleichermaßen manifestierenden Gesetzmäßigkeiten. // Für nicht-aktualistische und nicht-uniformitaristische Theorien ist die Erdvergangenheit dagegen nicht immerwährende Gegenwart, sondern die Geschichte 'anderer Zeiten' und 'gegenwartsfremder' Ereignisse. Neben den Gesetzmäßigkeiten, die Gegenwärtiges, Vergangenes und Zukünftiges in gleicher Weise betreffen, ist die Einmaligkeit singulärer Ereignisse, Zustände und Epochen und deren Aufeinanderfolgen in der Zeit das eigentliche Thema der Forschung, sofern sie nicht-aktualistisch oder nicht-uniformitarisch orientiert ist. In Theorien mit diesem Tenor entfernt sich die Geowissenschaft am weitesten von Physik und Chemie, konstituiert sie sich als historische Wissenschaft und wird der Humanhistorie verwandt, die an den einzelnen und spezifischen Ereignissen und am Neuen, das in der Zeit erscheint, mehr interessiert ist als an allgemeinen Gesetzen' (Engelhardt and Zimmermann 1982: 355).

anticipates and allows for catastrophes: 'local, sudden and rare events, that stand out by the intensity of their effects from the calm course of history before and afterwards, and which, because of their rareness and the short duration given to human observation, have not been observed' (Engelhardt and Zimmermann 1982: 355).[2]

However, Engelhardt's and Zimmermann's (1982) view agrees neither with the historical data concerning Ries research, nor with common geological practise. Outside the ideological framework of 'German Geology' (Chapter 7), historical geology (stratigraphy) relies on observed or inferred present processes to explain past deposits and rock formations, and thus values the principle of actualism for its specifically historical research programme. On the other hand, the idea of a Ries and Steinheim double impact was experienced (especially by the historically minded Ries researchers) as deeply non-actualistic. The catastrophic event was in fact dismissed with the express argument that it violated the principle of actualism, and the stratigraphers and palaeontologists of the Munich School in particular preferred an explanation based on small, local volcanic eruptions to one based on the cataclysmic central explosion of the Ries Volcano (Chapter 3). Thus, historical/stratigraphic Ries research did not allow for catastrophes.

Contrary to Engelhardt's and Zimmermann's (1982) statement, the Ries and Steinheim impacts – these 'local, sudden and rare events, which stand out by the intensity of their effects from the calm course of history before and afterwards' (see above) – were more readily accepted by those who had already been concerned previously with 'ahistorical' laws of nature rather than deciphering Earth history from the geological record. Consequently, instead of separating geology from physics (as it should have done according to Engelhardt and Zimmermann's philosophical view quoted above), the impact theory brought geology much closer to physics, chemistry, crystallography and mineralogy than before.

Highlighting the Ries Basin area as a special location – predisposed for a *singular* type of volcanic activity, as had frequently been argued by Ries experts up to the last resistance against the impact theory – was certainly a factual argument. However, it also points to a more pervasive problem in the research history of the two German impact craters, a problem of a different quality than mere epistemic discussions: namely, a certain exaggerated patriotism that was readily prepared to accept a piece of local landscape as unique in time and space; an attitude that was heavily promoted by the ultranationalist propaganda of

---

2   ' ... punktuelle, plötzliche und seltene Ereignisse, die sich durch die Intensität ihrer Auswirkungen vom vorher und nachher ruhigeren Ablauf des Geschehens abheben und die wegen ihrer Seltenheit in der kurzen Zeitspanne, die für die menschliche Beobachtung zur Verfügung steht, nicht beobachtet wurden' (Engelhardt and Zimmermann 1982: 355).

'German Geology' (Chapter 7), and which precluded process-oriented research in local geology rather effectively.

The chauvinistic excesses of 'German Geology' were the ugly peak of a more prevalent, prejudiced and unreflected-upon local patriotism. This attitude became possible due to the general political and social factors that were responsible for detaching German geological research from the international scientific community – the post-World War I depression and subsequent propaganda and political framework of Nazi Germany. The discomfort and pain of this detachment was compensated or masked by a special sort of mental narcotics, unquestioningly cherishing the own and familiar while shunning the other and alien. This, of course, was a deeply unscientific attitude, because instead of judging the quality of data and interpretations it became increasingly important to know who said something instead of what had actually been said.

This chauvinism, in a very broad sense of the word, acted on both a global and on a smaller scale. It secretly fuelled the rivalry between the 'knights' of the Kranz–Löffler model, which centred on a single huge central explosion to blast the Steinheim and Ries Basins respectively, in Tübingen, and the Munich school of Ries research, which for theoretical reasons supported an only slightly modified version of Branca's and Fraas's theory (Chapter 1). It also stood behind the ceremonies on the occasion of field trips crossing the Bavarian/Württemberg border (Chapter 3), which clearly voiced the traditional animosity between the two provinces – with the consequence that the Steinheim and Ries Basins had never been treated as a single structural unit. It discriminated as a matter of course between insiders (the local experts) and outsiders who, by definition, could not have anything interesting to say (those who could not boast to have spent sufficient time within the Steinheim and Ries Basins, and whose expertise in other fields was denigrated as irrelevant for the problem in question: Chapter 3).

This chauvinism consequently also separated 'German Geology' from foreign geologists' methods, surfacing not only in specific verbal attacks against foreign colleagues, but also in an unfocused disinterest concerning foreign research. Within this nationalistic framework local geology seemed self-sufficient, and comparison on a global scale seemed to make no sense because the effects of any process would always be unique due to the specific history of a given locality. This uniqueness was in turn quite acceptable, notwithstanding any debates about actualism or non-actualism, whereas generalisation seemed unnecessary (Chapter 7).

Because 'German Geology' had categorised research on 'ahistorical' processes as non-holistic, and thus rejected it by definition as deficient, chauvinism also lingered behind the rivalry between geologists and mineralogists, which became apparent especially in the on-going resistance against the impact theory throughout the early 1960s. Geology as defined by Nazi ideologists was, because of its historical perspective, styled as The Science *per se* and thus considered superior to the so-called 'ahistorical' sciences such as physics and even mineralogy (Becksmann 1939; Beurlen 1935). The distinct scientific cultures of geology and mineralogy (Laudan 1982), and their different methodologies, precluded a common language between the two sciences and led to serious doubts about the expertise of mineralogists among the 'old guard' of Ries experts, while their own expertise was insufficient to judge the mineralogists' data and methods. The stratigraphic focus of geology at that time, on the other hand, strengthened the geologists' routine method of viewing rock sequences as testimony to the passing of time rather than as evidence of processes (Chapter 10).

Chauvinism in the broad sense of the word even stood guard to defend the Earth's status as a geologically closed system. Meteorites were seen as insignificant freak events, possibly not even present in the geological past (Chapter 3). As in Aristotle's time, things outside the Earth were considered to be of a different nature and thus not affecting us here; a notion that had already been present in the national-romantic precursors of 'German Geology', who like the ancient Greek philosopher assumed the existence of a special celestial matter distinguishing the Earth from space (Fritscher 2002: 120; Chapter 7).

Caught in a chauvinistic web, it seemed unnecessary to look beyond one's own personal horizon: Why visit international meetings, why bother to learn another living language and read foreign literature, if by definition there was nothing worthwhile out there that would concern us here?

In contrast, intrusion from outside was countered by a number of rhetorical arguments (cf. Schultz 1988). The argument based on experience condescendingly highlighted the lack of expertise in Ries matters among impactists. It was argued by discipline that mineralogy was a method 'alien' to the geological Ries problem and by methodology that the impactist idea violated the actualistic principle. The restriction to terrestrial causes seemed a dogma, whereas Schröder and Dehm (1950) argued by sheer authority that they had considered the impact idea and found it wanting.

The influence of chauvinism or local patriotism becomes even clearer when we ask ourselves what would have happened if the conditions had been otherwise (Radick 2008; Henry 2008). It seems clear that Walter Kranz and Heinrich Ludwig Quiring together had the key (Chapter 4). They could have worked together, perhaps talking the geophysicist Herrmann Reich into joining them and testing their various theories (something that happened only after 1960; Chapter 11). They could also have invited mineralogists to study the petrology of suevite. Doubt would have arisen among volcanists, while impactists would have had nothing to fear, even though a *final* replacement of the volcanic explosion theory would not have been realistically possible before the discovery of coesite/stishovite. Then we would have had a gradual shift from volcanic to impactist interpretations and a hearty welcome to the discovery of coesite rather than a sharp shift of paradigm including stubborn resistance from the old elite.

On the other hand, had the Third Reich continued, 'German Geology' would most probably have caused a further conservation of the volcanic theory, because 'German Geology' was guided by ideology rather than facts and did not wonder about the singular and unrepeatable. Rather, under the cloak of 'holistic thinking' it prevented generalisations, instead furthering 'fragmentation' of explanations, welcoming and applauding a view of a singular German landscape as a fit mirror for a People who had styled themselves as equally unique.

Without isolation and the chauvinism to which it led, the impact theory would have been discussed continuously as a possible, albeit unloved, alternative. Perhaps drilling programmes such as the one led by Gerold Wagner and Rudolf Hüttner (Chapter 11) or petrological/mineralogical investigations (Chapters 11 and 12), would have been undertaken earlier. International conferences would have exposed Ries researchers to scientists investigating similar structures elsewhere around the globe.

It was arrogant ideology, which had been so strong as to have a tangible adverse effect on scientific performance and precluded any serious reflection about a conceptual alternative to the prevalent volcanic paradigm. Ideology and resulting disinterest detached the Ries experts and their colleagues from alternative views within the scientific community, building up a tension between a locally confined culture of science and the global scientific mainstream.

Chauvinism and disregard for outside factors prevented open-mindedness and scientific creativity. What remained was constantly rehashing the same old Ries problem again and again. Stasis prevailed despite all sorts of cover-up activity, and the protagonists were unaware of their ideological corset.

'[T]he starting point of induction, naïve observation, innocent observation, is a mere philosophical fiction. There is no such thing as unprejudiced observation. Every act of observation we make is biased. What we see or otherwise sense is a function of what we have seen or sensed in the past' (Medawar 1964: 9). As such, there is nothing wrong with speculation and preconceived ideas, as 'hypotheses arise by guesswork. ... I should say rather that they arise by inspiration' (Medewar 1964: 10), but what follows then distinguishes science from fantasy: 'hypotheses can be tested rigorously' (Medewar 1964: 11) by way of experiments or other predicted observations, and both honesty and open-mindedness are required to devise such tests and to accept the outcome, once every possible error in the experimental or observational setup has been addressed. The volcanist theories – whether the central explosion of Kranz and Löffler or the local 'outbursts' of the Munich School under Schröder and Dehm – had, however, never really been tested before the pressing necessity to do so arose in the 1960s (Chapter 11).

It would be all too simpleminded to denounce the Ries researchers of the 1930s to 1950s – including contributors to the volcanist 'resistance' of the

1960s – as bad scientists. However, things are more complicated than that. The Ries experts were often prominent members of their scientific community, and generally highly regarded among colleagues. They were widely read and widely published, and quite a number of them could boast of all sorts of honours and medals as tokens of the respect and regard of their peers. Although, as would be expected, a number of their papers have become obsolete over the passing decades, others are still readable 'classics'. Thus, the issue is not a question of good or bad scientists, because their eventual 'defeat' in the specific case of the Ries problem was unpredictable from within their contemporary community.

It was an intrusion, a transfer of methods and knowledge from outside – from foreign countries as well as from mineralogy and physics – that finally overcame the local stagnation, and moreover made clear that there had been something missing previously. The turnaround led to much more than just a solution to the age-old Ries problem. It achieved a new culture of science, including new interdisciplinary cooperation; and, importantly, it led to new questions, which lead to new results elsewhere. It was this creative fertility that distinguished the new paradigm sharply from the old.

When Eugene Shoemaker and Edward Chao successfully introduced 'foreign' mineralogical methods to Ries research in the early 1960s, they suddenly reconnected their German colleagues to the international scientific community, after some twenty-five years of scientific isolation from the impact crater research of their global contemporaries during the build-up to World War II and its aftermath.

While the most vocal of the 'old guards' were still occupied with resistance (Chapter 10), others prepared to test the volcanic and impact theories against new field evidence (Chapter 11). A full research programme was established, including drilling projects supported by an interdisciplinary working group. All of these activities, which also involved numerous workshops and conferences, brought together a critical mass of well-informed people, some of whom also took the time to explain physics and mineralogy to the older geologists. Eventually, after 1975, with evidence mounting and resistance dying away, considerable effort began to be invested in visualising and popularising the impact event for the general public in the form of museums and exhibits (Chapters 11 and 12).

In addition to the paradigm shift regarding the crater hypothesis, which concerned a limited sub-problem of geological investigation, the years between 1960 and 1975 had brought another shift of overarching importance: a shift from simple, field-based geology to more chemically and physically oriented geoscience; from self-sufficient stratigraphy to a process-oriented view; from local to global and even inter-planetary interests; and also from chauvinism to internationality. The lingering and often unconscious chauvinism of the 1930s to 1950s in German geology was supplanted by a new scientific culture, one with interdisciplinary interests and international scope.

Glossary

allochthonous: The rocks or materials in question have not been formed in their locality, but have been transported from elsewhere.

aragonite: A mineral consisting of calcium carbonate ($CaCO_3$) but having a different crystal symmetry from the more common calcite (also $CaCO_3$).

bentonite: A type of clay which commonly forms by weathering of volcanic ash.

bolide: A large impacting body capable of forming an impact crater; from the Greek word for missile.

breccia: A coarse-grained rock composed of angular rock fragments.

brecciation: The process of fracturing rocks.

caldera: A type of volcano which forms when, during large eruptions, the roof of the magma chamber – like a huge piston – sinks into the chamber along ring-fractures. A caldera is a more or less circular volcanic depression, largely filled with volcanic pumice and ash deposits, which also cover the surrounding area.

coesite: A high-pressure mineral consisting of Silica (SiO_2) but having a different, denser crystal structure from common quartz (also SiO_2).

diapir: A rising dome of magma or of soft, plastic rock (such as salt) of lower density than the surrounding rocks, rupturing and uplifting the overlying rocks.

diaplectic glasses: Glass produced by shock waves, which destroy the crystal structure of the minerals without melting them first.

diatreme: A steep-walled volcanic funnel filled with a breccia that was formed by an explosion of hot water vapour or volcanic gases.

doline: A sinkhole caused by the dissolution of limestone or gypsum.

downfaulted: Said of the rock formation, which moved downwards along a fault-line.

epirogenic: An old-fashioned term to characterise processes leading to the long-term elevation and subsidence of larger portions of the Earth's crust. The term belongs to a pre-plate tectonic paradigm.

eutectic: A term to characterise the melt composition which forms at the lowest possible temperature for a given mineral composition.

fulgurite: An irregular, sometimes branching, tube of melted and then fused sand grains produced by lightning strike.

fumaroles: Holes or vents which emit volcanic gases (most commonly hot water vapour).

Habilitation: In Germany, in addition to the doctorate a second thesis is required to qualify for professorship. This qualification is called *Habilitation*.

horst: An uplifted block of the Earth's crust, which is bordered by faults.

ignimbrite: A rock formed by volcanic ash flows.

isotropic: Said of substances whose properties are the same in all directions such as glasses or minerals crystallising in cubic symmetry.

juvenile gases: Gases, which are derived directly from the magma and not, for example, from rain water.

juvenile volcanic material: Volcanic material which is derived directly from the fresh magma and not reworked from other, older, volcanic rocks.

kimberlite pipe: A steep-walled volcanic funnel containing the rare and unusual rock-type of kimberlite, which is derived from a gas-rich magma that has been propelled directly from deep within the Earth's mantle.

laccolith: A roughly lentil-shaped, but flat-bottomed, magmatic intrusion which has domed up the overlying rock strata.

lahar: A volcanic mud-flow.

lapilli: Pyroclastics with a grain size of between 2 and 64 mm.

lechatelierite: SiO_2-glass usually formed by lightning strike or by meteorite impact.

liparite: An old-fashioned synonym of rhyolite, an SiO_2-rich volcanic rock.

listric faults: Curved fault plains which are steep in the upper part and near horizontal in the lower part, commonly associated with landslides.

mafic: Igneous rocks or minerals rich in magnesium and iron.

Miocene: A geological time period within the Tertiary from about 23 to 5.3 million years ago.

modifications: In mineralogy, this term means minerals which share the same chemical composition but are distinct by their different crystal structure. Quartz, coesite and stishovite are modifications of SiO_2 (Figure 8.3).

molasse: An extensive and diverse sedimentary rock formation derived from the erosion of a mountain range.

moldavites: Green tektites derived from the Ries Crater.

monomictic breccia: A breccia composed of fragments of a single rock type.

nappe: A sheet-like rock unit that has moved horizontally for quite some distance by thrust-faulting during the formation of mountain ranges.

orogenesis: The processes leading to the formation of a mountain range.

PDF: Short for planar deformation features.

pegmatite: Igneous rock formed in veins and fissures consisting of large crystals.

pegmatitic stage: Late and volatile-rich stage in the crystallisation sequence of a magma chamber. The residual fluid is concentrated at the magma chamber top and injected into fissures, where it crystallises forming large crystals.

phreatic eruption: A volcanic eruption due to the explosion of heated water.

planar deformation features: Single or multiple sets of parallel plains of deformation of the crystal structure in quartz and other minerals caused by a shock wave.

plutogenesis: Old-fashioned term meaning all processes of magma generation and magma chamber formation.

pluton: A body of rock derived from a crystallised magma chamber.

plutonism: All processes involved in the formation of a pluton.

polymictic breccia: A breccia composed of fragments of various different rock types.

pyroclastic flow: A hot density current consisting of gases and pyroclastic materials, which is produced during an explosive volcanic eruption either by collapse of the eruption column or when the explosive force is insufficient to create an eruption column in the first place.

pyroclastics: Short for pyroclastic rocks. These consist of volcanic ashes, lapilli and/or bombs, which are formed during a volcanic eruption when the magma is fragmented by explosive processes instead of forming a lava flow.

Reuter's blocks: Reworked distal Ries ejecta (Upper Jurassic limestone) within sandy Molasse sediments of the northern Alpine foreland (Figure 1.11).

shatter cone: Conical fragments of mostly fine-grained rock characterised by striations that radiate from the apex which are caused by shock waves (Figure 8.2).

sill: A sheet-like magma intrusion that parallels the bedding planes of the surrounding rocks.

spindle bombs: Spindle-shaped pyroclastics with a grain size larger than 64 mm, which acquired their characteristic form due to the lava batches from which they formed still being fluid when expelled from the volcanic vent.

stishovite: A high-pressure mineral consisting of Silica (SiO_2) but having a different, considerably denser crystal structure from quartz (also SiO_2). Stishovite is also denser than coesite.

stratigraphy: The science of rock strata including their formation and dating via their fossil content. While all rock types, in principle, fall under the general scope of stratigraphy, the main concerns are traditionally the sedimentary rocks.

suevite: An impact breccia with an appreciable amount of impact melt particles. Suevite is a deposit of the hot fallout from the impact plume (Figure 1.13).

syn-orogenic: Contemporary with the formation of a mountain range.

tectonic: A term to characterise processes and forces responsible for large scale structural and deformational features of the Earth's crust including their relationship, their origin and development through time.

tektites: Black, green or brown and mostly aerodynamically shaped glass bodies, which originate from terrestrial rocks located close to the point of first contact during large meteorite impacts.

Tertiary: A geological time period from about 65 to 2 million years ago.

thin-section: A rock fragment mounted on a glass slide and then mechanically ground to the thickness of about 0.03 mm, making it possible to study the optical properties of the composing minerals with the aid of a polarisation microscope.

trass: An old-fashioned term for volcanic tuff.

travertine: A rock formed by the precipitation of calcium carbonate around thermal springs.

tuff: Consolidated volcanic ash.

Widmanstätten figures: A pattern formed by nickel-iron crystals in some meteorites, which can be made visible by etching of the polished surface.

X-ray (powder) diffraction: A method used to analyse the mineral content of a rock sample (which has been ground to a fine powder) from characteristic interference patterns obtained when X-rays are diffracted by the crystals' lattices.

References

Interviews

Helmut Hölder (1915–2014), Stuttgart, 4 March 2000
Rudolf Hüttner, Suggenthal near Freiburg i. Br., 28 May 2001
Hans Pichler, Mössingen, 11 May 2000
Jean Pohl, München, 24 May 2000
Eugen Seibold (1918–2013), Freiburg i. Br., 29 May 2001
Peter Signer, CH-Grüt/Wetzikon, 5 May 2000
Rudolf Trümpy (1921–2009), CH-Küsnacht, 4 May 2000
Helmut Vidal (1919–2002), Starnberg, 8 March 2000
All interviews have been conducted by the author.

Archive Sources

Archiv der TU Bergakademie Freiberg (Nachlass Otto Stutzer)
Archiv des Deutschen Museums München
Bayerisches Hauptstaatsarchiv: Kriegsarchiv (Munich)
BGIB: Bibliothek des Geographischen Instituts der Universität Bonn (Nachlass Carl Troll)
Fürst zu Oettingen-Spielberg'sches Archiv, Harburg (Schwaben).
GA: Geologenarchiv, Universitätsbibliothek Freiburg i. Br.
JME: Jura-Museum Eichstätt (off-print collection)
NASA: Apollo Lunar Surface Journal (<http://www.hq.nasa.gov/alsj/>)
Niels Bohr Library and Archives, American Institute of Physics, College Park, MD USA
Universitätsarchiv München
UAT: Universitätsarchiv Tübingen (Nachlass Georg Wagner)
ZERIN: Zentrum für Rieskrater- und Impaktforschung Nördlingen (Engelhardt-Nachlass)

Printed Sources

Ackermann, W., 1958. Geologisch-petrographische Untersuchungen im Ries. *Geologisches Jahrbuch*, 75, pp.135–82.

Adam, K.D., 1980. Vom Wandel der Anschauungen in den Erdwissenschaften am Beispiel des Steinheimer Beckens (Württemberg). *Annalen des Naturhistorischen Museums Wien*, 83, pp.13–23.

AG Ries, 1963. *Besprechung der Arbeitsgemeinschaft Ries am Montag, den 25.III.63 in München, veranlaßt von der Deutschen Forschungsgemeinschaft.* Unpublished minutes of the meeting.

Ahrens, W., 1929. Die Tuffe des Nördlinger Rieses und ihre Bedeutung für das Gesamtproblem. *Zeitschrift der Deutschen Geologischen Gesellschaft*, 81, pp.94–9.

Ahrens, W., 1936. Bemerkung zum Vortrag von O. Stutzer. *Zeitschrift der Deutschen Geologischen Gesellschaft*, 88, pp.590–91.

Aken, P.A. van, 2001. In honour of the 60th birthday of Wolfgang Friedrich Müller. *European Journal of Mineralogy*, 13, pp.219–20.

Alvarez, W., 1997. *T. rex and the Crater of Doom.* London: Penguin Books.

Amstutz, G.C., 1964. Impact, Cryptoexplosion, or Diapiric Movements? (A discussion of the origin of polygonal fault patterns in the Precambrian and overlying rocks in Missouri and elsewhere). *Transactions of the Kansas Academy of Science*, 67, pp.343–56.

Amstutz, G.C., 1965. A morphological comparison of diagenetic cone-in-cone structures and shatter cones. *Annals of the New York Academy of Science*, 123, pp.1050–56.

Angenheister, G. and Pohl, J., 1969. Die seismischen Messungen im Ries von 1948–1969. *Geologica Bavarica*, 61, pp.304–26.

Arndt, J., 1986. [Untitled laudation on Wolf von Engelhardt]. Unpublished manuscript.

Arp, G., 2006. *Sediments of the Ries Crater Lake (Miocene, Southern Germany). Sediment 2006: 21st Meeting of Sedimentologists / 4th Meeting of SEPM Central European Section: Field Trip F2.* Available through: The University of Göttingen <http://goedoc.uni-goettingen.de/goescholar/bitstream/handle/goescholar/3181/Exkursionsfuehrer-Sediment2006-Ries.pdf?sequence=1> [Accessed 30 May 2012].

Bailey, N.G. and Ulrich, G.E., 1975. *Apollo 14 Voice Transcript pertaining to the geology of the landing site. U.S. Geological Survey, Branch of Astrogeology, Flagstaff, Arizona.* Available through: The Lunar and Planetary Institute, Houston <http://lpi.usra.edu/lunar/documents/Apollo14VoiceTranscript-Geology.pdf> [Accessed 16 January 2012].

Baldwin, R.B., 1949. *The Face of the Moon.* Chicago: The University of Chicago Press.

Barber, B., 1961. Resistance by Scientists to Scientific Discovery. *Science*, 134, pp.569–602.

Barsig, W. and Kavasch, W.-D., 1983. *Das Ries und der Kulturpreis des Vereins Rieser Kulturtage 1983.* Nördlingen: Verlag F. Steinmeier.

Barthel, W., 1964. Das Ries und sein Werden. Eine geologische Skizze Band 1. *Rieser Schriften*, 3, 55pp.

Becksmann, E., 1939. Erdgeschichtliche Gestalten – Grundsätzliches zur erdgeschichtlichen Fragestellung. *Zeitschrift der Deutschen Geologischen Gesellschaft*, 91, pp.734–56.

Bentz, A., 1929. Der heutige Stand des Riesproblems. *Zeitschrift der Deutschen Geologischen Gesellschaft*, 81, pp.71–5.

Bentz, A., 1936. Bemerkungen zum Vortrag von O. Stutzer. *Zeitschrift der Deutschen Geologischen Gesellschaft*, 88, pp.589–90.

Bernauer, F., 1939. Die Italienreise der Deutschen Geologischen Gesellschaft 1939 und einige dabei behandelte Einzelfragen. *Zeitschrift der Deutschen Geologischen Gesellschaft*, 91, pp.450–65.

Bernoulli, D., 2009. Rudolf Trümpy (1921–2009). *International Journal of Earth Sciences (Geologische Rundschau)*, 98, pp.1557–9.

Beurlen, K., 1935. Sinn und Ziel geologischer Forschung – Antrittsrede gehalten an der Christian-Albrechts-Universität am 1. Dezember 1934 unter dem Rektorat von K.L. Wolf. *Kieler Universitätsreden*, Neue Folge 3, 16pp.

Beurlen, K., 1937. Bedeutung und Aufgabe der Deutschen Geologischen Gesellschaft. *Zeitschrift der Deutschen Geologischen Gesellschaft*, 89, pp.52–8.

Beurlen, K., 1939. Einige Bemerkungen zur Geschichte der Geologie. *Zeitschrift der Deutschen Geologischen Gesellschaft*, 91, pp.236–52.

Beurlen, K. and Lehmann, U., 1944. Osborn und seine Stellung in der modernen Paläontologie. *Zeitschrift der Deutschen Geologischen Gesellschaft*, 96 (printed in 1947), pp.229–36.

Beyerchen, A.D., 1982. *Wissenschaftler unter Hitler. Physiker im Dritten Reich.* Frankfurt am Main: Ullstein.

BGL [Bayerisches Geologisches Landesamt] (ed.), 1969. Das Ries – Geologie, Geophysik und Genese eines Kraters – Bericht der Arbeitsgemeinschaft Ries. *Geologica Bavarica*, 61.

BGL [Bayerisches Geologisches Landesamt] (ed.), 1977. Ergebnisse der Ries-Forschungsbohrung 1973: Struktur des Kraters und Entwicklung des Kratersees. *Geologica Bavarica*, 75.

BGL [Bayerisches Geologisches Landesamt] (ed.), 1999. *Geologische Karte Ries 1:50 000 mit Kurzerläuterung a.d. Rückseite + Multimedia-CD*, München: Bayerisches Geologisches Landesamt.

Birzer, F., 1969. Molasse und Ries-Schutt im westlichen Teil der Südlichen Frankenalb. *Geologische Blätter von Nordostbayern*, 19, pp.1–28.

Böhme, M., Gregor, H.-J. and Heißig, K., 2001. The Ries and Steinheim meteorite impacts and their effect on environmental conditions in time and space. In: Buffetaut, E. and Koeberl, C., (eds), 2001. *Geological and biological effects of impact events*. Berlin: Springer-Verlag, pp.217–35.

Bolten, R. and Müller, D., 1969. Das Tertiär im Nördlinger Ries und in seiner Umgebung. *Geologica Bavarica*, 61, pp.87–130.

Boon, J.D. and Albritton, C.C., 1937. Meteorite scars in ancient rocks. *Field and Laboratory*, 5, pp.53–64.

Bourgeois, J. and Koppes, St., 1998. Robert S. Dietz and the Recognition of Impact Structures on Earth. *Earth Sciences History*, 17, pp.139–56.

Boyd, F.R. and England J.L., 1960. The Quartz–Coesite Transition. *Journal of Geophysical Research*, 65(2), pp.749–56.

Branco, W. and Fraas, E., 1905. Das kryptovulcanische Becken von Steinheim. *Abhandlungen der königlich preußischen Akademie der Wissenschaften*, phys. Abh. 1, pp.1–64.

Bucher, W.H., 1963a. Cryptoexplosion structures caused from without or from within the Earth? ("Astroblemes" or "Geoblemes?"). *American Journal of Science*, 261, pp.597–649.

Bucher, W.H., 1963b. Are cryptovolcanic structures due to meteorite impact? *Nature*, 4874, pp.1241–5.

Buchner, E., Schweigert, G. and Seyfried, H., 1998. Revision der Stratigraphie der süddeutschen Brackwassermolasse. *Zeitschrift der Deutschen Geologischen Gesellschaft*, 149(2), pp.305–20.

Buchner, E., Seyfried, H. and Hische, R., 1996. Die Graupensandrinne der süddeutschen Brackwassermolasse: ein Incised Valley-Fill infolge des Ries-Impaktes. *Zeitschrift der Deutschen Geologischen Gesellschaft*, 147(2), pp.169–81.

Buchner, E., Seyfried, H. and Van den Bogaard, P., 2003. ^{40}Ar/^{39}Ar laser probe age determination confirms the Ries impact crater as the source of glass particles in Graupensand sediments (Grimmelfingen Formation, North Alpine Foreland Basin). *International Journal of Earth Sciences (Geologische Rundschau)*, 92, pp.1–6.

Bülow, K. von, 1937. Sendboten aus dem Weltenraum: Riesenmeteoriten und ihre Spuren im Antlitz der Erde. *Kosmos*, 34(1), pp.21–4.

Bunch, T.E. and Cohen, A.J., 1964. Shock deformation of quartz from two meteorite craters. *Geological Society of America Bulletin*, 75, pp.1263–6.

Bunch, T.E. and Cohen, A.J., 1968. Shock-induced structural disorder in plagioclase and quartz. In: French, B.M. and Short, N.M., (eds), 1968. *Shock metamorphism of natural materials*. Baltimore: Mono Book Corp., pp.509–18.

Bunch, T.E., Cohen, A.J, and Dence, M.R., 1967. Natural terrestrial maskelynite. *American Mineralogist*, 52, pp.244–53.

Carlé, W., 1972. Georg Wagner 1885 bis 1972. *Jahresheft der Gesellschaft für Naturkunde in Württemberg*, 127, pp.35–42.

Carlé, W., 1987. Die Wende am Ries 1961. *Jahresheft der Gesellschaft für Naturkunde in Württemberg*, 142, pp.73–98.

Carter, N.L., 1965. Basal quartz deformation lamellae – a criterion for recognition of impacts. *American Journal of Science*, 263, pp.786–806.

Carter, N.L., 1968. Dynamical deformation of quartz. In: French, B.M. and Short, N.M., (eds), 1968. *Shock metamorphism of natural materials*. Baltimore: Mono Book Corp., pp.453–74.

Cas, R.A.F. and Wright, J.V., 1988. *Volcanic successions – modern and ancient*. London: Unwin Hyman.

Cassidy, W.A., 1993. Memorial of Alvin Jerome Cohen 1918–1991. *American Mineralogist*, 78, pp.1340–41.

Chao, E.C.T., 1964. Selective mineral transformation as evidence for meteorite impact. *US Geological Survey Astrogeological Studies Annual Progress Report*, July 1, 1963–July 1, 1964, Part B, pp.39–570.

Chao, E.C.T., 1966. Impact metamorphism. *US Geological Survey Astrogeological Studies Annual Progress Report*, July 1, 1965–July 1, 1966, Part B, pp.135–80.

Chao, E.C.T., 1967a. Impact metamorphism. In: Abelson, P.H., (ed.), 1967. *Researches in Geochemistry*. New York: John Wiley & Sons, pp.204–33.

Chao, E.C.T., 1967b. Shock effects of certain rock-forming minerals. *Science*, 156, pp.192–202.

Chao, E.C.T., 1968. Pressure and temperature histories of impact metamorphosed rocks – based on petrographic observations. In: French, B.M. and Short, N.M., (eds), 1968. *Shock metamorphism of natural materials*. Baltimore: Mono Book Corp., pp.135–58.

Chao, E.C.T., 1977a. The Ries Crater of Southern Germany, a Model for Large Basins on Planetary Surfaces. *Geologisches Jahrbuch*, Reihe A, 43, 85pp.

Chao, E.C.T., 1977b. Preliminary interpretation of the 1973 Ries research deep drill core and a new Ries cratering model. *Geologica Bavarica*, 75, pp.421–41.

Chao, E.C.T., Fahey, J.J., Littler, J. and Milton, D.J., 1962. Stishovite, SiO_2, a very high pressure new mineral from Meteor Crater, Arizona. *Journal of Geophysical Research*, 67, pp.119–21.

Chao, E.C.T. and El Goresy, A., 1977. Shock attenuation and the implantation of Fe-Cr-Ni veinlets in the compressed zone of the 1973 Ries research deep drill core. *Geologica Bavarica*, 75, pp.289–304.

Chao, E.C.T., Hüttner, R. and Schmidt-Kaler, H., 1978. *Principal exposures of the Ries meteorite crater in Southern Germany*. München: Bayerisches Geologisches Landesamt.

Chao, E.C.T., Hüttner, R. and Schmidt-Kaler, H., 1992. *Aufschlüsse im Ries-Meteoriten-Krater*. 4th ed. München: Bayerisches Geologisches Landesamt.

Chao, E.C.T. and Littler, J., 1963. Additional evidence for the impact origin of the Ries basin, Bavaria, Germany. *Geological Society of America Special Paper*, 73, p.127.

Chao, E.C.T., Shoemaker, E.M. and Madsen, B.M., 1960. First Natural Occurrence of Coesite. *Science*, 132, pp.220–22.

Cloos, H., 1947. *Gespräch mit der Erde.* München: Piper Verlag.

Cohen, A.J., 1961. A semi-quantitative asteroid impact hypothesis of tektite origin. *Journal of Geophysical Research,* 66, p.2521.

Daniels, M. and Michl, S., 2010. Strukturwandel unter ideologischen Vorzeichen – Wissenschafts- und Personalpolitik an der Universität Tübingen 1933–1945. In: Wiesing, U., Britzinger, K.-R., Grün, B., Junginger, H. and Michl, S., (eds), 2010. Die Universität Tübingen im Nationalsozialismus. *Contubernium, Tübinger Beiträge zur Universitäts- und Wissenschaftsgeschichte,* 73, pp.13–73.

David, E., 1969. Das Ries-Ereignis als physikalischer Vorgang. *Geologica Bavarica,* 61, pp.350–78.

David, E., 1977a. Anmerkungen zur Bruchmechanik der shatter-cone-Bildung. *Geologica Bavarica,* 75, pp.285–7.

David, E., 1977b. Abschätzung von impaktmechanischen Daten aufgrund von Ergebnissen der Forschungsbohrung Nördlingen 1973. *Geologica Bavarica,* 75, pp.459–70.

Davis, W.M., 1922. Biographical Memoir of G.K. Gilbert. *Biographical Memoirs of the National Academy of Sciences,* 21(5), pp.1–300.

Deecke, W., 1925. Das innere System im west- und süddeutschen Thermalsystem. *Zeitschrift der Deutschen Geologischen Gesellschaft,* 77, pp.96–111.

Deffner, C., 1870. Der Buchberg bei Bopfingen. *Jahreshefte des Vereins für vaterländische Naturkunde in Württemberg,* 26, pp.95–142.

Dehm, R., 1951. Zur Gliederung der jungtertiären Molasse in Südbayern nach Säugetieren. *Neues Jahrbuch für Geologie und Paläontologie,* Monatshefte 1951, pp.140–52.

Dehm, R., 1962. Das Nördlinger Ries und die Meteortheorie. *Mitteilungen der Bayerischen Staatssammlung für Paläontologie und historische Geologie,* 2, pp.69–87.

Dehm, R., 1966. Gümbel, Carl Wilhelm von. *Neue Deutsche Biographie,* 7, p.259. Available through <http://www.deutsche-biographie.de/pnd116909943. html> [Accessed 25 February 2011].

Dehm, R., 1969. Geschichte der Riesforschung. *Geologica Bavarica,* 61, pp.25–35.

Dehm, R., 1977. Vorgeschichte der Geologischen Karte des Rieses 1:50 000. *Geologica Bavarica,* 76, pp.8–11.

Dehm, R., 1978. Geschichte der Riesforschung. *Rieser Kulturtage Dokumentation,* 2, pp.9–24.

Dence, M.R., 1964. A comparative structural and petrographic study of probable Canadian meteorite craters. *Meteoritics,* 2, pp.249–70.

Dence, M.R., 1965. The extraterrestrial origin of Canadian craters. *Annals of the New York Academy of Science,* 123, pp.941–69.

Dence, M.R., 1968. Shock zoning at Canadian craters: Petrography and structural implications. In: French, B.M. and Short, N.M., (eds), 1968. *Shock metamorphism of natural materials*. Baltimore: Mono Book Corp., pp.169–84.

Dietz, R.S., 1946. The meteoritic impact origin of the Moon's surface features. *The Journal of Geology*, 54, pp.359–75.

Dietz, R.S., 1959. Shatter Cones in Cryptoexplosive Structures (Meteorite Impact?). *The Journal of Geology*, 67, pp.496–505.

Dietz, R.S., 1960. Meteorite Impact Suggested by Shatter Cones in Rock. *Science*, 131, pp.1781–4.

Dietz, R.S., 1968. Shatter cone in cryptoexplosion structures. In: French, B.M. and Short, N.M., (eds), 1968. *Shock metamorphism of natural materials*. Baltimore: Mono Book Corp., pp.267–85.

Dorn, C., 1942. *Beiträge zur Geologie des Rieses. Neues Jahrbuch für Mineralogie und Centralblatt für Mineralogie*, B (1942–1944), pp.1–7.

Dorn, P., 1950. Ein Jahrhundert Riesgeologie. *Zeitschrift der Deutschen Geologischen Gesellschaft*, 100, pp.348–65.

El Goresy, A., 1964. Die Erzmineralien in den Ries und Bosumtwi-Krater-Gläsern und ihre genetische Deutung. *Geochimica et Cosmochimica Acta*, 28, pp.1881–91.

El Goresy, A., 1965. Baddeleyite and its significance in impact glasses. *Journal of Geophysical Research*, 70, pp.3453–6.

El Goresy, A., 1968. The opaque minerals in impactite glasses. In: French, B.M. and Short, N.M., (eds), 1968. *Shock metamorphism of natural materials*. Baltimore: Mono Book Corp., pp.531–3.

El Goresy, A. and Chao, E.C.T., 1977. Discovery, origin, and significance of Fe-Cr-Ni veinlets in the compressed zone of the 1973 Ries research drill core. *Geologica Bavarica*, 75, pp.305–21.

Engelhardt, W. von, 1967. Chemical composition of Ries glass bombs. *Geochimica et Cosmochimica Acta*, 31, pp.1677–89.

Engelhardt, W. von, 1982. Hypotheses on the origin of the Ries Basin, Germany, from 1792 to 1960. *Geologische Rundschau*, 71, pp.475–85.

Engelhardt, W. von, 1987. Das Nördlinger Ries im Wandel geologischer Theorien. *Der Daniel – Nordschwaben – Zeitschrift für Landschaft, Geschichte, Kultur und Zeitgeschehen*, 15 (23. Daniel), pp.303–8.

Engelhardt, W. von, 1994. Von der Geologie zur Planetologie – Geowissenschaften im Widerstreit regulativer Ideen. In: Pfeiffer, J. and Fichtner, G., (eds), 1994. *Erlebte Geschichte – Zeitzeugen berichten in einer Tübinger Ringvorlesung*. Tübingen: Verlag Schwäbisches Tagblatt, pp.51–64.

Engelhardt, W. von, Arndt, J., Stöffler, D., Müller, W.F., Jeziorkowski, H. and Gubser, R.A., 1967a. Diaplektische Gläser in den Breccien des Ries von

Nördlingen als Anzeichen für Stoßwellenmetamorphose. *Contributions to Mineralogy and Petrology*, 15, pp.93–102.

Engelhardt, W. von and Graup, G., 1977. Stoßwellenmetamorphose im Kristallin der Forschungsbohrung Nördlingen 1973. *Geologica Bavarica*, 75, pp.255–71.

Engelhardt, W. von, Bertsch, W., Stöffler, D., Groschopf, P. and Reiff, W., 1967b. Anzeichen für den meteoritischen Ursprung des Beckens von Steinheim. *Naturwissenschaften*, 54, pp.198–9.

Engelhardt, W. von and Hölder, H., 1977. *Mineralogie, Geologie und Paläontologie an der Universität Tübingen von den Anfängen bis zur Gegenwart*. Tübingen: J.C.B. Mohr (Paul Siebeck).

Engelhardt, W. von and Hörz, F., 1964. Hochdruckgläser im Nördlinger Ries. *Die Naturwissenschaft*, 51, p.264.

Engelhardt, W. von, Hörz, F., Stöffler, D. and Bertsch, W., 1968. Observations of Quartz Deformation in the Breccias of West Clearwater Lake, Canada, and the Ries Basin, Germany. In: French, B.M. and Short, N.M., (eds), 1968. *Shock metamorphism of natural materials*. Baltimore: Mono Book Corp., pp.475–82.

Engelhardt, W. von and Stöffler, D., 1965. Spaltflächen im Quarz als Anzeichen für Einschläge großer Meteoriten. *Die Naturwissenschaft*, 52, pp.489–90.

Engelhardt, W. von and Stöffler, D., 1968. Stages of Shock Metamorphism in the Crystalline Rocks of the Ries Basin, Germany. In: French, B.M. and Short, N.M., (eds), 1968. *Shock metamorphism of natural materials*. Baltimore: Mono Book Corp., pp.159–68.

Engelhardt, W. von, Stöffler, D. and Schneider, W., 1969. Petrologische Untersuchungen im Ries. *Geologica Bavarica*, 61, pp.229–95.

Engelhardt, W. von and Zimmermann, J., 1982. *Theorie der Geowissenschaft*. Paderborn: Verlag Ferdinand Schöningh.

Engelmann, G., 1977. Keilhack, Konrad. *Neue Deutsche Biographie*, 11, pp.408–9. Available through: <http://www.deutsche-biographie.de/pnd116094583.html> [Accessed 5 April 2011].

Erismann, Th., Heuberger, H. and Preuss, E., 1977. Der Bimsstein von Köfels (Tirol), ein Bergsturz"Friktionit". *Mineralogy and Petrology*, 24, pp.67–119.

Fahlbusch, V., 1996. Richard Dehm. *Jahreshefte der Gesellschaft für Naturkunde in Württemberg*, 152, pp.297–9.

Focke-Museum, (ed.), 2013. *Graben für Germanien: Archäologie unterm Hakenkreuz*. Darmstadt: Theiss-Verlag.

Förstner, U., 1967. Petrographische Untersuchungen des Suevit aus den Bohrungen Deiningen und Wörnitzostheim im Ries von Nördlingen. *Contributions to Mineralogy and Petrology*, 15, pp.281–8.

Fraas, E., 1900. Der geologische Aufbau des Steinheimer Beckens. *Jahreshefte des Vereins für Vaterländische Naturkunde in Württemberg*, 56, pp.47–59.

Fraas, E., 1903. I. Die geologischen Verhältnisse im Ries. *Jahresberichte und Mitteilungen des Oberrheinischen Geologischen Vereins*, 36, pp.8–18.

Fraas, E., 1914. Erwiderung auf W. Kranz, "Das Problem des Steinheimer Beckens". *Jahresberichte und Mitteilungen des Oberrheinischen Geologischen Vereins*, Neue Folge 4, pp.113–15.

Franke, H., 1983. *Die Entwicklung der wissenschaftlichen Tektitforschung im Zeitraum von 1787 bis 1965*. University of Jena: Unpublished doctoral thesis.

French, B.M., 1968a. Shock metamorphism as geological process. In: French, B.M. and Short, N.M., (eds), 1968. *Shock metamorphism of natural materials*. Baltimore: Mono Book Corp., pp.1–17.

French, B.M., 1968b. Sudbury structure, Ontario: Some petrographic evidence for an origin by meteorite impact. In: French, B.M. and Short, N.M., (eds), 1968: *Shock metamorphism of natural materials*. Baltimore: Mono Book Corp., pp.383–412.

French, B.M., 1998. *Traces of Catastrophe: A Handbook of Shock-Metamorphic Effects in Terrestrial Meteorite Impact Structures*. LPI Contribution No. 954, Lunar and Planetary Institute, Houston. Available through: The Lunar and Planetary Institute, Houston <http://www.lpi.usra.edu/publications/books/CB-954/CB-954.pdf> [Accessed 26 June 2012].

French, B.M. and Short, N.M., (eds), 1968. *Shock metamorphism of natural materials*. Baltimore: Mono Book Corp.

Freyberg, B. von, 1962. Dehm, Richard: Das Nördlinger Ries und die Meteortheorie. – Mitteilungen der Bayerischen Staatssammlung für Paläontologie und Historische Geologie 2: 69–87. München 1962. *Geologische Blätter von Nordostbayern*, 12, pp.237.

Freyberg, B. von, 1974. Das geologische Schrifttum über Nordost-Bayern (1476–1965). Teil II Biographisches Autoren-Register. *Geologica Bavarica*, 71, 177pp.

Fritscher, B., 2002. Erdgeschichte zwischen Natur und Politik: Lorenz Okens "Zeugungsgeschichte" der Erde. In: Engelhardt, D. von and Nolte, J., (eds), 2002. Von Freiheit und Verantwortung in der Forschung. Symposium zum 150. Todestag von Lorenz Oken (1779–1851). *Gesellschaft Deutscher Naturforscher und* Ärzte, 9, pp.110–29.

Fritscher, B., 2012. Erdgeschichtsschreibung als montanistische Praxis: Zum nationalen Stil einer ‚preußischen Geognosie'. In: Schleiff, H. and Konečný, P., (eds), 2012. Staat, Bergbau und Bergakademie im 18. und frühen 19. Jahrhundert. *Vierteljahresschrift für Sozial- und Wirtschaftsgeschichte*, Beihefte, 223, pp.205–29.

Fritschl, E. and Hubmann, B., 2003. *Robert Schwinner (1878–1953), ein Vorkämpfer der Plattentektonik*. Available through The University of Graz

<http://www.uni-graz.at/ubwww/aktuelles-ub/ausstellungen/ausstellungen-archiv/ausstellungen-2003-schwinner.htm> [Accessed 15 February 2011].

Fromm, H., (ed.), 1985. *Kalevala – Das finnische Epos des Elias Lönnrot.* Stuttgart: Philipp Reclam jun.

Gall, H., 1976. *Das Nördlinger Ries – Ein Meteoritenkrater.* München: Freunde der Bayerischen Staatssammlung für Paläontologie und Historische Geologie e.V.

Gall, H., Hüttner, R. and Müller, D., 1977. Erläuterungen zur Geologischen Karte des Rieses 1:50000. *Geologica Bavarica,* 76, 191pp. + 1 map.

Gehlen, K. von, 1961. Shoemaker, E. M. und E. C. T. Chao: New evidence for the impact origin of the Ries basin, Bavaria, Germany. – Fortschr. Miner. 39, 359, Stuttgart 1961. *Geologische Blätter von Nordostbayern,* 11, p.229.

Gehlen, K. von, 1962. Shoemaker, E. M. und Chao, E. C. T.: New evidence for the impact origin of the Ries basin, Bavaria, Germany. – Journ. Geophysical Research 66, 3371–3378. *Geologische Blätter von Nordostbayern,* 12, pp.236–7.

Gentner, W. and Wagner, G.A., 1969. Altersbestimmungen an Riesgläsern und Moldaviten. *Geologica Bavarica,* 61, pp.296–303.

Gifford, A.C., 1924. The Mountains of the Moon. *New Zealand Journal of Science and Technology,* 7, pp.129–42.

Gilbert, G.K., 1893. The Moon's Face. *Bulletin of the Philosophical Society of Washington,* 12, p.241.

Glen, W., 1994. What the Impact/Volcanism/Mass-Extinction Debates Are About. In: Glen, W., (ed.), 1994. *The Mass-Extinction Debates: How Science Works in a Crisis.* Stanford: Stanford University Press, pp.7–38.

Grau, W., 1978. Das Nördlinger Ries und seine Entstehungstheorien. *Geographische Rundschau,* 30, pp.144–7.

Grau, W. and Höfling, R., 1978. *Das Nördlinger Ries – ein Objekt geowissenschaftlicher Forschung.* München: Paul List Verlag.

Greene, M.T., 1998. Alfred Wegener and the Origin of Lunar Craters. *Earth Sciences History,* 17, pp.111–38.

Greenough, G.B., 1819. *A Critical Examination of the First principles of Geology in a Series of Essays.* London: Longman.

Groschopf, P. and Reiff, W., 1966. Ergebnisse neuerer Untersuchungen im Steinheimer Becken (Württemberg). *Jahreshefte des Vereins für vaterländische Naturkunde in Württemberg,* 121, pp.155–68.

Groschopf, P. and Reiff, W., 1969. Das Steinheimer Becken. Ein Vergleich mit dem Ries. *Geologica Bavarica,* 61, pp.400–412.

Groschopf, P. and Reiff, W., 1971. Es war ein Meteoreinschlag. Ergebnis der Bohrungen im Steinheimer Becken. *Kosmos,* 12, pp.520–25.

Groschopf, P. and Reiff, W., 1993. *Der geologische Wanderweg im Steinheimer Becken.* Steinheim am Albuch: Bürgermeisteramt.

Grüttner, M., 2004. *Biographisches Lexikon zur nationalsozialistischen Wissenschaftspolitik. Studien zur Wissenschaftsgeschichte 6.* Heidelberg: Synchron Publishers.

Gudden, H., 1974. Die Forschungsbohrung Nördlingen 1973 – Durchführung und erste Befunde. *Geologica Bavarica,* 72, pp.11–31.

Guest, J.E. and Greely, R., 1979. *Geologie auf dem Mond.* Stuttgart: Enke-Verlag.

Guntau, M., 2002. Zu einigen Aspekten der Geologie in der Zeit des Nationalsozialismus in Deutschland (1933–1945). *Geohistorische Blätter,* 5, pp.125–50.

Hahn, A., 1969. Deutung der magnetischen Anomalie in der Umgebung der Bohrung Wörnitzostheim. *Geologica Bavarica,* 61, pp.343–7.

Hantzsch, V., 1904. Fraas, Oskar von. *Allgemeine Deutsche Biographie,* 48, pp.671–4. Available through <http://www.deutsche-biographie.de/pnd116678364.html> [Accessed 5 April 2011].

Haunschild, H., 1969. Die Trias im Ries und Vorries. *Geologica Bavarica,* 61, pp.43–58.

Heide, F., 1933. Meteoritenkrater. *Forschungen und Fortschritte,* 9, pp.379–81.

Heide, F., 1934. Über Riesenmeteoriten. *Chemie der Erde,* 1934, pp.224–51.

Heide, F., 1957. *Kleine Meteoritenkunde.* 2nd ed. Berlin: Springer-Verlag.

Heißig, K., 1986. No effect of Ries impact event on the local mammal fauna. *Modern Geology,* 10, pp.171–9.

Heller, F., 1974. Hummel, Karl. *Neue Deutsche Biographie,* 10, pp.55–6. Available through <http//www.deutsche-biographie.de/pnd117062332.html> [Accessed 5 April 2011].

Hennig, E., 1936. Bemerkungen zum Vortrag von O. Stutzer. *Zeitschrift der Deutschen Geologischen Gesellschaft,* 8, pp.588–9.

Hennig, E., 1969. Hennig, Richard. *Neue Deutsche Biographie,* 8, pp.544–5. Available through: <http//www.deutsche-biographie.de/pnd116716002.html> [Accessed 4 October 2011].

Hennig, R., 1934. Meteorkrater auf der Erde, eine neuentdeckte Naturerscheinung. *Geographische Wochenschrift,* 2, pp.730–34.

Henry, J., 2008. Ideology, Inevitability, and the Scientific Revolution. *Isis,* 99, pp.552–9.

Herold, R., 1969. Eine Malmkalk-Trümmermasse in der Oberen Süßwassermolasse Niederbayerns. *Geologica Bavarica,* 61, pp.413–27.

Heybrock, W., 1934. Zur Kenntnis von Meteoriten und Meteorkratern. *Die Sterne,* 14, pp.51–7.

Hölder, H., 1962. Zur Geschichte der Ries-Forschung. *Jahresheft des Vereins für vaterländische Naturkunde Württemberg,* 117, pp.10–17.

Hölder, H., 1971. Die Rolle der Wissenschaftsgeschichte für die meteoritische Deutung des Nördlinger Rieses, einer in statu nascendi zerstörten Lagerstätte. *Geologie,* 20, pp.591–6.

Hölder H., 1976. Reinhold Seemann – Geologe. *Jahreshefte der Gesellschaft für Naturkunde Württemberg*, 131, pp.203–6.

Hölder, H., 1989. *Kurze Geschichte der Geologie und Paläontologie – Ein Lesebuch*. Berlin: Springer-Verlag.

Hollaus, E., 1969. Kurze Übersicht der bisherigen Kenntnisse des Pleistozäns im Nördlinger Ries. *Geologica Bavarica*, 61, pp.131–41.

Hooykaas, R., 1959. *Natural Law and Divine Miracle: A Historical-critical Study of the Principle of Uniformity in Geology, Biology, and Theology*. Leiden: Brill.

Horn, P., 1991. Marginalien zur wissenschaftlichen Literatur über irdische Meteoriten-Krater, den Nördlinger Ries-Krater und die Tektitfrage von 1890 bis 1988. *Acta Albertina Ratisbonensia*, 47, pp.27–36.

Hörz, F., 1965. Untersuchungen an Riesgläsern. *Beiträge zur Mineralogie und Petrographie*, 11, pp.621–61.

Hörz, F., 1968. Statistical Measurements of Deformation Structures and Refractive Indices in Experimentally Shock Loaded Quartz. In: French, B.M. and Short, N.M., (eds), 1968. *Shock metamorphism of natural materials*. Baltimore: Mono Book Corp., pp.243–53.

Hörz, F., Gall, H., Hüttner, R. and Oberbeck, V.R., 1977. Shallow Drilling in the 'Bunte Breccia' Impact Deposits, Ries Crater, Germany. LSC VIII abstracts, 1977, pp.874–6. Available through: The Lunar and Planetary Institute, provided by the NASA Astrophysics Data System <http://www.lpi.usra. edu/meetings/lsc1977/pdf/1157.pdf> [Accessed 13 February 2014].

Hough, R.M., Gilmour, I., Pillinger, C.T., Arden, J.W., Gilkes, K.W.R., Yuan, J. and Milledge, H.J., 1995. Diamond and silicon carbide in impact melt rock from the Ries impact crater. *Nature*, 378, pp.41–4.

Hoyt, W.G., 1987. *Coon Mountain Controversies: Meteor Crater and the Development of Impact Theory*. Tucson: The University of Arizona Press.

Hummel, K., 1940a. Das Fremdwort im geologischen Schrifttum. *Zeitschrift der Deutschen Geologischen Gesellschaft*, 92, pp.128–32.

Hummel, K., 1940b. Geochemie und Erdgeschichte. *Zeitschrift der Deutschen Geologischen Gesellschaft*, 92, pp.459–68.

Hüttner, R., 1969. Bunte Trümmermassen und Suevit. *Geologica Bavarica*, 61, pp.142–200.

Hüttner, R., 1974. Das Ries als geologisches Problem. *Der Aufschluß*, 25, pp.381–94.

Hüttner, R., 1977. Makroskopische Beobachtungen zur Deformation des Kristallins in der Forschungsbohrung Nördlingen 1973. *Geologica Bavarica*, 75, pp.273–83.

Hüttner, R. and Schmidt-Kaler, H., 1999. *Meteoritenkrater Nördlinger Ries. Wanderungen in die Erdgeschichte (10)*. München: Pfeil-Verlag.

Hüttner, R., Schmidt-Kaler, H. and Treibs, W., 1969. Anmerkungen zur geologischen Übersichtskarte. *Geologica Bavarica*, 61, pp.142–454.

Hüttner, R. and Wagner, G., 1965a. Bericht über Bohrungen in Suevittuffen des württembergischen Riesgebietes. *Jahreshefte des geologischen Landesamts Baden-Württemberg*, 7, pp.223–8.

Hüttner, R. and Wagner, G., 1965b. Über Lagerung und Herkunft einiger Suevitvorkommen. *Neues Jahrbuch für Mineralogie*, Monatshefte 1965, pp.316–21.

Ives, H.E., 1919. Some Large-Scale Experiments Imitating the Craters of the Moon. *Astrophysical Journal*, 50, pp.245–50.

Jung, C.G., 1959. *Flying saucers: a northern myth*. New York: Hartcourt Brace.

Jung, K., Schaaf, H. and Kahle, H.G., 1969. *Ergebnisse gravimetrischer Messungen im Ries*. Geologica Bavarica, 61, pp.337–42.

Jurasky, K.A., 1937. Otto Stutzer zum Gedächtnis! *Blätter der Bergakademie Freiberg*, 16, pp.2–6.

Kaljuvee, J.O., 1933. *Die Großprobleme der Geologie*. Talinn: F. Wassermann.

Kavasch, J., 1997. *Meteoritenkrater Ries – ein geologischer Führer*. 11th ed. Donauwörth: Auer-Verlag.

Keller, G., Smit, J. et al., 2004. *The Geological Society of London Great Online Chicxulub Debate 2004*. Available through: The Geological Society <http://www.geolsoc.org.uk/chicxulub> [Accessed 14 November 2012].

Keller, J., Brey, G., Lorenz, V. and Sachs, P., 1990. *Urach, Hegau, Kaiserstuhl. Excursion 2A, August 27 to September 2, 1990*. Mainz: IAVCEI International Volcanological Congress Mainz (FRG) 1990.

King, E.A., 1976. *Space Geology – An Introduction*. New York: John Wiley and Sons, Inc.

Klarmann, J., 1896. *Offiziers-Stammliste des Bayerischen Ingenieur-Corps 1744–1894*. München: Hübschmann'sche Buchdruckerei.

Knoblauch, H., 1991. *Die Welt der Wünschelrutengänger und Pendler – Erkundungen einer verborgenen Wirklichkeit*. Frankfurt: Campus.

Kockel, T., 2005. Deutsche Ölpolitik 1928–1938. *Jahrbuch für Wirtschaftsgeschichte*, Beiheft 7, 393pp.

Koeberl, C., 1998. Identification of meteoritic components in impactites. In: Grady, M.M., Hutchison, R., McGall, G.J.H. and Rothery, D.A., (eds), 1998. *Meteorites: Flux with Time and Impact Effects*. Geological Society, London, Special Publications, 140, pp.133–53.

Kohring, R. and Kreft, G., (eds), 2003: *Tilly Edinger. Leben und Werk einer jüdischen Wissenschaftlerin*. Stuttgart: Schweizerbart'sche Verlagsbuchhandlung.

Kohring, R. and Schlüter, Th., 2009. Auf der Spur zur Wiege der Menschheit: Hans Reck. *Fossilien*, 6(9), pp.367–71.

Kölbl, L., 1945 [printed in 1949]. *Franz Eduard Sueß. Mitteilungen der Geologischen Gesellschaft in Wien*, 38, pp.267–84.

Kölbl-Ebert, M., 2003. From volcano to impact crater: a history of the impact hypothesis at Ries Crater and Steinheim Basin from 1900 to 1970. *Neues Jahrbuch für Geologie und Paläontologie*, Monatshefte, 2003(10), pp.591–602.

Kölbl-Ebert, M., 2013. Vom Riesvulkan zum Mondlabor – Die geologische Riesforschung im Wandel der öffentlichen Wahrnehmung. *Geohistorische Blätter*, 23, pp.141–62.

Kranz, W., 1911. Das Nördlinger Riesproblem. *Jahresberichte und Mitteilungen des Oberrheinischen Geologischen Vereins*, 1, pp.32–5.

Kranz, W., 1912. Das Nördlinger Riesproblem II. *Jahresberichte und Mitteilungen des Oberrheinischen Geologischen Vereins*, 2(1), pp.54–65.

Kranz, W., 1914a. Das Problem des Steinheimer Beckens. *Jahresberichte und Mitteilungen des Oberrheinischen Geologischen Vereins*, N. F. 4(2), pp.92–112.

Kranz, W., 1914b. Aufpressung und Explosion oder nur Explosion im vulkanischen Ries bei Nördlingen und im Steinheimer Becken? *Zeitschrift der Deutschen Geologischen Gesellschaft*, Monatsberichte 66, pp.9–25.

Kranz, W., 1915–16. Das Problem des Steinheimer Beckens II. *Jahresberichte und Mitteilungen des Oberrheinischen Geologischen Vereins*, N. F. 5(2), pp.125–8.

Kranz, W., 1926. Zum Problem des Rieses und des Steinheimer Beckens. In: Oberrheinischer Geologischer Verein, (ed.), 1926. *Das Problem des Rieses*. Nördlingen: Verlag der Stadt Nördlingen, pp.84–98.

Kranz, W., 1928. Vulkanexplosionen, Sprengtechnik, praktische Geologie und Ballistik. *Zeitschrift der Deutschen Geologischen Gesellschaft*, 80, pp.257–307.

Kranz, W., 1937a. "Krater von Sall" auf Ösel, wahrscheinlich "Meteorkrater". *Gerlands Beiträge zur Geophysik*, 51(1), pp.50–55.

Kranz, W., 1937b. Steinheimer Becken, Nördlinger Ries und 'Meteorkrater'. *Petermanns Mitteilungen*, 83, pp.198–202.

Kranz, W., 1939. Beitrag zum Köfels-Problem: Die 'Bergsturz-Hebungs- und Sprengtheorie'. *Neues Jahrbuch für Mineralogie, Geologie und Paläontologie*, Abt. B, Beilagen 80, pp.113–38.

Kranz, W. and Gottschick, F., 1925. Zur Tektonik des Steinheimer Beckens. *Zeitschrift der Deutschen Geologischen Gesellschaft*, 77, pp.37–65.

Kraus, E., Meyer, R. and Wegener, A., 1928. Untersuchungen über den Krater von Sall auf Ösel. *Gerlands Beiträge zur Geophysik*, 20, pp.312–78.

Kraut, F., 1969. Über ein neues Impaktit-Vorkommen im Gebiete von Rochechouart-Chassenon (Départements Haute Vienne und Charente, Frankreich). *Geologica Bavarica*, 61, pp.428–50.

Kuhn, O., 1964. *Geologie von Bayern*. 3rd ed. München: BLV Verlagsgesellschaft.

Kuhn, T.S., 1962. *The Structure of Scientific Revolutions*. Chicago: University of Chicago Press.

Künstner, E., 1981. Zur 100. Wiederkehr des Geburtstages von Prof. Dr. phil. habil. Otto Stutzer (1881–1936). *Neue Bergbautechnik*, 11(12), pp.711–12.

Küppers, A., 2003. Quiring, Heinrich Ludwig. *Neue Deutsche Biographie*, 21, pp.50–51. Available through <http://www.deutsche-biographie.de/pnd116318562.html> [Accessed 21 February 2011].

Küppers, H., [2007]. *Die Geschichte der Mineralogie in Kiel*. Available through: The University of Kiel <http://www.ifg.uni-kiel.de/AGs/Depmeier/MinKiel-2007.pdf> [Accessed 22 September 2011].

Lakatos, I., 1970. Falsification and the methodology of scientific research programmes. In: Lakatos, I. and Musgrave, A., (eds), 1970. *Criticism and the growth of knowledge*. Cambridge: Cambridge University Press.

Lakatos, I., 1971. History of science and its rational reconstructions. *Boston Studies in the Philosophy of Science*, 8, pp.91–136.

Lang, A., 1887. Oken, Lorenz. *Allgemeine Deutsche Biographie*, 24, pp.216–26. Available through <http://www.deutsche-biographie.de/pnd118589717.html> [Accessed 5 April 2011].

Lang, H.-J., 1995. Eine Art kopernikanische Wende. *Schwäbisches Tagblatt*, 9 February 1995, p.31.

Laudan, R., 1982. Tensions in the concept of geology: natural history or natural philosophy? *History of Geology [Earth Sciences History]*, 1(1), pp.7–13.

Lemcke, K., 1997. Spuren eines präriesischen Meteoriteneinschlags in Nordostbayern. *Mitteilungen der Bayerischen Staatssammlung für Paläontologie und Historische Geologie*, 37, pp.135–8.

Löffler, R., 1912. Die Zusammensetzung des Grundgebirges im Ries. *Jahreshefte des Vereins für vaterländische Naturkunde in Württemberg*, 68, pp.107–54.

Löffler, R., 1926. Der Eruptionsmechanismus im Ries. *Zeitschrift der Deutschen Geologischen Gesellschaft*, B Monatshefte, 78, pp.177–8.

Lofgren, G., 2008. Teaching Geology to Apollo Astronauts. *ASK Magazine*, 32, pp.18–21.

Lorenz, V., 1982. Zur Vulkanologie der Tuffschlote der Schwäbischen Alb. *Jahresberichte und Mitteilungen des Oberrheinischen Geologischen Vereins*, N.F. 64, pp.167–200.

Mark, K., 1995. *Meteorite Craters*. Tucson: The University of Arizona Press.

Martin, H., 1999. *Wenn es Krieg gibt, gehen wir in die Wüste*. 3rd ed. Hamburg: Abera-Verlag.

Martini, H.J., 1965. Ansprache anläßlich der Trauerfeier für Alfred Bentz am 10. 7. 1964 in Hannover. *Geologisches Jahrbuch*, 83, pp.XXI–XXVIII.

Mattmüller, C.R., 1994. *Ries und Steinheimer Becken. Geologischer Führer und Einführung in die Meteoritenkunde*. Stuttgart: Enke-Verlag.

Marvin, U.B., 1986. Meteorites, the Moon and the History of Geology. *Journal of Geological Education*, 34, pp.140–65.

Marvin, U.B., 1996. Ernst Florens Friedrich Chladni (1756–1827) and the origins of modern meteorite research. *Meteoritics and Planetary Science*, 31, pp.545–88.

Marvin, U.B., 1999. Impacts from space: the implications for uniformitarian geology. In: Craig, G.Y. and Hull, J.H., (eds), 1999. James Hutton – Present and Future. *Geological Society, London, Special Publications*, 150, pp.89–117.

Marvin, U.B., 2002. Geology: from an Earth to a planetary science in the twentieth century. In: Oldroyd, D.R., (ed.), 2002. The Earth Inside and Out: Some Major Contributions to Geology in the Twentieth Century. *Geological Society, London, Special Publications*, 192, pp.17–57.

Mäussnest, O., 1974. Die Eruptionspunkte des Schwäbischen Vulkans. *Zeitschrift der Deutschen Geologischen Gesellschaft*, 125, pp.23–54 and 277–352.

Mäussnest, O., 1978. *Karte der vulkanischen Vorkommen der Mittleren Schwäbischen Alb und ihres Vorlandes (Schwäbischer Vulkan) 1:100.000.* Stuttgart: Landesvermessungsanstalt Baden-Württemberg.

Mayr, H., 2003. Reck, Hans. *Neue Deutsche Biographie*, 21, pp.232–3. Available through <http://www.deutsche-biographie.de/pnd116373555.html> [Accessed 21 February 2011].

McIntyre, D.B., 1962. Impact metamorphism at Clearwater Lakes, Quebec. *Journal of Geophysical Research*, 67, p.1647.

Medawar, P.B., 1964. Is the Scientific Paper a Fraud? In: Edge, D., (ed.), 1964. *Experiment – A Series of Scientific Case Histories first broadcast in the BBC Third Programme.* London: British Broadcasting Corporation, pp.7–12.

Meinecke, F., 1909. Der Meteorkrater von Canyon Diablo in Arizona und seine Bedeutung für die Entstehung der Mondkrater. *Naturwissenschaftliche Wochenschrift*, 8, pp.801–10.

Melosh, H.J., 1989. *Impact cratering: a geologic process.* Oxford monographs on geology and geophysics 11, Oxford: Oxford University Press.

Metz, K., [undated]. *Univ.-Prof. Dr. Robert Schwinner †.* Available through: Naturwissenschaftlicher Verein für Steiermark, <http://www.landesmuseum.at/pdf_frei_remote/MittNatVer St_84_0007–0014.pdf> [Accessed 15 February 2011].

Metz, R. and Treibs, W., 1970. Bericht über die 91. Tagung des Oberrheinischen Geologischen Vereins e.V. in Nördlingen vom 31. März bis 4. April 1970. *Jahresbericht und Mitteilungen des Oberrheinischen Geologischen Vereins*, Neue Folge 52, pp.5–21.

Meyer, M.W., ca.1906. *Kometen und Meteore.* Stuttgart: Francksche Verlagsbuchhandlung – Kosmos Gesellschaft der Naturfreunde.

Mohr, B.A., 2010. Wives and daughters of early Berlin geoscientists and their work behind the scenes. *Earth Sciences History*, 29(2), pp.291–310.

Mehrtens, H. and Richter, St., (eds), 1980. *Naturwissenschaft, Technik und NS-Ideologie. Beiträge zur Wissenschaftsgeschichte des Dritten Reichs.* Frankfurt: Suhrkamp.

Mosebach, R., 1964. Das Nördlinger Ries, vulkanischer Explosionskrater oder Einschlagstelle eines Großmeteoriten? *Berichte der Oberhessischen Gesellschaft für Natur- und Heilkunde zu Gießen,* Neue Folge, Naturwissenschaftliche Abteilung 33, pp.165–204.

Neufeld, M.J., 1995. *The Rocket and the Reich. Peenemünde and the Coming of the Ballistic Missile Era.* New York: The Free Press.

Nölke, F., 1937. Astronomie und Geologie. *Zeitschrift der Deutschen Geologischen Gesellschaft,* 89, pp.167–75.

Nyhart, L., 2009. *Modern Nature: The Rise of the Biological Perspective in Germany.* Chicago: The University of Chicago Press.

Oberdorfer, R., 1904. Die vulkanischen Tuffe des Ries bei Nördlingen. Inaugural-Dissertation Universität Tübingen. *Jahreshefte des Vereins für Vaterländischen Naturkunde in Württemberg,* Jahrgang 1905, 61, pp.1–40.

Öpik, E., 1936. Researches on the physical theory of meteor phenomena. I.: Theory of the formation of meteor craters. *Publication de l'observatoire astronomique de l'université de Tartu (Dorpat),* 28, pp.1–12.

Park, W.C., Zimmerman, R. and Schot, E.H., 2008. Memorial to Gerhard Christian Amstutz (1922–2005). *Geological Society of America Memorials,* 37, pp.13–15.

Pecora, W.T., 1960. Coesite craters and space geology. *Geotimes,* 5, pp.16–19 and 32.

Pflugmacher, K., 1970a. Das sollen Mondfahrer sein ... Staunen in Stuttgart: Die Apollo-14-Leute sehen ja aus wie Sommerfrischler aus Texas – Heute ins Ries. *Augsburger Allgemeine,* 11 August 1970. Available through NASA <http://next.nasa.gov/alsj/a14/RiesNews1.jpg> [Accessed 16 January 2012].

Pflugmacher, K., 1970b. Mondfahrer ruhen in der Sonne – Die Astronauten der Apollo-14-Besatzung sammeln Mondgestein und irdische Steinkrüge. *Augsburger Allgemeine,* 13 August 1970. Available through NASA <http://next.nasa.gov/alsj/a14/RiesNews2.jpg> [Accessed 16 January 2012].

Pflugmacher, K., 1970c. Im Rieser Riesenkrater die Romantik entdeckt. Mondfahrer von ihrem Trainingsaufenthalt in Nördlingen begeistert – Weltall-Witze am Biertisch – Weißwürste bleiben den Astronauten unbekannt. *Augsburger Allgemeine,* 13 August 1970. Available through NASA <http://next.nasa.gov/alsj/a14/RiesNews3.jpg> [Accessed 16 January 2012].

Pichler, H., 2000. *Laudatio auf Wolf von Engelhardt (12.05.2000).* Unpublished manuscript.

Plotkin, H. and Tait, K., 2011. V. Ben Meen and the Riddle of Chubb Crater. *Earth Sciences History*, 30(2), pp.240–66.

Pohl, J., 1965. Die Magnetisierung der Suevite des Rieses. *Neues Jahrbuch für Mineralogie*, Monatshefte, 1965, pp.268–76.

Pohl, J. and Angenheister, G., 1969. Anomalien des Erdmagnetfeldes und Magnetisierung der Gesteine im Nördlinger Ries. *Geologica Bavarica*, 61, pp.327–36.

Pohl, J., Bader, K., Berktold, A., Blohm, E.K., Bram, K., Ernstson, K., Friedrich, H., Haak, V., Hänel, R., Homilius, J., Knödel, K., Schmidt-Kaler, H., Rodemann, H., Wagner, G.A. and Wiesner, H., 1977. The research drill hole Nördlingen 1973 in the Ries crater – a summary of geophysical investigations. *Geologica Bavarica*, 75, pp.323–8.

Porstendorfer, G., 2003. Reich, Georg Hermann. *Neue Deutsche Biographie*, 21, p.289. Available through <http://www.deutsche-biographie.de/pnd116394536.html> [Accessed 21 February 2011].

Pösges, G. and Schieber, M., 1994. *Das Rieskrater-Museum Nördlingen*. München: Verlag Dr. Friedrich Pfeil.

Preuss, E., 1964. Das Ries und die Meteoritentheorie. *Fortschritte der Mineralogie*, 41(2), pp.271–312.

Preuss, E., 1965. Das Rätsel um das Ries. *Kosmos*, 61, pp.59–66.

Preuss, E., 1969a. Einführung in die Ries-Forschung. *Geologica Bavarica*, 61, pp.12–24.

Preuss, E., 1969b. Kennzeichen von Meteoritenkratern mit Bezug auf das Ries. *Geologica Bavarica*, 61, pp.389–99.

Proctor, R.A., 1873. *The Moon*. Manchester: Alfred Brothers.

Quenstedt, W., 1955. Branca, Karl Wilhelm Franz von. *Neue Deutsche Biographie*, 2, pp.514–15. Available through <http://www.deutsche-biographie.de/pnd116396423.html> [Accessed 21 February 2011].

Quenstedt, W., 1961a. Fraas, Oscar Friedrich von. *Neue Deutsche Biographie*, 5, pp.307–8. Available through <http://www.deutsche-biographie.de/sfz16730.html> [Accessed 21 February 2011].

Quenstedt, W., 1961b. Fraas, Eberhard. *Neue Deutsche Biographie*, 5, p.308. Available through <http://www.deutsche-biographie.de/pnd116678364.html> [Accessed 5 April 2011].

Quiring, H.L., 1948a. Gedanken über Alter, Zusammensetzung und Entstehung des Mondes. *Zeitschrift der Deutschen Geologischen Gesellschaft*, 98, pp.172–87.

Quiring, H.L., 1948b. Die irdische Mondnarbe. *Forschung und Fortschritt*, 24, pp.211–14.

Quiring, H.L., 1949a. Das Loch im Pazifik. Mondentstehung und Laurentische Revolution. *Orion*, 4, pp.569–74.

Quiring, H.L., 1949b. Die exäquatoriale Mondabschleuderung als Ursache der Mondbahnneigung und Ekliptikschiefe. *Petermanns geographische Mitteilungen*, 1949, pp.125–9.

Quiring, H.L., 1955a. *Vorträge und Schriften 1911–1955.* Berlin.

Quiring, H.L., 1955b. Bernauer, Ferdinand. *Neue Deutsche Biographie*, 2, p.104. Available through <http://www.deutsche-biographie.de/pnd116137134. html> [Accessed 27 March 2012].

Quiring, H.L., 1965. Eiszeit und Rassentrennung. *Deutsche Hochschullehrer-Zeitung*, 11(4), no pagination on off-print.

Rabinowitz, C.B., undated. Daniel Moreau Barringer and the battle for the impact theory. Available through: The Barringer Crater Company <http://www.barringercrater.com/adventure/> [Accessed 30 March 2011].

Radick, G., 2008. Why what if? *Isis*, 99, pp.547–51.

Rasmussen, K.L., Aaby, B. and Gwozdz, R., 2000. The age of the Kaalijärv meteorite craters. *Meteoritics and Planetary Science*, 35, pp.1067–71.

Reichenbacher, B., Böttcher, R., Bracher, H., Doppler, G., Engelhardt, W. von, Gregor, H.-J., Heißig, K., Heizmann, E.P.J., Hofmann, F., Kälin, D., Lemcke, K., Luterbacher, H., Martini, E., Pfeil, F., Reiff, W., Schreiner, A. and Steininger, F.F., 1998a. Graupensandrinne – Ries Impakt: Zur Stratigraphie der Grimmelfinger Schichten, Kirchberger Schichten und Oberen Süßwassermolasse (nördliche Vorlandmolasse, Süddeutschland). *Zeitschrift der Deutschen Geologischen Gesellschaft*, 149(1), pp.127–61.

Reichenbacher, B., Doppler, G., Schreiner, A., Böttcher, R., Heißig, K. and Heizmann, E.P.J., 1998b. Lagerungsverhältnisse von Grimmelfinger Schichten und Kirchberger Schichten: Kommentar zur "Revision der Stratigraphie der süddeutschen Brackwassermolasse". *Zeitschrift der Deutschen Geologischen Gesellschaft*, 149(2), pp.321–6.

Reif, W.-E., 1976. Die Erforschung des Steinheimer Beckens – Ein Beitrag zur Geschichte der Erdwissenschaften in Süddeutschland. In: Akermann, M., (ed.), 1976. *75 Jahre Heimat- und Altertumsverein Heidenheim 1901–1976.* Heidenheim: Heimat- und Altertumsverein, pp.65–86.

Reimold, W.U., Koeberl, Ch., Partridge, T.C. and Kerr, S.J., 1992. Pretoria Saltpan crater: Impact origin confirmed. *Geology*, 20, pp.1079–82.

Reinwaldt, I., 1928. Bericht über geologische Untersuchungen am Kaalijärv (Krater von Sall) auf Ösel. *Sitzungsberichte der Naturforscher-Gesellschaft bei der Universität Tartu*, 25, pp.30–70.

Reuter, L., 1925. Die Verbreitung jurassischer Kalkblöcke aus dem Ries im südbayerischen Diluvial-Gebiet (Ulm – Augsburg – Neuburg a. D.) Beitrag zur Lösung des Riesproblems. *Jahresberichte und Mitteilungen des Oberrheinischen Geologischen Vereins*, 14, pp.191–218.

Richter, R., 1936. Nacht am Meteor-Krater. *Natur und Volk*, 66, pp.454–8.

Rieppel, O., 2012. Karl Beurlen (1901–1985); Nature Mysticism, and Aryan Paleontology. *Journal of the History of Biology*, 45, pp.253–99.

Rittmann, A., 1936. *Vulkane und ihre Tätigkeit*. Stuttgart: Ferdinand Enke Verlag.

Robertson, P.B., Dence, M.R. and Vos, M.A., 1968. Deformation in rock-forming minerals from Canadian craters. In: French, B.M. and Short, N.M., (eds), 1968. *Shock metamorphism of natural materials*. Baltimore: Mono Book Corp. pp.433–52.

Rohleder, H.P.T., 1933a. Meteor-Krater (Arizona) – Salzpfanne (Transvaal) – Steinheimer Becken. *Zeitschrift der Deutschen Geologischen Gesellschaft*, 85, pp.463–8.

Rohleder, H.P.T., 1933b. The Steinheim Basin and the Pretoria Salt Pan. Volcanic or Meteoritic Origin? *The geological magazine or monthly journal of geology*, 70, pp.489–98.

Rohleder, H.P.T., 1934. Über den Fund von Vergriesungserscheinungen und Drucksuturen am Kesselrand des kryptovulkanischen Bosumtwi-Sees, Ashanti. *Zentralblatt für Mineralogie, Geologie und Paläontologie*, 10, pp.316–18.

Rohleder, H.P.T., 1936. Lake Bosumtwi, Ashanti. *Geographical Journal (London)*, 87, pp.51–65.

Rose, E.P.F., Häusler, H. and Willig, D., 2000. Comparison of British and German applications of geology in world war. In: Rose, E.P.F. and Nathanail, C.P., (eds), 2000. *Geology and warfare, examples of the influence of terrain and geologists on military operations*. The Geological Society, London, pp.107–140.

Rudwick, M.J.S., 2005. *Bursting the Limits of Time: The Reconstruction of Geohistory in the Age of Revolution*. Chicago: The University of Chicago Press.

Sachs, O., 2011. Die Erforschung und Namensgebung von 'Suevit'. *Jahresberichte und Mitteilungen des Oberrheinischen Geologischen Vereins*, N.F. 93, pp.77–88.

Salomon-Calvi, W., (ed.), 1922–26. *Grundzüge der Geologie: Ein Lehrbuch für Studierende, Bergleute und Ingenieure*. 4 vols. Stuttgart: E. Schweizerbart'sche Verlagsbuchhandlung.

Schmid, H., 2003. Nachruf auf Helmut Vidal. *Geologica Bavarica*, 108, pp.225–37.

Schmidt-Kaler, H., 1969a. Versuch einer Profildarstellung für das Rieszentrum vor der Kraterbildung. *Geologica Bavarica*, 61, pp.38–40.

Schmidt-Kaler, H., 1969b. Der Jura im Ries und in seiner Umgebung. *Geologica Bavarica*, 61, pp.59–86.

Schnell, Th., 1926. Der bayerische Trass und seine Entstehung. In: Oberrheinischer Geologischer Verein, (ed.), 1926. *Das Problem des Rieses.* Nördlingen: Verlag der Stadt Nördlingen, pp.222–79.

Schröder, J., 1934. Die Deutung der vulkanischen Vorgänge im Nördlinger Ries. *Schwabenland*, 1, pp.93–102.

Schröder, J. and Dehm, R., 1950. Geologische Untersuchungen im Ries. Das Gebiet des Blattes Harburg. *Abhandlungen des Naturwissenschaftlichen Vereins Schwaben Augsburg*, 5, pp.1–147.

Schuberth, K., 2007. *Wilhelm Ahrens (1894–1968).* Available through: Preussische Geologische Landesanstalt, Biographien <http://pgla.de/ahren.htm> [Accessed 21 February 2011].

Schüller, A. and Ottemann, J., 1963. Vergleichende Geochemie und Petrographie meteoritischer und vulkanischer Gläser. *Neues Jahrbuch für Mineralogie, Abhandlungen*, 100, pp.1–26.

Schultz, P., 1998. Shooting the Moon: Understanding the History of Lunar Impact Theories. *Earth Sciences History*, 17(2), pp.92–110.

Schuster, M., 1926. Neues zum Problem des Rieses. In: Oberrheinischer Geologischer Verein, (ed.), 1926. *Das Problem des Rieses.* Nördlingen: Verlag der Stadt Nördlingen, pp.280–91.

Schwinner, R., 1933. Das Steinheimer Becken ein Meteor-Krater? *Zeitschrift der Deutschen Geologischen Gesellschaft*, 85, pp.801–2.

Schwinner, R., 1940. Zur Geschichte der Ostalpen-Tektonik. *Zeitschrift der Deutschen Geologischen Gesellschaft*, 92, pp.263–70.

Seemann, R., 1939. Versuch einer vorwiegend tektonischen Erklärung des Nördlinger Rieses. *Neues Jahrbuch für Mineralogie, Geologie und Paläontologie, Beilagen, Abteilung B*, 81, pp.70–214.

Seibold, E. and Seibold, I., 2000. Hans Cloos (1885–1951). Dokumente aus seinem Leben. *International Journal of Earth Sciences (Geologische Rundschau)*, 88, pp.853–67.

Seibold, E. and Seibold, I., 2002. Alfred Bentz – Erdölgeologe in schwieriger Zeit, 1938–1947. *International Journal of Earth Sciences (Geologische Rundschau)*, 91, pp.1081–93.

Seibold, E. and Seibold, I., 2008. Curt Teichert – Dokumente zu einer Emigration (Dänemark – Australien – USA). *International Journal of Earth Sciences (Geologische Rundschau)*, 97, pp.665–73.

Seidl, E., 1932. Nördlinger Ries, eine typische Zerreiß-Zone; entstanden durch tektonische Spannungen der Erdrinde. *Zeitschrift der Deutschen Geologischen Gesellschaft*, 84, pp.18–23.

Shea, J.H., 1982. Twelve fallacies of uniformitarianism. *Geology*, 10, pp.455–60.

Shoemaker, E.M., 1960. *Impact mechanics at Meteor Crater Arizona.* Princeton University, unpublished PhD Thesis, 55pp.

Shoemaker, E.M. and Chao, E.Ch.-T., 1961a. New evidence for the impact origin of the Ries basin, Bavaria, Germany. *Journal of Geophysical Research*, 66, pp.3371–8.

Shoemaker, E.M. and Chao, E.Ch.-T., 1961b. New evidence for the impact origin of the Ries basin, Bavaria, Germany. *Fortschritte der Mineralogie*, 39, p.359.

Shoemaker, E.M., Goddard, E.N., Mackin, J.H., Schmitt, H.M. and Waters, A.C., 1965. *Apollo manned lunar landing scientific experiment proposal, geological field investigation in early Apollo manned lunar landing missions, Abstract and Technical Section*. Available through: Lunar and Planetary Institute, Houston <http://www.lpi.usra.edu/lunar/documents/ap_geological_shoemaker_1965.pdf> [Accessed 26 June 2012].

Short, N.M., 1966a. Effects of shock pressures from a nuclear explosion on mechanical and optical properties of granodiorite. *Journal of Geophysical Research*, 71, pp.1195–1215.

Short, N.M., 1966b. Shock processes in geology. *Journal of Geological Education*, 14, pp.149–66.

Simon, Th., 1996. Walter Carlé. *Jahreshefte der Gesellschaft für Naturkunde in Württemberg*, 152, pp.289–96.

Snowman, D., 2002. *The Hitler Emigrés: The Cultural Impact on Britain of Refugees from Nazism*. London: Chatto & Windus.

Spencer, L.J., 1933. Meteorite craters as topographical features on the earth's surface. *The Geographical Journal*, 81, pp.227–48.

Sperling, Th., (ed.), 2001. *Carl Wilhelm von Gümbel (1823–1898). Leben und Werk des bedeutendsten Geologen Bayerns*. München: Verlag Dr. Friedrich Pfeil.

Steinert, H., 1973. *Der Ries-Krater wird angebohrt – Bohrung bis 1000 Meter Tiefe / Mondkrater auf der Erde?* Clipping from unknown newspaper, 1 August 1973 [Jura-Museum Eichstätt: Off-print collection].

Steinert, H., 1974. Meteoritenkrater Nördlinger Ries – Ergebnisse der 1206 m tiefen Forschungsbohrung / Kraterboden noch nicht erreicht. *Die Zeit*, 85, 1. April 1974, p.I.

Stishov, S.M., 1995. Memoir on the discovery of high-density silica. *High Pressure Research*, 13, pp.245–80.

Stöffler, D., 1965. Anzeichen besonderer mechanischer Beanspruchung an Mineralien der Kristallineinschlüsse des Suevits (Stosswellenmetamorphose). *Neues Jahrbuch für Mineralogie*, Monatshefte, 1965, pp.350–54.

Stöffler, D., 1966. Zones of impact metamorphism in the crystalline rocks of the Nördlingen Ries crater. *Contributions to Mineralogy and Petrology*, 12, pp.15–24.

Stöffler, D., 1967. Deformation und Umwandlung von Plagioklas durch Stoßwellen in den Gesteinen des Nördlinger Ries. *Contributions to Mineralogy and Petrology*, 16, pp.51–83.

Stöffler, D., 1977. Research drilling Nördlingen 1973: polymictic breccias, crater basement, and cratering model of the Ries impact structure. *Geologica Bavarica*, 75, pp.443–58.

Stöffler, D., 1987. Die Bedeutung des Rieskraters für die Planetengeologie. *Der Daniel Nordschwaben Zeitschrift für Landschaft, Geschichte, Kultur und Zeitgeschehen*, 15(4), pp.276–85.

Stöffler, D. and Ostertag, R., 1983. The Ries Impact Crater. *Fortschritte der Mineralogie*, 61 (Beiheft 2), pp.71–116.

Stutzer, O., 1936a. Der Meteor-Krater in Arizona. *Natur und Volk*, 66, pp.442–53.

Stutzer, O., 1936b. "Meteor Crater" (Arizona) u. Nördlinger Ries. *Zeitschrift der Deutschen Geologischen Gesellschaft*, 88, pp.510–23.

Stutzer, O., 1936c. Die Talerweitung von Köfels im Ötztal (Tirol) als Meteorkrater. *Zeitschrift der Deutschen Geologischen Gesellschaft*, 88, pp.523–5.

Suess, E., 1909. *Das Antlitz der Erde*. 3rd vol., 2nd part. Wien: Tempsky.

Sueß, F.E., 1937. Der Meteor-Krater von Köfels bei Umhausen im Ötztale, Tirol. *Neues Jahrbuch für Mineralogie, Geologie und Paläontologie*, Abhandlungen, Beilage-Band, Abt. A, 72, pp.98–155, plsVIII–XI.

Taylor, St.R., 1975. *Lunar Science: A Post-Apollo View. Scientific Results and Insights from the Lunar Samples*. New York: Pergamon Press Inc.

Tollmann, A., 1986. Karl Beurlen 17. April 1901 – 27. Dezember 1985. *Mitteilungen der* Österreichischen *Geologischen Gesellschaft*, 79, pp.373–4.

Torrens, H.S., 1998. "No Impact": René Gallant (1906–1985) and His Book of 1964 Bombarded Earth (An Essay on the Geological and Biological Effects of Huge Meteorite Impacts). *Earth Sciences History*, 17, pp.174–89.

Treibs, W., 1965. Beitrag zur Kenntnis der Geologie des Rieses und östlichen Vorrieses nach Beobachtungen im Rohrgraben der Rhein-Donau-Ölleitung. *Geologica Bavarica*, 55, pp.310–16.

Trusheim, F., 1940. Fliegerbomben und Geologie. *Natur und Volk*, 70(7), pp.317–21.

UA Freiburg, 1996. *Bestand B0015 Naturwissenschaftlich-Mathematische Fakultät 1882–1971*. Available through Universitätsarchiv Freiburg i.Br. <http://www.uniarchiv.uni-freiburg.de/bestaende/provenienzgerechte-bestaende/fakultaeten/b0015/findbuchb0015> [Accessed 30 May 2012].

Vand, V., 1963. The meteoritic craters of Ries Kessel and Steinheimer Basin and their relation to tektites. *Mineral Industries*, 32, pp.1–8.

Verne, J., 1966. *Reise um den Mond*. Frankfurt: Bärmeier & Nikel.

Vidal, H., 1969. Warum Ries-Forschung? *Geologica Bavarica*, 61, pp.9–11.

Vidal, H., 1974. Die Forschungsbohrung Nördlingen 1973 – Vorgeschichte, Verwirklichung und Organisation der wissenschaftlichen Bearbeitung. *Geologica Bavarica*, 72, pp.5–10.

Wagenbreth, O., 1999. *Geschichte der Geologie in Deutschland.* Stuttgart: Enke.

Wagner, Georg, 1931. *Einführung in die Erd- und Landschaftsgeschichte mit besonderer Berücksichtigung Süddeutschlands.* Öhringen: Verlag der Hohenlohenschen Buchhandlung F. Rau.

Wagner, Georg, 1960. *Einführung in die Erd- und Landschaftsgeschichte mit besonderer Berücksichtigung Süddeutschlands.* 3rd ed. Öhringen: Verlag der Hohenlohenschen Buchhandlung F. Rau.

Wagner, Georg, 1963. Riesproblem noch nicht gelöst! *Geologische Blätter NO-Bayern*, 13, pp.13–16.

Wagner, Georg and Löffler, R., 1949. Zum Riesproblem. *Geologische Rundschau*, 37, pp.90–91.

Wagner, Gerold H., 1964. Kleintektonische Untersuchungen im Gebiet des Nördlinger Rieses. *Geologisches Jahrbuch*, 81, pp.519–600.

Wagner, Gerold H., 1965. Über Bestand und Entstehung typischer Riesgesteine. *Jahreshefte des Geologischen Landesamts Baden-Württemberg*, 7, pp.199–222.

Waldmann, L., 1953. Das Lebenswerk von Franz Eduard Sueß. *Jahrbuch der Geologischen Bundesanstalt*, 96, pp.193–216.

Wegener, A., 1921. *Die Entstehung der Mondkrater.* Braunschweig: Friedrich Vieweg & Sohn.

Weiskirchner, W., 1962. Untersuchungen und Überlegungen zur Entstehung des Rieses. *Jahresberichte und Mitteilungen des Oberrheinischen Geologischen Vereins*, NF 44, pp.17–30.

Werner, E., 1904. Das Ries in der schwäbisch-fränkischen Alb. *Blätter des Schwäbischen Albvereins*, 16(5), columns 155–68.

Whewell, W., 1858. *History of the Inductive Sciences from the Earliest to the Present Time.* New York: Appleton.

Williams, H., 1941. *Calderas and their origin.* Berkeley: University of California Press.

Wutzke, U., 2000. Alfred Wegener (1880–1930) und die Entwicklung der Vorstellungen über die Entstehung der Erde – eine Einführung. *Berichte der Geologischen Bundesanstalt*, 51, pp.76–8.

Zauner, St., 2010. Die Entnazifizierung (Epuration) des Lehrkörpers – Von der Suspendierung und Entlassung 1945/46 zur Rehabilitierung und Wiedereinsetzung der Professoren und Dozenten bis Mitte der 1950er Jahre. In: Wiesing, U., Britzinger, K.-R., Grün, B., Junginger, H. and Michl, S., (eds), 2010. Die Universität Tübingen im Nationalsozialismus. *Contubernium, Tübinger Beiträge zur Universitäts- und Wissenschaftsgeschichte*, 73, pp.937–97.

ZDGG. *Zeitschrift der Deutschen Geologischen Gesellschaft.*

Zimmerman, R., 2003. Rhythmic layering in the suevite, and reworked "Bunte Bresche" of the Otting Quarry, Otting, Bavaria, Germany: Evidence for a diatremic origin for the Ries Basin, Bavaria, Germany. *Geological Society of America Abstracts with Programs*, 35(6), p.511. Available through GSA <http://gsa.confex.com/gsa/2003AM/finalprogram/abstract_60058. htm> [Accessed 10 May 2012].

Zinnstein, G., 2005. Schindewolf, Otto Heinrich Nikolaus. *Neue Deutsche Biographie*, 22, pp.786–7. Available through <http://www.deutsche-biographie.de/pnd118755099.html> [Accessed 5 April 2011].

Index

Images are listed in bold

Ackermann, Wilhelm 84–5, 91, 211, 278, 303
actualism 75–6, 162–3, 335–9; *see also* uniformitarianism
Ahrens, Wilhelm 49, 59, 83, 91–2, 113, 203, 282
air resistance 118, 125, 131, 133
Alberti, Friedrich August von 21
Albritton, C. C. 34
Aldrin, Edwin 308
Altenbürg Quarry 91, 257–8, **259**, 282
Alvarez, Luis W. 328
Alvarez, Walter 328
Amstutz, Christian 218–20
Angenheister, Gustav 64, 196, 273–4, 279–82
Apollo 11: 299, 307–8,
Apollo 14: **195**, **311**, 311–2, 314–8
Apollo 17: 312
Apollo expeditions 299; *see also* Apollo mission, Apollo programme, Apollo Space Program
Apollo mission 172, 260, 302, 308; *see also* Apollo expeditions, Apollo programme, Apollo Space Program
Apollo programme 307, 310, 324; *see also* Apollo expeditions, Apollo mission, Apollo Space Program
Apollo Space Program 309–13; *see also* Apollo expeditions, Apollo mission, Apollo programme
aragonite 43–4, 59, 111, 113, 293, 343
Aristotle 147, 340
Arizona Crater 32, 36, 110, 136, 321; *see also* Meteor Crater

Armstrong, Neil 308
Arndt, Jörg 307, 309
Astrogeology Group (of the USGS) 172, 244, 270, 300, 306, 312
astronauts 310, **311**, 312–18, 324
astronomers 29, 34, 36, 47, 54, 64, 103, 131–2, 134, 142, 171, 197–8, 238, 297–8, 319
Aumühle quarry **3**, 91, 212, 252–3, 256, 282

Baden-Württemberg Geological Survey 270, 291–2, 321; *see also* Geological Survey of Baden-Württemberg
Baldwin, Ralph B. 171–2, 175, 298–9
Bandai-San 15–7, 86–8, 336
Barringer, Daniel Moreau 33–6, 40, 47, 58, 111, 171
Barringer Crater 32; *see also* Arizona Crater, Meteor Crater
Barthel, Karl Werner 223
Bavarian Geological Survey 196, 268–70, 284–6; *see also* Geological Survey of Bavaria
Becksmann, Ernst 143, 155, 157, 160–62, 338
Bentz, Alfred 57–8, 66–8, 75, 78, 151, 172, 183–5, 187, 196, 279, 282, 287
Berckhemer, Hans 292
Beurlen, Karl 143, 148–9, 151–5, 157–60, 162–8, 336, 340
Bickerton, Alexander William 132
Birzer, Friedrich Georg Joseph 230, 265
Boon, J.D. 34
Boué, Ami 22

Branca, Karl Wilhelm Franz von 8, 12,
      14–5, 19, 25, 27, 65–6, 68, 75, 82,
      89, 172, 177, 187–8, 281, 339
Branco, *see* Branca, Karl Wilhelm Franz von
brecciation 1, 10, 13–4, 19, 30, 36, 45, 58,
      80, 90, 120, 187, 198, 219, 276,
      282, 290, 292
Bridgman, Percy Williams 35
Brinkmann, Roland 250–51
Brückner, Eduard 36
Bucher, Walter H. 23, 192, 213, **214**,
      215–8, 227, 235, 237–8, 251, 256,
      263–4
Bülow, Kurd von 37–8, 45, 297
Bunte Brekzie, see polymictic breccia

caldera 22, 36, 43, 59, 68, 86–8, 90, 94–7,
      112, 122, 172, 219, 285, 343
caldera collapse 87, 94; *see also* caldera
Campo del Cielo 36, 138
Canyon Diablo iron meteorites 29, 32, 55,
      104
Carlé, Walter 199, 201–2, 221
Caspers, Carl von 9
central explosion (theory by W. Kranz)
      14–5, 18–9, **20**, 24, 65, 73, 120,
      122, 124, 127, 172, 208, 251–2,
      266, 287, 338–9, 341; *see also*
      Kranz-Löffler model
central mountain 1, **2**, 4–5, 22–3, 34–5, 42,
      50, 56, 99–101, **100**, 284, 291, 298
Cernan, Eugene **311**, 312
Chao, Edward C.T. 64–5, 84, 168, 179,
      181, **191**, 192–3, 196, 200, 205,
      210–12, 218, 230–34, 240, 244–5,
      247, 260, 263–4, 270, 273, 289–90,
      303–7, 321–2, 342
chauvinism xix, 156, 202, 230–2, 339–42
Chubb Crater 178
Cloos, Hans 9, 149–50
Coes, Loring 178
coesite 45, 143, 178–9, 181–2, **182**, 191–3,
      194, 196, 202, 204, 211, 215–6,
      218–9, 230, 232, 234–7, 240, 242,
      248, 252, 255–6, 258, 261, 263–4,

      273, 275–7, 292, 300, 303, 305,
      319, 321–2, 340–1, 343
Coon Mountain 32, 60; *see also* Arizona
      Crater, Meteor Crater
craters of the Moon 27, 29, 35–6, 47, 171,
      178, 181, 295, 298; *see also* lunar
      craters
cryptovolcanic explosion/structure 22–3,
      43, 90, 173, 175, 192, 216–9

David, Erwin 284, 287–9, 320–22
Decaturville 215, 218
Deffner, Carl 10, 27
Dehm, Richard 15, 19, 24–5, 62, **67**,
      68–71, 74–5, 82, 85, 166, 168,
      187, 199–201, 208, 211, 223–31,
      240–41, 251, 265–6, 274–5, 310,
      327, 340–1
Deluge 31, 76, 157
denazification 70, 102, 117, 165–6, 193,
      205, 210
Deutsche Erdöl AG (DEA) 185, 279–80
Deutsche Forschungsgemeinschaft (DFG)
      116, 260, 274, 279, 286
Deutsche Geologische Gesellschaft, *see*
      DGG
Deutsche Mineralogische Gesellschaft, *see*
      DMG
DFG, *see* Deutsche Forschungsgemeinschaft
      (DFG)
DGG 42, 46, 55, 61, 148, 151
diamonds 33, 38, 103–5, 303
diaplectic glass **194**, 303, 306, 309, 343
diatremes **6**, 7–9, **8**, 72, 93, 95–6, 111, 172,
      180, 192, 216, 224, 275, 285, 291,
      343
Dietz, Robert Sinclair 45, 171, 173, **174**,
      175–8, 192, 215, 217, 220, 227,
      303, 328
DMG 193, 210, 303
drilling 23, 40, 42, 47, 49, 58, 112–3, 116,
      196, 216, 222, 252–3, 255, 258,
      268, 271, 279, 281, 285–7, 290,
      292–5, 309–10, 324–5, 341–2

Earth historian 107, 240, 243
Earth history 107, 146, 152, 154, 159, 161, 178, 204, 239–40, 243, 338
Edinger, Tilly 149
Eifel 9, 59, 95, 103, 192, 216
El Goresy, Ahmed 289–90, 304, 306
Elwerath Company 151, 255, 330, 333
energy, *see* kinetic energy
Engel, Theodor 22
Engelhardt, Wolf von xix, 61, 193, **195**, 195–8, 201–3, 205, 207, 212, 220, 229, 234, 253, 260–1, 268, 278–80, 282, 287, 300–303, 305–12, **311**, 314–7, 324, 330–34, 336–8
Engle, Joe **311**, 312
eruptive plutons 96
eruptive sill 96

Federal Geological Survey (of the Federal Republic of Germany) 196, 257, 269, 279
Fischer, Georg 84, 278
Fisher, Osmond 102
Förstner, Ulrich 280
Fraas, Eberhard 8, 12, 22, 25, 27, 82, 122
Fraas, Oscar 10, 12, 27
French, Bevan 217, 227, 300, 306, 328
Freyberg, Bruno von 82, 210–12, 265–6
Frickhinger, Albert 9
Frickhinger, Hans Walter 183

Gallant, René 185, 328
'gas volcanoes' 93
Gehlen, Kurt von 210–1
Gentner, Wolfgang 212–3, 277, 282, 285
Geologica Bavarica 268, 271, 281, 284, 286; *see also* 'thick blue book'
geological myths 86, 88, 336
Geological Survey of Baden-Württemberg 23, 63, 160, 202, 244, 257; *see also* Baden-Württemberg Geological Survey
Geological Survey of Bavaria 65, 72; *see also* Bavarian Geological Survey

Geological Survey of Württemberg 105; *see also* Württemberg Geological Survey, Württemberg Office of Statistics
geophysics/geophysical investigations 24, 38, 171, 148, 294; *see also* seismic survey/seismic investigations
German Geological Society, *see* DGG
'German Geology' xix, 75–6, 107, 143, 145–7, 155, 157, 160, 163, 169, 201, 204, 231, 234, 239, 338–41
German Mineralogical Society, *see* DMG
Gifford, Algernon Charles 104, 131–2, 276, 298
Gilbert, Grove Karl 29, 297–8
Goebbels, Josef 152, 218
Goethe, Johann Wolfgang von 80–81, 147, 158, 160
Göhring, Hermann 57, 148
Goldschmidt, Victor Moritz 128, 149, 193
Goresy, Ahmed El, *see* El Goresy, Ahmed
Gosses Bluff 330–32
gradualism 75, 78, 158, 326, 336
Gran Chaco 47
Graupensand 329–34
Gries/Griesbildungen/Vergriesung 31, 90–91, 104, 109, 188, 277–8
Grieß, see Gries/Griesbildungen/Vergriesung
Grimm, Wolf–Dieter 269
Groschopf, Paul 23, 291–5
Gümbel, Carl Wilhelm von 9–10

Haunschild, Hellmut 269
Heide, Hermann Wilhelm Friedrich 'Fritz' 125, 127, 134, 138–42, **139**, 197, 223, 275
Henbury Craters 30, 36, 47, 132–3, 138, 330–31
Hennig, Edwin 55–7, 165–6, 221–2
Hennig, Richard 36–7
Herold 284–5
Heybrock 132–3
Hoba meteorite 30, 140
Hölder, Helmut 56–7, 81, 185, 197–8, 200–201, 203–4, 257–8, 336

Horrix, Wilhelm 278
Hörz, Friedrich [Fred] 253–6, 278,
    300–301, 303, 306, 310–12
Hummel, Karl 152–3, 162–3
Hüttner, Rudolf 63–4, 66, 81, 89, 178, 195,
    198, 244–5, 257–8, 262, 266–7,
    270, 282–3, 288, 341
Hyde, Herbert né Rohleder, *see* Rohleder,
    Herbert P.T.
hypervelocity 15, 142–3, 173, 306, 308,
    321

impact physics 131, 136; *see also* physics
impactists 27, 32–3, 51, 57, 59, 78–9,
    85, 104, 120, 133, 143, 145, 182,
    217–8, 220, 234, 281, 294, 309,
    340–1
International Geological Congress 33, 36,
    181, 192, 207
Ives, Herbert Eugene 35, 104, 132, 291, 298

Jeptha Knob 215

Kaalijärv Crater 29–30, 36, **115**, 115–22,
    124, 133, 136, 145
Kaljuvee, Julius Osvald 24, 28–32, 44, 55,
    59, 113, 120, 133–4, 221, 227
karst, karst water 50, 96, 111
Katmai 96–7
Kavasch, Julius 238
Kennedy, John F. 172, 297, 299
Kentland disturbance/structure 173, 215
kinetic energy 49, 94, 103–4, 107, 118, 121,
    124–5, 127, 132–4, 136–8, 142,
    215, 298
Klosterberg 4, 22–3, 42, 56, 99, 101
Knebel, Walter von 13
Köfels landslide 90, 122, **123**, 124–9, 133,
    145, 192
Koken, Ernst 10–12, 22, 83
Korn, Hermann 150
Krakatoa 15, 19, 59, 86, 336
Kranz, Walter 15–7, 19, 22–5, 43, 47–8,
    50, 65–6, 72–4, 89–91, 102–5,
    106, 107–14, 115, 117–8, 120–4,
    128, 133, 137, 172, 177, 183–4,
    189, 208, 213, 266–7, 269, 340–1
Kranz-Löffler model 19, 65–6, 96, 185–6,
    339, 341; *see also* central explosion
Kraus, Ernst Carl 116–8, 222
Kraut, François [Ferencz] 261
K/T-Boundary 325, 329
Kuhn, Oskar 222–3
Kühne, Georg Walter 149

Laacher See 59, 91, 95, 228, 325
laccolith 14–6, 22, 27–8, 68, 74, 96, 185,
    187, 344
Lake Bosumtwi 30, 36, 44–6, 138, 175,
    194, 261
Lehmann, Ulrich 166–8
Lemcke, Kurt 329–33
'local outbursts' 14, 65–6, 68, 72, 74–5,
    185, 208, 251–2, 266, 341
Löffler, Richard 18–9, 65–6, 82, 88–9,
    176–7, 184, **186**, 187–9, 197–9,
    208, 215, 256, 263–4, 266–7, 278,
    341
lunar craters 27–9, 104, 131, 136, 145, 171,
    298–9, 309, 319; *see also* craters of
    the Moon
Lyell, Charles 154, 159–62

maar volcanoes 7–8, 93, 95–7, 117, 192,
    216
McEwen, Michael 312
Maclaren, James Malcolm 45
Madsen, Beth M. 181, 191
Martin, Henno 150
Maucher, Albert 164, 269–70
Max Planck Institute for Nuclear Physics
    (Heidelberg) 212, 253, 277, 282,
    289, 329
Meinecke, Franz 298
Merrill, George Perkins 118–9
Meteor Crater (Arizona) 24, 28–9, **32**,
    32–8, 40, 42, 46–8, 53–6, 58–60,
    105, 111–12, 117, 120, 136, 138,
    171, 176,179–81, 191, 193, 217,
    235, 238, 242, 276, 288, 298, 306;

see also Arizona Crater, Barringer
  Crater, Coon Mountain
Meteorkratermuseum (Sontheim) 4, 324
Meyer, Max Wilhelm 36
Meyer, Rudolf 116–8
MICE project 180–81, 192
mineralogy xix, 82–3, 147–8, 180–81, 202,
  208, 234, 239–41, 336, 338, 340
Mitchell, Edgar **311**, 312, 318
moldavites 2, 31, 277, 282, 284, 329, 344
monomictic breccia 277, 346; *see also Gries/*
  Griesbildungen/Vergriesung
Moon 27, 35, 102, 107, 110, 131, 171–2,
  181, 260, 270, 295, 297–300, 302,
  307–19, 324–5; *see also* craters of
  the Moon; lunar craters
Moore, Patrick 171
Moos, August 75, 150–51, 187–8
Moos, Beate 150–51
Müller, Wolfgang 307
Munich model 19, 89; *see also* 'Munich
  School'
'Munich School' 15, 19, 71–2, 74, 188, 208,
  251, 269, 281, 338–9, 341; *see also*
  Munich model
myths, *see* geological myths

NASA 173, 217, 245, 253, 270, 299–302,
  306–7, 310, 312, 314–7
National Geological Survey 58, 102,
  155, 201; *see also Reichsamt für*
  Bodenforschung
nationalism 155, 168, 230, 232
nickel (Ni) 33–5, 38, 40, 47–9, 58, 128,
  193, 197, 200, 211–2, 224, 242,
  247, 252, 254, 290, 303
Niedertrennbach 284–5
Niggli, Paul 218
Nininger, Harvey H. 179, 181
Nölke, Friedrich 53–4
Nördlinger Ries 1, 33, 48, 50–1, 53–4, 93,
  103, 107–8, 192, 224, 228, 248,
  302, 309–10, 317, 319; *see also* Ries
  Basin, Ries Crater
Novarupta 97

nuclear reaction as trigger for Ries
  explosion 183–4
nuclear testing 175, 179–80, 185, 230, 300,
  304

Oberdorfer, Richard 21–2, 82–4, 303
Oberrheinischer Geologischer Verein 17–8,
  22, 27, 185, 271–2, 319
Odessa (Texas) 36, 47, 138
Oettingen-Spielberg, Therese zu 164
Oken, Lorenz 146–7
Öpik, Ernst Julius 103, 125–6, 132, 134,
  135, 136–8, 140, 142
Osborn, Henry Fairfield 167–8
Ösel 29–30, 37–8, 47, 115–7, 124
Ottemann, Joachim 212, 247–8, 251, 304
Otting quarry 192, 252–3, 256
oxyhydrogen gas 50, 121

palaeontologists (and Ries research) 19, 64,
  72, 152, 265, 326, 329–32, 336,
  338
PDFs, *see* planar deformation features
  (PDFs)
percussion pits 117, 131
Pestalozzi, Johann Heinrich 272
phreatic eruption 7, 172, 256, 344
physics xix, 29, 33, 60, 83, 120, 125, 131,
  134, 136, 143, 155, 183–4, 212,
  233, 337–8, 340, 342
Pichler, Hans 61, 71, 86, 241, 332–4
picric acid 103–4
planar deformation features (PDFs) 261,
  292, 295, 303, **304**, 309, 344–5
Pohl, Jean 64, 230, 260, 281–2, 289
polymictic breccia **3**, 23, 58, 74, 90–91,
  242, 245, 258, 275, 277, 280, 282,
  284, 288, 292, 310, 330–31, 345
Popova, Svetlana 277
Preuss, Ekkehard 223–4, 227, 230, 238,
  251, 275–8, 281, 291, 297
Proctor, Richard Antony 34, 297
Prussian Geological Survey 12, 37, 44, 57,
  59, 72

Quenstedt, Friedrich August 22, 24
Quiring, Heinrich Ludwig 73, 101–5, 107–10, 114, 128, 133, 142, 145, 276, 340

racism 41, 58, 60, 102, 147, 149, 153, 155, 166, 232
Reck, Hans 89–90, 122, 172
Regelmann, C. 24
regolith 307, 310
Reich, (Georg) Hermann 72–5, 108–9, 185, 196, 278, 282, 340
Reichsamt für Bodenforschung 37, 58, 102; *see also* National Geological Survey
Reiff, Winfried 292–5
Reinwaldt, Ivan 29, 116–20, 134
Reuter, Lothar 17, 269
Reuter's blocks 17, **18**, 265, 269, 285, 345
Richter, Rudolf 36, 149
Ries Basin xvii–xix, **1**, 1–5, 9–19, 21–5, 27–8, 30–1, 35–6, 42–50, 54–5, 57–60, 62, 65–9, 72–3, 78, 80, 82–4, 87–9, 91, 93–4, 97, 99, 101, 104–5, 107, 111–3, 120–21, 128, 131, 138, 141, 143, 164, 171–2, 176, 178, 182, 185, 193–4, 196–7, 199, 201–2, 211, 213, 215–7, 223–4, 226, 228–9, 235, 247, 251, 256–7, 264, 267, 274–7, 281–2, 285, 287, 289, 291, 303, 310, 314, 316, 319, 325, 330, 335–6, 338–9; *see also* Nördlinger Ries, Ries Crater
Ries catastrophe 254–6, 270, 273–4; *see also* Ries event, Ries explosion
Ries Crater xviii, xx, 31, 63, 66, 76, 79, 89, 91, 169, 192, 196, 221, 229–31, 234, 247, 258, 266, 268, 281–2, 286, 291, 295, 300, 303, 305–7, 309–13, 317, 324, 332; *see also* Nördlinger Ries, Ries Basin
Ries event 19, 57, 92, 202–3, 227, 241, 273, 275, 281–2, 284–6, 330–31, 333; *see also* Ries explosion, Ries catastrophe

Ries experts 46, 51, 53–4, 57, 60, 75, 101, 110, 114, 169, 196, 230, 244, 264, 338, 340–2
Ries explosion **20**, 79, 84, 96 102, 172, 183, 203, 223, 252, 264, 270, 281; *see also* Ries catastrophe, Ries event
Ries genesis 24, 27, 65–6, 71, 85, 89–90, 203, 241, 248, 267
Ries mountain 10, 12–4, 90
Ries problem 208, 242, 74
Ries research 62, 64, 168, 202, 205, 220, 229–31, 234, 240–41, 244, 247, 268–70, 273, 292, 297, 301–2, 317–8, 320, 323–5, 336, 338–9, 341–2
Ries volcano/Ries Basin interpreted as volcano 5, 9–10, 55, 93–4, 112, 124, 172, 211, 269, 335, 338
Ries Working Group 196, 273, 275, 278–9, 292
Rieskratermuseum Nördlingen 4, 317, 325
Rittmann, Alfred 61, 87–8, 92–7, 172
Rochechouart 205, 261
Rohleder, Herbert P.T. 24, 28, 41–6, 49–50, 53–5, 59, 62, 111, 133, 156, 172, 175, 177, 184, 199, 221

Saaremaa, *see* Ösel
Salomon-Calvi, Wilhelm 149–50, 154
Santorini 86, 89, 237
Sauer, Adolf 82–3
Schädel, Karl 321
Schafhäutl, Karl von 21
Schindewolf, Otto 176–7
Schmidt-Kaler, Hermann 269
Schmidt-Thomé, Paul 270, 275
Schmitt, Harrison 312
Schnitzlein, Adalbert 9
Schröder, Joachim 15, 24–5, 65, 68–70, 74–5, 82, 85, 224, 340–1
Schüller, Arno 212–3, 247–8, 251, 304
Schwäbischer Vulkan, see Swabian Volcano
Schwinner, Robert Gangolf 54–5, 155–7

Seemann, Reinhold 24–5, 73–4, **77**, 78–82, 185, 188–9, 198–9, 203, 209–10, 220, 336
Seibold, Eugen 175–7
Seidl, Armin 24
seismic survey/seismic investigations 72–4, 185, 226, 282, 293; *see also* geophysics/geophysical investigations
Serpent Mound 23, 215
shatter cones 43, 45, 111, 173, **175**, 175, 177–8, 219, 219–20, 261, 282, 289, 303, 330–31, 345
Shepard, Alan **311**, 312, 315–7
shock metamorphism 267, 282, 287, 299–300, 304–6, 330
shock waves 29–30, 136, 175, 185, 254, 266, 269, 284, 287–90, 292, 294, 299–300, 303, 310, 320, 343, 345
Shoemaker, Eugene Merle 64–5, 84, 179–81, **191**, 191–3, 211–2, 217–8, 223–4, 230–34, 244, 260, 275, 277, 299, 303, 312–3, 342
Siedentopf, Heinrich Friedrich 198
Signer, Peter 260–61
'Southwest-German Mega-Plate' 224, 227
Spencer, Leonard James 127, 142
Spiegler, Georg 163–4
Steinheim Basin xvii–xviii, 1, 2, 4–5, 9, 12, 22–5, 28, 30–31, 42–6, 49–50, 55–7, 59–60, 62, 65, 87, 97, 99, 101, 107, 110–13, 121, 131, 138, 141, 145, 169, 171–3, 175–7, 182, 192, 201, 213, 215–6, 221, 224–8, 231, 234, 275, 284, 290–95, 324–5, 335
Steinheimer Becken, *see* Steinheim Basin
Steno, Nicolaus 158
Stille, Hans 166
Stishov, Sergey 276
stishovite **182**, 218, 236, 252, 255–6, 276, 278, 300, 303, 305, 319, 339, 344–5
Stöffler, Dieter 260, 288–9, 300–303, 305–7, **311**, 312

stratigraphy xix, 63, 70, 72, 145–6, 239, 251, 263, 284, 294, 327, 329, 331–4, 338, 342, 345
Stutzer, Otto 24, 28, 38, **39**, 40, 46–51, 53–9, 62, 85–6, 105, 107–8, 110–13, 121, 128–9, 133, 142, 199, 221
Sueß, Eduard 15, 24, 41, 122, 156–7
Sueß, Franz Eduard 122–8, 133, 137, 156
suevite 1, **3**, 4, 10, 12, 16–7, **21**, 21–3, 31, 42, 45, 47–9, 59, 68, 83, 86, 89, 91–3, 96, 103–4, 109, 113, 172, 176–7, 185, 192–3, 195, 198, 200, 203, 211–2, 216, 221–2, 230, 235–6, 238–40, 247–8, 252–8, **259**, 261, 266–7, 270, 274–5, 277–82, 284, 286–9, 291, 299, 307, 310, 329–30, 334, 335, 340, 345
Swabian Lineament **225**, 226
Swabian Volcano 5, **6**, 7, **8**, 8, 12, 57, 72, 96, 111, 192, 216, **225**, 226–8, 236, 275, 291
tectonic Ries theories 23–5, 73–4, 78–80, 187–8, 203, 210

Teichert, Curt 149
tektites 2, 31, 132–3, 224, 238, 260, 277, 282, 345; *see also* moldavites
'thick blue book' 271, 281, 284, 286, 323; *see also* Geologica Bavarica
Thomas-conference 196, 203–4, 208, 220, 234, 260
Tilghman, Benjamin Chew 33
Troll, Carl 41, 184
Trümpy, Rudolf 236–7, 327–32, 334
Trusheim, Ferdinand 99–101, 291
tuff (suevite interpreted as volcanic tuff) xvii, 1, 9–10, 12–4, 42–3, 47–8, 58, 82, 91–2, 252–3, 256–7, **259**, 279, 335, 346
Tunguska event 30, 36–7, 47, 116–7, 138

uniformitarianism 75–6, 78, 154, 160, 336–7; *see also* actualism
Unterneul 326

Upheaval Dome 215
U.S. Geological Survey 171, 179, 299; *see also* USGS
USGS 33, 171–2, 179–81, 191, 193, 244, 270, 300, 306, 312; *see also* U.S. Geological Survey

Valley of the 10,000 smokes 97
Vand[t], Vladimir 238
Verne, Jules 132
Verein für Vaterländische Naturkunde 196
Vidal, Helmut 196, 285–6, 319–20, 323
volcano-tectonic depression 93–6, 172
volcanology 9, 59, 61, 82, 88, 92, 96–7, 263, 335–6
volcanist 78–9, 84, 89, 234, 254, 280, 340–1
Vredefort Dome 217

Wabar 36, 47, 132, 138, 303
Wagner, Georg 19, 62, 65–6, 74, 76, 96, 143, 166, 168–9, 176–7, 187–9, 196–200, 202, 205, **206**, 207–13, 215, 217–8, 220, 229, 231–43, 248–51, 256, 260–68, 270–73, 280–81, 292, 319–24
Wagner, Gerold Heinrich 65–6, 205, 208, 215, 220, 235, 248, **249**, 250–54, 256, 258, 260–67, 270, 274, 282, 341

water vapour explosion 7, 15, 19, 22, 29–30, 43, 47–8, 50, 80, 87–8, 111, 119, 121–2, 125, 133–4
Waterston, John James 132
Weber, Emil 187–8
Wegener, Alfred 29, 31, 37–8, 102, 116–8, 171, 298
Wells Creek 215, 217
Werner, Abraham Gottlob 159–60
Werner, Ernst 24, 27–8, 35, 105, 113, 230, 321–2
Whipple, Fred L. 103
Widmanstätten figures 32, 346
Wörnitzostheim 216, 235, 271, 279–81, 286
Wurster, Paul 265
Württemberg Geological Survey 65, 201; *see also* Geological Survey of Württemberg, Württemberg Office of Statistics
Württemberg Office of Statistics 24, 105; *see also* Geological Survey of Württemberg, Württemberg Geological Survey
Wylie, C.C. 103

Zähringer, Josef 277, 329
Zeller, [Heinrich?] 105, 107–10
ZERIN 325

Printed and bound by CPI Group (UK) Ltd, Croydon, CR0 4YY

24/10/2024

01778283-0012